MODULES AND RINGS

L.M.S. MONOGRAPHS

Editors: D. A. Edwards and P. M. Cohn

1. Surgery on Compact Manifolds *by* C. T. C. Wall, F.R.S.
2. Free Rings and Their Relations *by* P. M. Cohn
3. Abelian Categories with Applications to Rings and Modules *by* N. Popescu
4. Sieve Methods *by* H. Halberstam and H.-E. Richert
5. Maximal Orders *by* I. Reiner
6. On Numbers and Games *by* J. H. Conway
7. An Introduction to Semigroup Theory *by* J. M. Howie
8. Matroid Theory *by* D. J. A. Welsh
9. Subharmonic Functions, Volume 1 *by* W. K. Hayman and P. B. Kennedy
10. Topos Theory *by* P. T. Johnstone
11. Extremal Graph Theory *by* B. Bollobás
12. Spectral Theory of Linear Operators *by* H. R. Dowson
13. Rational Quadratic Forms *by* J. W. S. Cassels, F.R.S.
14. C^*-Algebras and their Automorphism Groups *by* G. K. Pedersen
15. One-Parameter Semigroups *by* E. B. Davies
16. Convexity Theory and its Applications in Functional Analysis *by* L. Asimow and A. J. Ellis
17. Modules and Rings *by* F. Kasch (*translated by* D. A. R. Wallace)

Published for the London Mathematical Society
by Academic Press Inc. (London) Ltd.

MODULES AND RINGS

A translation of

MODULN UND RINGE

German text by

F. KASCH

Ludwig-Maximilian University, Munich, Germany

Translation and editing by

D. A. R. WALLACE

University of Stirling, Stirling, Scotland

1982

ACADEMIC PRESS

A Subsidiary of Harcourt Brace Jovanovich, Publishers

LONDON NEW YORK

PARIS SAN DIEGO SAN FRANCISCO

SÃO PAULO SYDNEY TOKYO TORONTO

ACADEMIC PRESS INC. (LONDON) LTD.
24/28 Oval Road
London NW1 7DX

United States Edition published by
ACADEMIC PRESS INC.
111 Fifth Avenue
New York, New York 10003

British Library Cataloguing in Publication Data
Kasch, F.
Modules and rings.—(London Mathematical Society monograph; no. 17; ISSN 0076-0560).
1. Modules (Algebra) 2. Rings (Algebra)
I. Title II. Series III. Moduln und Ringe. *English*
512′.522 QA247.3 LCCCN 78-18028

ISBN 0-12-400350-8

Printed by J. W. Arrowsmith Ltd., Bristol BS3 2NT

Preface

This book has two predominant objectives. On the one hand, the fundamental concepts of the theory of modules and rings are presented, for which the presentation is set out in detail so that the book is suitable for private study. On the other hand, it is my intention to develop, in an easily comprehensible manner, certain themes which so far have not been presented conveniently in a text book, but which however occupy an important place in this area. In particular rings with perfect duality and quasi-Frobenius rings (*QF*-rings) are considered.

In summary the book aims to put the reader in the position of advancing from the most basic concepts up to the posing of questions and of considerations which are of topical interest in the development of mathematics. For this purpose numerous exercises of varying degrees of difficulty are provided. Here the intention is not merely to give practice in the material of the text, but also to touch upon concepts and lines of development not otherwise considered in the book.

The structure of the book is determined by the conviction that the concepts of projective and injective modules are among the most important fundamental concepts of the theory of rings and modules and consequently should be placed as far as possible at its very beginning. These concepts can then also be used in the treatment of the classical parts of the theory. For the same reason I have developed the fundamental concepts of generator and cogenerator as early as possible in order to have them always available. If different finiteness conditions are added, then one has the main theme of the book. This culminates, accordingly, in the theory of rings which are injective cogenerators resp. injective cogenerators with finiteness conditions (*QF*-rings).

In order to prevent the size of the book becoming excessive it was only possible to take up categorical concepts as far as absolutely necessary. Since there are numerous good books on categories the reader can easily broaden his knowledge in this respect. In other areas too a choice of the themes to be considered was obviously necessary. The basic principle in such a selection was first to cover the fundamental concepts which are absolutely necessary but beyond that to focus as directly as possible on the material of the last three chapters.

This book has resulted from lectures and seminars which I have given in different universities. The teaching experience which has been so gained is incorporated in the book. Thus the expert in the subject will easily recognize that I have not always chosen the "shortest" version of a proof, occasionally calculating with elements where this might be avoidable. Also I have not been deterred in places from repetitions or from the presentation of a second proof. All of this is done to render the book more intelligible, and in so doing I am aware that from a teaching point of view there can be very different opinions.

It is my belief that in a textbook—as opposed to a scholarly monograph—one is not obliged to state the authorship of all results in detail. I have made extensive use of this freedom and have only provided a name in places of particular significance. In many developments which derive from several authors, precise assignment of responsibilities is often difficult. From experience with other books it therefore seemed better to me to make no statement rather than to risk introducing false attributions.

As well as a selection of textbooks on modules and rings some original literature is given as suggestions for further reading in connection with the last three chapters. This is very much a matter of individual choice which does not imply any evaluation of the authors.

To numerous colleagues, collaborators and students I owe suggestions and critical remarks for this book. To all I express my profound thanks. I owe very particular thanks to W. Müller, W. Zimmermann and H. Zöschinger for their assistance. In particular, the later chapters have arisen from discussion with H. Zöschinger who has also contributed numerous exercises. Without the keen interest of those named in the ensuing mathematical and didactic questions, the book would almost certainly not have attained its present draft.

To the editors and the publisher I have to express my thanks for their helpful and unbureaucratic co-operation.

Munich, Autumn 1976 F. KASCH

Translator's Preface

The translator, in undertaking the task of translation, was initially motivated by his belief that an edition in English would be very worthwhile and was subsequently encouraged to embark upon the task by two reviews of the German edition indicating that a translation would be of considerable value. This English edition is a direct translation of the German text which has been essentially unaltered with the exception of Lemma 5.2.4, Lemma 5.2.5 and Corollary 11.1.4, for which more succinct proofs are now provided, and with the addition of Section 11.7, which is entirely new.

In preparing this edition the translator is much indebted to Professor Kasch for a list of the (few) corrections and for a careful over-seeing of the translation. Thanks are also owing to various members of the Department of German of the University of Stirling for their willingness to be consulted and to offer advice. A profound debt is owing to Mrs. M. Abrahamson, Secretary of the Department of Mathematics of the University of Stirling, for her unfailing cheerfulness and for the consummate skill with which she produced a beautifully typed manuscript with the many displayed formulae neatly inserted. Finally thanks are owing to Professor P. M. Cohn for many suggestions towards an improvement of the final draft.

Stirling, Spring 1981 D. A. R. WALLACE

Contents

Symbols

$\wedge$	and
$\vee$	or (in the inclusive sense)
$\forall$	quantifier ("for all" resp. "for every")
$\exists$	quantifier ("there exists")
$\Rightarrow$	implication
$\Leftrightarrow$	equivalence
$:\Leftrightarrow$ $:=$	definitions
$\lightning$	contradiction
$\subset$	subset
$\subsetneqq$	proper subset
$\not\subset$	not a subset
$\hookrightarrow$	sub-object in the sense of the relevant structure
$\overset{\neq}{\hookrightarrow}$	proper sub-object
$\not\hookrightarrow$	not a sub-object
$\overset{\circ}{\hookrightarrow}$	is small in
$\overset{*}{\hookrightarrow}$	is large (essential) in
$\mid$	divides ($a \mid b$ means "a divides b")
$\setminus$	complementary set ($A \setminus B := \{a \mid a \in A \wedge a \notin B\}$
$\square$	end of a proof
$\mathbb{N}$	set of natural numbers ($\mathbb{N} = \{1, 2, 3, \ldots\}$)
$\mathbb{Q}$	field of rational numbers
$\mathbb{R}$	field of real numbers
$\mathbb{Z}$	ring of integers
$\mathrm{Le}(M)$	composition length of module M
$\mathrm{Ord}(G)$	order of group G

[Note the difference between $A, B, C, \ldots$ and $\mathsf{A}, \mathsf{B}, \mathsf{C}, \ldots$ (e.g. $M \in \mathsf{M}_{\mathbb{R}}$)]

Chapter 1

Fundamental Ideas of Categories

The theory of categories has developed, since the year 1945, as a new branch of Mathematics. This theory is not only of interest in itself, as having produced essentially new ideas and methods, but it is also contributing to an overall understanding of mathematics. Its significance rests on the possibility that important concepts and considerations from different parts of mathematics may be brought together and be developed uniformly. In particular it furnishes the possibility of formulating and investigating common properties of different structures.

In this way it has given rise to new points of view and to the posing of questions which are not only themselves of interest in the theory of categories, but which have revealed new avenues for investigation in various concrete categories. This analysis arises in the particular case of module categories which have given, in their turn, the motivation for the development of categories.

Finally it is evident that, increasingly, fundamental concepts from the theory of categories are being accepted into the everyday jargon of mathematics and are being employed in formulating concepts and in assembling the relevant facts in other areas of mathematics. Such categorical modes of expression are essential for module categories.

In the following, knowledge of such categorical language will be provided. However we shall confine ourselves as much as possible to developing the concepts only as far as it appears absolutely necessary for their understanding.

1.1 DEFINITION OF CATEGORIES

We assume here the idea of set and class. To a first approximation a class is understood to be a "very big set", in which no operations, capable of leading to an antinomy, may be performed. For example, and in contrast to the set of all subsets, it is not permissible to form the class of all subclasses. In an axiomatic theory of classes and sets, the sets are exactly the classes which appear as elements of some classes. A class can also be conceived intuitively as the "totality of all objects with a certain property". The relevant text books can be recommended for a more thorough treatment of classes. Here, the intuitive concept of a class is enough for the understanding of what follows.

1.1.1 *Definition.* A *category* K is given by means of:

I. A class $\mathrm{Obj}(K)$, which is called the *class of objects* of K, whose elements are to be called *objects* (of K) and to be denoted by $A, B, C, \ldots$.

II. For every pair (A, B) of objects there is a set $\mathrm{Mor}_K(A, B)$ such that for different pairs of objects $(A, B) \neq (C, D)$

$$\mathrm{Mor}_K(A, B) \cap \mathrm{Mor}_K(C, D) = \varnothing.$$

The elements of $\mathrm{Mor}_K(A, B)$ are called *morphisms* from A to B and are denoted by $\alpha, \beta, \gamma, \ldots$.

III. To every triple (A, B, C) of objects there is a mapping

$$\mathrm{Mor}_K(B, C) \times \mathrm{Mor}_K(A, B) \ni (\beta, \alpha) \mapsto \beta\alpha \in \mathrm{Mor}_K(A, C)$$

which is called *multiplication* and for which we have:

(1) *Associative law*: $\gamma(\beta\alpha) = (\gamma\beta)\alpha$ for all $\alpha \in \mathrm{Mor}_K(A, B)$, $\beta \in \mathrm{Mor}_K(B, C)$, $\gamma \in \mathrm{Mor}_K(C, D)$.

(2) *Existence of identities*: To every object $A \in \mathrm{Obj}(K)$ there exists a morphism $1_A \in \mathrm{Mor}_K(A, A)$, called the *identity* of A so that for all $\alpha \in \mathrm{Mor}_K(A, B)$, $\alpha 1_A = 1_B \alpha = \alpha$.

We may now indicate some notations and simple properties. If no confusion is possible we write

$$\mathrm{Mor}(A, B) := \mathrm{Mor}_K(A, B).$$

We write further

$$\mathrm{Mor}(K) := \bigcup_{A, B \in \mathrm{Obj}(K)} \mathrm{Mor}_K(A, B)$$

to denote the class of morphisms of K. We use also the abbreviated notation

$$A \in K :\Leftrightarrow A \in \mathrm{Obj}(K)$$

$$\alpha \in K :\Leftrightarrow \alpha \in \mathrm{Mor}(K).$$

Now let $\alpha \in \mathrm{Mor}(A, B)$, then as in the case of a mapping we define

$$\textit{Domain of } \alpha := \mathrm{Dom}(\alpha) := A$$

$$\textit{Codomain of } \alpha := \mathrm{Cod}(\alpha) := B$$

Since the sets $\mathrm{Mor}(A, B)$ are disjoint for different pairs (A, B), $\mathrm{Dom}(\alpha)$ and $\mathrm{Cod}(\alpha)$ are uniquely determined by α.

Instead of writing $\alpha \in \mathrm{Mor}(A, B)$ we also write

$$\alpha : A \to B \qquad \text{or} \qquad A \overset{\alpha}{\to} B.$$

The symbol

$$A \to B$$

denotes an element from $\mathrm{Mor}(A, B)$ and an arrow $\to$ an element from $\mathrm{Mor}(K)$.

The commutativity of the diagram

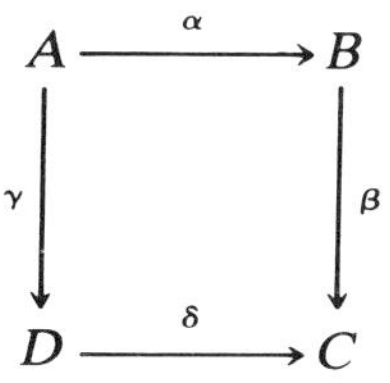

indicates that $\beta\alpha = \delta\gamma$.

If $\alpha, \beta \in \mathrm{Mor}(K)$ we write $\beta\alpha$ for the product, thereby incorporating the assumption $\mathrm{Cod}(\alpha) = \mathrm{Dom}(\alpha)$ which is required in Definition 1.1.1 for multiplication.

1.1.2 Proposition. *The identity* 1_A *(by virtue of the property given by* III(2)) *is uniquely determined.*

Proof. Let e_A be another identity of A. Then there follows

$$e_A = e_A 1_A = 1_A. \qquad \square$$

1.1.3 *Definition.* Let K be a category and let $\alpha : A \to B$ be a morphism of K. Then the following nomenclature applies.

(1) α is a *monomorphism* $:\Leftrightarrow$
$\forall C \in K\ \forall \gamma_1, \gamma_2 \in \mathrm{Mor}(C, A)\ [\alpha\gamma_1 = \alpha\gamma_2 \Rightarrow \gamma_1 = \gamma_2]$.

(2) α is an *epimorphism* $:\Leftrightarrow$
$\forall C \in K\ \forall \beta_1, \beta_2 \in \text{Mor}(B, C)\ [\beta_1\alpha = \beta_2\alpha \Rightarrow \beta_1 = \beta_2]$.

(3) α is a *bimorphism* $:\Leftrightarrow$
α is a monomorphism $\wedge\ \alpha$ is an epimorphism

(4) α is an *isomorphism* $:\Leftrightarrow$
$\exists \beta \in \text{Mor}(B, A)\ [\beta\alpha = 1_A \wedge \alpha\beta = 1_B]$

(5) α is an *endomorphism* $:\Leftrightarrow$
$\text{Dom}(\alpha) = \text{Cod}(\alpha)$

(6) α is an *automorphism* $:\Leftrightarrow$
α is an isomorphism $\wedge\ \alpha$ is an endomorphism.

1.1.4 PROPOSITION. *α is an isomorphism $\Rightarrow \alpha$ is a bimorphism.*

Proof. Let $\beta\alpha = 1_A$ and $\alpha\beta = 1_B$. It follows then from $\alpha\gamma_1 = \alpha\gamma_2$ that

$$\gamma_1 = 1_A\gamma_1 = \beta\alpha\gamma_1 = \beta\alpha\gamma_2 = 1_A\gamma_2 = \gamma_2.$$

It follows analogously from $\beta_1\alpha = \beta_2\alpha$ that

$$\beta_1 = \beta_1 1_B = \beta_1\alpha\beta = \beta_2\alpha\beta = \beta_2 1_B = \beta_2. \qquad \square$$

We observe that the converse of 1.1.4 does not hold in general (examples in exercises). Of course the converse is valid in several important categories, e.g. in module categories, the proof of which we give later.

1.2 EXAMPLES FOR CATEGORIES

In each of these examples we understand by (I) the class of the objects, by (II) the sets $\text{Mor}(A, B)$ and by (III) the multiplication $\beta\alpha$ for $\alpha \in \text{Mor}(A, B)$, $\beta \in \text{Mor}(B, C)$. The axioms are easily verified in each case.

1.2.1 S = CATEGORY OF SETS

(I) $\text{Obj}(S)$ = class of all sets.
(II) $\text{Mor}(A, B)$ = set of all mappings of A into B.
(III) $\beta\alpha$ = composition of the mappings α and β, α being followed by β.

1.2.2 G = CATEGORY OF GROUPS

(I) $\text{Obj}(G)$ = class of all groups.
(II) $\text{Mor}(A, B) = \text{Hom}(A, B)$ = set of all group homomorphisms of A into B.
(III) Usual composition.

1.2.3 A = CATEGORY OF ABELIAN GROUPS

(I) $\mathrm{Obj}(A)$ = class of all abelian groups.

(II) and (III) as for G.

In this case $\mathrm{Hom}(A, B)$ can itself be made further into an abelian group.

Definition. Let the group operations in B be written additively and let $\alpha_1, \alpha_2 \in \mathrm{Hom}(A, B)$. Then we define $\alpha_1 + \alpha_2$ by

$$\mathrm{Dom}(\alpha_1+\alpha_2) := A, \qquad \mathrm{Cod}(\alpha_1+\alpha_2) := B,$$

$$\forall a \in A[(\alpha_1+\alpha_2)(a) := \alpha_1(a)+\alpha_2(a)].$$

From the definition we see immediately that $\mathrm{Hom}(A, B)$ is in fact an abelian group. In particular the zero mapping of A into B is the zero element of this group and for $\alpha \in \mathrm{Hom}(A, B)$, $-\alpha$ is defined by

$$\mathrm{Dom}(-\alpha) := A, \qquad \mathrm{Cod}(-\alpha) := B, \qquad \forall a \in A[(-\alpha)(a) := -\alpha(a)].$$

1.2.4 R = CATEGORY OF RINGS WITH UNIT ELEMENT

(I) $\mathrm{Obj}(R)$ = class of all rings with unit element.

(II) $\mathrm{Mor}(R, S)$ = set of all unitary ring homomorphisms of R into S (Definition, see 3.2.1).

(III) Usual composition.

1.2.5 M_R = CATEGORY OF UNITARY RIGHT R-MODULES OVER A RING R WITH A UNIT ELEMENT

(I) $\mathrm{Obj}(M_R)$ = class of unitary right R-modules.

(II) $\mathrm{Mor}(A, B) := \mathrm{Hom}_R(A, B)$ = set of module homomorphisms of A into B (Definition, see 3.1.1).

(III) Usual composition.

As in the case of the category of abelian groups $\mathrm{Hom}_R(A, B)$ by the same definition as in 1.2.3 turns into an abelian group, in general however not again into an R-module! Relevant details follow later.

If S is also a ring with a unit element, we denote by ${}_SM$ and ${}_SM_R$ the categories of unitary left S-modules and unitary S–R bimodules respectively (Definition, see 2.1.1, etc.).

1.2.6 T = CATEGORY OF TOPOLOGICAL SPACES

(I) $\mathrm{Obj}(T)$ = class of all topological spaces.

(II) $\mathrm{Mor}(A, B)$ = set of continuous mappings of A into B.

(III) Usual composition.

In all of the categories so far considered the objects were sets with or without (in $\mathbf{S}$) an additional structure, and the morphisms were structure-preserving mappings. We exhibit now some examples in which other conditions are present.

1.2.7 $\mathbf{S}$ = A GROUP AS A CATEGORY

Let G be an arbitrary group and let $*$ be an object. Then we obtain a category $\hat{G}$ by

(I) $\operatorname{Obj}(\hat{G}) = \{*\}$.

(II) $\operatorname{Mor}(*, *) = G$.

(III) Group operations in G.

Obviously, 1_* is then the neutral element of G.

1.2.8 AN ORDERED SET AS A CATEGORY

Let $(M, \leqslant)$ be an ordered set. A category $\hat{M}$ is then defined by the following statements:

(I) $\operatorname{Obj}(\hat{M}) = M$.

(II) $\operatorname{Mor}(A, B) := \begin{cases} \varnothing & \text{for } A \not\leqslant B \\ \{(A \leqslant B)\} & \text{for } A \leqslant B. \end{cases}$

This means, in the case $A \leqslant B$, that $\operatorname{Mor}(A, B)$ signifies the set whose single element is the symbol $(A \leqslant B)$.

(III) $(B \leqslant C)(A \leqslant B) :\!= (A \leqslant C)$.

The identity of A is now $1_A = (A \leqslant A)$.

1.2.9 THE DUAL CATEGORY

Let $\mathbf{K}$ be a given category. The category $\mathbf{K}^\circ$ dual to the category $\mathbf{K}$ is defined by:

(I) $\operatorname{Obj}(\mathbf{K}^\circ) = \operatorname{Obj}(\mathbf{K})$.

(II) $\forall A, B \in \operatorname{Obj}(\mathbf{K}^\circ)[\operatorname{Mor}_{\mathbf{K}^\circ}(A, B) = \operatorname{Mor}_{\mathbf{K}}(B, A)]$.

(III) $\operatorname{Mor}_{\mathbf{K}^\circ}(B, C) \times \operatorname{Mor}_{\mathbf{K}^\circ}(A, B) \ni (\gamma, \beta) \mapsto \beta\gamma \in \operatorname{Mor}_{\mathbf{K}^\circ}(A, C)$,

where $\beta\gamma$ is to be formed in $\operatorname{Mor}(\mathbf{K})$.

1.3 FUNCTORS

Functors play the same role for categories as do structure-preserving mappings (= homomorphisms) for the usual algebraic structures or as do continuous mappings for topological structures. A functor is accordingly (in our definition) a pair of structure-preserving mappings of one category into another (possibly the same as the first).

1.3.1 *Definition.* A *covariant*, respectively *contravariant*, *functor* F of a category K into a category L is a pair $F = (F_O, F_M)$ of mappings satisfying:

(I) $F_O : \mathrm{Obj}(K) \to \mathrm{Obj}(L)$,

(II) $F_M : \mathrm{Mor}(K) \to \mathrm{Mor}(L)$,

with the following properties

(1) $\forall \alpha \in \mathrm{Mor}(K)[\alpha \in \mathrm{Mor}(A, B) \Rightarrow F_M(\alpha) \in \mathrm{Mor}(F_O(A), F_O(B))]$ resp. $[\alpha \in \mathrm{Mor}(A, B) \Rightarrow F_M(\alpha) \in \mathrm{Mor}(F_O(B), F_O(A))]$;

(2) $\forall A \in \mathrm{Obj}(K)[F_M(1_A) = 1_{F_O(A)}]$;

(3) $\forall \alpha, \beta \in \mathrm{Mor}(K)[\mathrm{Cod}(\alpha) = \mathrm{Dom}(\beta) \Rightarrow F_M(\beta\alpha) = F_M(\beta)F_M(\alpha)]$ resp. $[\mathrm{Cod}(\alpha) = \mathrm{Dom}(\beta) \Rightarrow F_M(\beta\alpha) = F_M(\alpha)F_M(\beta)]$.

In place of F_O and F_M we write also simply F, thus $F(A) := F_O(A)$, $F(\alpha) := F_M(\alpha)$. Condition (1) can then also be formulated as follows:

(1) $\alpha : A \to B \Rightarrow F(\alpha) : F(A) \to F(B)$ resp. $\alpha : A \to B \Rightarrow F(\alpha) : F(B) \to F(A)$

or

(1) $\mathrm{Dom}(F(\alpha)) = F(\mathrm{Dom}(\alpha)) \wedge \mathrm{Cod}(F(\alpha)) = F(\mathrm{Cod}(\alpha))$ resp. $\mathrm{Dom}(F(\alpha)) = F(\mathrm{Cod}(\alpha)) \wedge \mathrm{Cod}(F(\alpha)) = F(\mathrm{Dom}(\alpha))$.

In order to indicate that F is a functor from K to L we also write $F : K \to L$. If $G : L \to P$ is also a functor then the composition $GF : K \to P$ is obviously also a functor. If both functors F and G are covariant or both functors are contravariant then GF is covariant, if F and G are of different "variance" then GF is contravariant. We indicate now some examples of functors.

1.3.2 FORGETFUL FUNCTORS

The *forgetful functor* F from M_R into the category A of abelian groups is defined by:

$$F_O : \mathrm{Obj}(M_R) \ni A \mapsto A \in \mathrm{Obj}(A)$$

$$F_M : \mathrm{Mor}(M_R) \ni \alpha \mapsto \alpha \in \mathrm{Mor}(A).$$

This covariant functor "forgets" the R-module structure; it preserves only the additive structure of a module. If the additive structure is also "forgotten" then we obtain the forgetful functor F from M_R into the category S of sets

$$F_O : \mathrm{Obj}(M_R) \ni A \mapsto A \in \mathrm{Obj}(S)$$

$$F_M : \mathrm{Mor}(M_R) \ni \alpha \mapsto \alpha \in \mathrm{Mor}(S).$$

The functorial rules are, in any given instance, trivially satisfied. Further examples of forgetful functors are easily indicated.

1.3.3 REPRESENTABLE FUNCTORS

Let now K be an arbitrary category and let $A \in K$. Then we define

$$\mathrm{Mor}_K(A, -)\colon \mathrm{Obj}(K) \ni X \mapsto \mathrm{Mor}_K(A, X) \in \mathrm{Obj}(S)$$

$$\mathrm{Mor}_K(A, -)\colon \mathrm{Mor}(K) \ni \xi \mapsto \mathrm{Mor}_K(A, \xi) \in \mathrm{Mor}(S),$$

in which for $X := \mathrm{Dom}(\xi)$, $Y := \mathrm{Cod}(\xi)$, $\mathrm{Mor}_K(A, \xi)$ is given by

$$\mathrm{Mor}_K(A, \xi)\colon \mathrm{Mor}_K(A, X) \ni \alpha \mapsto \xi\alpha \in \mathrm{Mor}_K(A, Y).$$

It is easy to verify that $\mathrm{Mor}_K(A, -)$ is a covariant functor of K into S.

Analogously we define for a fixed object $B \in K$:

$$\mathrm{Mor}_K(-, B)\colon \mathrm{Obj}(K) \ni X \mapsto \mathrm{Mor}_K(X, B) \in \mathrm{Obj}(S)$$

$$\mathrm{Mor}_K(-, B)\colon \mathrm{Mor}(K) \ni \xi \mapsto \mathrm{Mor}_K(\xi, B) \in \mathrm{Mor}(S),$$

in which with $X := \mathrm{Dom}(\xi)$, $Y := \mathrm{Cod}(\xi)$ we put

$$\mathrm{Mor}_K(\xi, B)\colon \mathrm{Mor}_K(Y, B) \ni \gamma \mapsto \gamma\xi \in \mathrm{Mor}_K(X, B).$$

It is easy to verify that $\mathrm{Mor}_K(-, B)$ is a contravariant functor of K into S.

So far we have considered functors of one argument, i.e. of one category into another. Often, however, functors of more arguments also occur. These can, indeed, with the use of product categories (and dual categories) be reduced to (covariant) functors of one argument; nevertheless it is convenient for our purpose if we indicate functors of two arguments.

1.3.4 *Definition.* Let K, K', L be categories. A *functor F of two arguments,* that is *covariant,* respectively *contravariant,* in the first and *covariant* in the second argument of $K \times K'$ into L is a pair of mappings $F = (F_O, F_M)$ for which we have

(I) $F_O\colon \mathrm{Obj}(K) \times \mathrm{Obj}(K') \to \mathrm{Obj}(L)$

(II) $F_M\colon \mathrm{Mor}(K) \times \mathrm{Mor}(K') \to \mathrm{Mor}(L)$

with the following properties

(1) For $\alpha \in \mathrm{Mor}(K) \wedge \alpha' \in \mathrm{Mor}(K')$
with $\alpha\colon A \to B \wedge \alpha'\colon A' \to B'$
we have $F_M(\alpha, \alpha')\colon F_O(A, A') \to F_O(B, B')$
resp. $F_M(\alpha, \alpha')\colon F_O(B, A') \to F_O(A, B')$

(2) $F_M(1_A, 1_{A'}) = 1_{F_O(A,A')}$

(3) $F_M(\beta\alpha, \beta'\alpha') = F_M(\beta, \beta')F_M(\alpha, \alpha')$
resp. $F_M(\beta\alpha, \beta'\alpha') = F_M(\alpha, \beta')F_M(\beta, \alpha')$

Correspondingly we define functors which are contravariant in the first and second argument or which are covariant in the first and contravariant in the second argument.

1.3.5 THE FUNCTOR Mor

Associated with every category K there is the functor $\mathrm{Mor} = \mathrm{Mor}_K$ of $K \times K$ into S which is contravariant in the first argument and covariant in the second. It is defined by:

$$\mathrm{Mor}\colon \mathrm{Obj}(K) \times \mathrm{Obj}(K) \ni (A, B) \mapsto \mathrm{Mor}(A, B) \in \mathrm{Obj}(S)$$

$$\mathrm{Mor}\colon \mathrm{Mor}(K) \times \mathrm{Mor}(K) \ni (\alpha, \gamma) \mapsto \mathrm{Mor}(\alpha, \gamma) \in \mathrm{Mor}(S),$$

in which $\mathrm{Mor}(\alpha, \gamma)$ for $\alpha : A \to B$, $\gamma : C \to D$ is defined as follows

$$\mathrm{Mor}(\alpha, \gamma)\colon \mathrm{Mor}(B, C) \ni \beta \mapsto \gamma\beta\alpha \in \mathrm{Mor}(A, D).$$

It is easy to establish the validity of the functorial rules. As a special case of the above we have the Hom-functor

$$\mathrm{Hom}_R : M_R \times M_R \to S.$$

1.4 FUNCTORIAL MORPHISMS AND ADJOINT FUNCTORS

Let F and G be two given functors of the category K into L. In numerous important examples these functors are not "independent" of one another; there exists between them a functorial morphism which we now wish to define.

1.4.1 *Definition.* Let $F : K \to L$ and $G : K \to L$ be two co-, respectively contravariant, functors. A *functorial morphism* $\Phi : F \to G$ is a family of morphisms

$$\Phi = (\Phi_A \mid \Phi_A \in \mathrm{Mor}_L(F(A), G(A)) \wedge A \in K),$$

so that for all morphisms $\alpha : A \to B$ from K we have:

$$G(\alpha)\Phi_A = \Phi_B F(\alpha),$$

so that the diagram

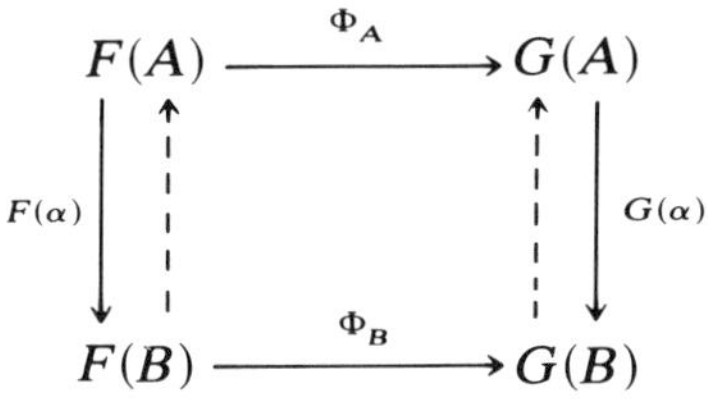

is commutative, where the vertical complete arrows denote the covariant and the vertical dotted arrows the contravariant case. It is important moreover that Φ_A depends indeed on F, G and A, and not, however, on α.

A trivial example of a functorial morphism is the identity $F \to F$ with $\Phi_A = 1_{F(A)}$ for every $A \in K$. Further it is clear that the composition of two functorial morphisms $\Phi: F \to G$ and $\Psi: G \to H$ is again such a functorial morphism. If $\Psi = (\Psi_A | A \in K)$, then we define $\Psi\Phi := (\Psi_A \Phi_A | A \in K)$.

Except for set-theoretical difficulties we can now define for two categories K and L a new category Func(K, L), the functor category of K into L, whose objects are the functors of K into L and whose morphisms are the functorial morphisms of functors of K into L. According to our notation $\mathrm{Mor}_{\mathrm{Func}(K,L)}(G, F)$ would be for instance the "set" of functorial morphisms of F into G. Certainly we must here exercise caution since, for an arbitrary category, this need not be a set. If we assume, however, that the object class of K is a set (K is then called a *small category*) then for arbitrary functors F and G—as we realise easily—$\mathrm{Mor}_{\mathrm{Func}(K,L)}(F, G)$ is again a set and Func(K, L) is in fact a category. Functor categories of this sort play an important role in category theory. They are not, however, of significance for us, so that we shall not consider them here any further.

1.4.2 *Definition* (notation as in 1.4.1). The functorial morphism $\Phi: F \to G$ is called a *functorial isomorphism* when Φ_A is an isomorphism for all $A \in K$. If a functorial isomorphism exists between the functors $F: K \to L$ and $G: K \to L$, then we write briefly $F \cong G$.

All that we have so far established for functorial morphisms of functors of one argument holds also, with appropriate modifications, for functors of more arguments. For instance let $F: K \times K' \to L$ and $G: K \times K' \to L$ be two functors in two arguments, being contravariant in the first and covariant in the second. A family of morphisms

$$\Phi = (\Phi_{(A,A')} | \Phi_{(A,A')} \in \mathrm{Mor}_L(F(A, A'), G(A, A')) \wedge (A, A') \in K \times K')$$

is then a *functorial morphism* of F into G if for all morphisms $\alpha: B \to A$ from K and $\alpha': A' \to B$ from K' the diagram

$$\begin{array}{ccc} F(A, A') & \xrightarrow{\Phi_{(A,A')}} & G(A, A') \\ \downarrow{\scriptstyle F(\alpha,\alpha')} & & \downarrow{\scriptstyle G(\alpha,\alpha')} \\ F(B, B') & \xrightarrow{\Phi_{(B,B')}} & G(B, B') \end{array}$$

is commutative. Φ is again called a *functorial isomorphism* between F and G, $F \cong G$, if all $\Phi_{(A,A')}$ are isomorphisms.

We can now introduce the concept of adjoint functors which plays a fundamental role in category theory. It is also convenient to have the concept at our disposal in module theory since only with its help can the connection between the functor Hom and the tensor product be properly understood, in which connection it is a question of adjoint functors.

Let $F: \mathsf{K} \to \mathsf{L}$ and $G: \mathsf{L} \to \mathsf{K}$ be two given functors, for which therefore G has the opposite direction to F. Let us consider now the "compound" functor

$$\mathrm{Mor}_{\mathsf{L}}(F-, -): \mathsf{K} \times \mathsf{L} \to \mathsf{S}.$$

Here we are dealing with the case of a functor in two arguments of $\mathsf{K} \times \mathsf{L}$ into the category S of sets, contravariant in the first argument and covariant in the second. This holds similarly for the functor

$$\mathrm{Mor}_{\mathsf{K}}(-, G\,-): \mathsf{K} \times \mathsf{L} \to \mathsf{S}.$$

Under these assumptions the following definition holds.

1.4.3 *Definition.* The functors F and G are said to be a *pair of adjoint functors*, of which G is said to be *right adjoint* to F and F *left adjoint* to G, if there exists a functorial isomorphism between $\mathrm{Mor}_{\mathsf{L}}(F-, -)$ and $\mathrm{Mor}_{\mathsf{K}}(-, G-)$.

1.4.4 EXAMPLE OF A FUNCTORIAL MORPHISM

Let M_K be the category of vector spaces over the field K.

As is well known with regard to a vector space V_K, there are associated two spaces, the dual

$${}_K V^* := \mathrm{Hom}_K(V, K)$$

and the bidual

$$V^{**}_K := \mathrm{Hom}_K(V^*, K).$$

If

$$\alpha: V \to W$$

is a linear mapping, and thus a morphism from M_K, then

$$\alpha^*: W^* \ni \psi \mapsto \psi\alpha \in V^*$$

is the dual and

$$\alpha^{**}: V^{**} \ni \xi \mapsto \alpha^*\xi \in W^{**}$$

is the bidual linear mapping to α (notice above the application of the linear mapping is on the opposite side from K).

PROPOSITION

(1) *By means of the definition*

$$\Delta(V) := V^{**}, \qquad \Delta(\alpha) := \alpha^{**}$$

a functor

$$\Delta : \mathsf{M}_K \to \mathsf{M}_K$$

is obtained.

(2) *For $v \in V$, let $\hat{v} \in V^{**}$ be defined by*

$$\hat{v} : V^* \ni \varphi \mapsto \varphi(v) \in K,$$

likewise let $\Phi_V \in Hom_K(V, V^{**})$ be defined by

$$\Phi_V : V \ni v \mapsto \hat{v} \in V^{**}.$$

$\Phi := (\Phi_V | V \in \mathsf{M}_K)$ is then a functorial morphism between the identity functors of M_K and Δ.

(3) *If Φ is restricted to the category of finite-dimensional vector spaces over K then Φ is a functorial isomorphism.*

Proof. The simple proof may be left to the reader as an exercise. □

1.5 PRODUCTS AND COPRODUCTS

In the investigation of modules two distinct possibilities arise. On the one hand we can analyse a given module by means of its submodules and factor modules and from the knowledge of these we can make inferences upon the structure of the module itself. On the other hand we attempt to construct a new module out of given module in order to obtain information about the category of modules. In connection with this second possibility the formation of products and coproducts is of the greatest interest. In order to make their significance more intelligible, we formulate these concepts here for arbitrary categories.

1.5.1 *Definition.* Let K be a category.

(1) Let $(A_i | i \in I)$ be a family of objects from K. A pair $(P, (\varphi_i | i \in I))$ is called a *product of the family* $(A_i | i \in I)$:⇔

(I) $P \in \mathrm{Obj}(\mathsf{K})$.

(II) $(\varphi_i | i \in I)$ is a family of morphisms from K such that

$$\varphi_i : P \to A_i, \qquad i \in I.$$

(III) For every family $(\gamma_i | i \in I)$ of morphisms $\gamma_i : C \to A_i$, $i \in I$ from K, there exists exactly one morphism $\gamma : C \to P$ from K such that

$$\gamma_i = \varphi_i \gamma, \qquad i \in I.$$

(2) Let $(A_i | i \in I)$ be a family of objects from K. A pair $(Q, (\eta_i | i \in I))$ is called a *coproduct of the family* $(A_i | i \in I)$:$\Leftrightarrow$

(I) $Q \in \mathrm{Obj}(K)$.

(II) $(\eta_i | i \in I)$ is a family of morphisms from K such that

$$\eta_i : A_i \to Q, \qquad i \in I.$$

(III) For every family $(\alpha_i | i \in I)$ of morphisms $\alpha_i : A_i \to B$, $i \in I$ from K, there exists exactly one morphism $\alpha : Q \to B$, from K such that

$$\alpha_i = \alpha \eta_i, \qquad i \in I.$$

If $(P, (\varphi_i | i \in I))$ is a product of the family $(A_i | i \in I)$, then we put

$$\prod_{i \in I} A_i := P$$

and let $\prod_{i \in I} A_i$ denote the product. This can lead to misunderstanding, since the brief notation $\prod_{i \in I} A_i$ creates the impression that the product is uniquely determined and because, moreover, the reference to the family $(\varphi_i | i \in I)$ is omitted. Caution is therefore needed in the use of $\prod_{i \in I} A_i$!

If $(Q, (\eta_i | i \in I))$ is a coproduct of the family $(A_i | i \in I)$, then we put

$$\coprod_{i \in I} A_i := Q$$

and let this denote the coproduct. The warning, mentioned above for the product, is also applicable here. The requirement mentioned in (III) for the product can, in the case of $I = \{1, 2\}$, be characterized by means of the following commutative diagram:

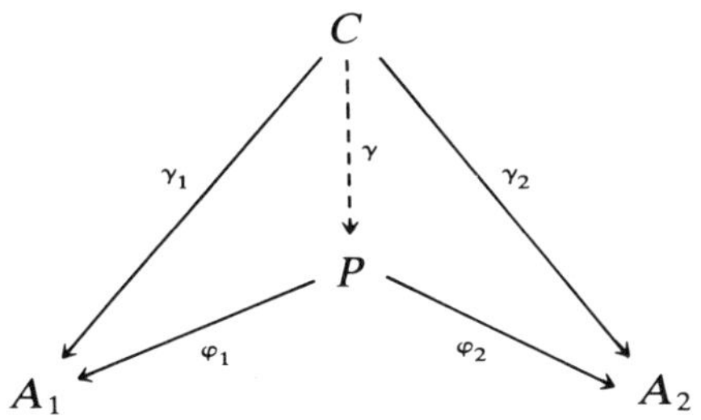

Correspondingly we obtain for the coproduct the commutative diagram:

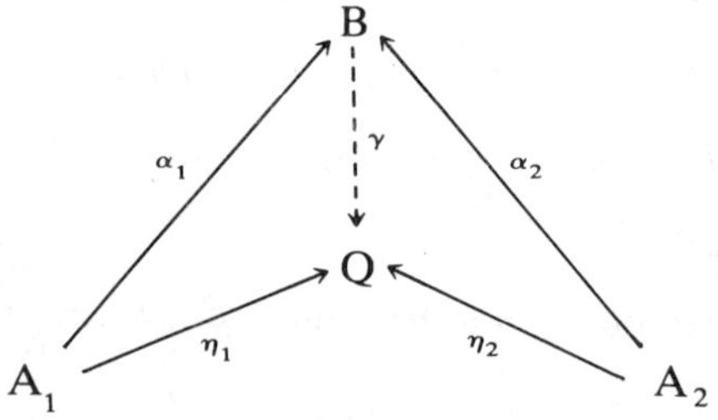

In a given category K products and coproducts do not necessarily exist. If, in the event, they exist for an arbitrary family $(A_i|i \in I)$, then K is called a *category with products*, respectively *coproducts*. If these exist at least for all finite index sets I, then K is called a *category with finite products*, respectively *finite coproducts*.

Products and coproducts—if they happen to exist—are uniquely determined up to isomorphism. More precisely the following theorem holds.

1.5.2 THEOREM. *Let K be an arbitrary category.*

(1) *If $(P, (\varphi_i|i \in I))$ and $(P', (\varphi_i'|i \in I))$ are products of the family $(A_i|i \in I)$, then there is an isomorphism $\sigma: P \to P'$ with*

$$\varphi_i = \varphi_i'\sigma, \qquad i \in I.$$

(2) *If $(Q, (\eta_i|i \in I))$ and $(Q', (\eta_i'|i \in I))$ are coproducts of the family $(A_i|i \in I)$, then there is an isomorphism $\tau: Q \to Q'$ with*

$$\eta_i' = \tau\eta_i, \qquad i \in I.$$

Proof. (1) If we replace $(\gamma_i|i \in I)$ of 1.5.1 (III) by the family $(\varphi_i'|i \in I)$ and replace C by P' then we obtain from the definition a

$$\sigma': P' \to P \quad \text{with} \quad \varphi_i' = \varphi_i\sigma'.$$

Analogously there exists a $\sigma: P \to P'$ with $\varphi_i = \varphi_i'\sigma$. From this it follows that

$$\varphi_i = \varphi_i\sigma'\sigma, \qquad \varphi_i' = \varphi_i'\sigma\sigma'.$$

If in the definition of the product we put $(\varphi_i|i \in I)$ for $(\gamma_i|i \in I)$ then $\gamma = 1_P$ yields the desired result: $\varphi_i = \varphi_i 1_P$. Since γ is uniquely determined, and on the other hand $\varphi_i = \varphi_i\sigma'\sigma$ holds, it follows that $1_P = \sigma'\sigma$ and analogously $1_{P'} = \sigma'\sigma$, as was to be shown.

(2) The proof for the coproduct results from dualizing (= reversal of the arrow) and is left to the reader as an exercise. □

We shall meet examples of products and coproducts in the category M_R.

EXERCISES

(1)

Let K be a category. Prove:

$\beta\alpha$ is a monomorphism $\Rightarrow \alpha$ is a monomorphism.
α, β are monomorphisms $\wedge \operatorname{Cod}(\alpha) = \operatorname{Dom}(\alpha) \Rightarrow \beta\alpha$ is a monomorphism.
$\beta\alpha$ is an epimorphism $\Rightarrow \beta$ is an epimorphism.
β, α are epimorphisms $\wedge \operatorname{Cod}(\alpha) = \operatorname{Dom}(\beta) \Rightarrow \beta\alpha$ is an epimorphism.

(2)

(a) Show for the category S of sets and for the category T of topological spaces: if α is a morphism, then we have

α is a monomorphism $\Leftrightarrow \alpha$ is injective as a mapping of sets,

α is an epimorphism $\Leftrightarrow \alpha$ is surjective as a mapping of sets.

(b) Let T_2 be the category of Hausdorff spaces. Investigate whether (a) also holds for T_2.

(3)

An abelian group A is called *divisible* $:\Leftrightarrow \forall n \in \mathbb{N}[nA = A]$. Let $A_\circ$ be the category of divisible abelian groups. Give an example of a monomorphism in $A_\circ$ which is not injective as a mapping of sets.
(Hint: use $\mathbb{Q}$ and $\mathbb{Q}/\mathbb{Z}$).

(4)

Let G be a group with more than one element and let $\hat{G}$ be the associated category in the sense of 1.2.7. Determine exactly the sets I for which products and coproducts exist on the index set I.

(5)

Let M be an ordered set and let $\hat{M}$ be the associated category in the sense of 1.2.8.

(a) By use of the ordering on M give a necessary and sufficient condition so that finite, respectively arbitrary, products and coproducts exist.

(b) Which morphisms from $\hat{M}$ are bimorphisms and which bimorphisms are isomorphisms?

(6)

Define a category K such that $\operatorname{Obj}(K) = \mathbb{N} = \{1, 2, 3, \ldots\}$ and in which also the product of the family $(A_i | i = 1, 2, \ldots, n)$ with $A_i \in K$ is the greatest common divisor of $A_1, \ldots, A_n$ and the coproduct of $(A_i | i = 1, 2, \ldots, n)$ is the least common multiple of $A_1, \ldots, A_n$.

Chapter 2

Modules, Submodules and Factor Modules

2.1 ASSUMPTIONS

The reader is expected to have some familiarity with the simplest ideas of rings and modules. At the least he should have already become familiar with two special cases of modules: linear vector spaces and abelian groups. Although the definitions of most of the basic ideas are here presented once again—above all, in order to fix notation—yet in view of the expected prerequisites these ideas are not especially motivated.

In consequence we shall be very brief. Motivations and examples are then best exhibited whenever we pass beyond the basic ideas and whenever the issue is not immediately concerned with a direct generalization of the ideas of linear vector spaces.

In the following all rings, which are mostly denoted by R, S or T, are to possess a unit element 1.

2.1.1 *Definition.* Let R be a ring. A *right R-module M* is

(I) an additive abelian group M together with

(II) a mapping

$$M \times R \to M \text{ with } (m, r) \mapsto mr,$$

called module multiplication, for which we have

(1) Associative law: $(mr_1)r_2 = m(r_1r_2)$.

(2) Distributive laws: $(m_1 + m_2)r = m_1r + m_2r$, $m(r_1 + r_2) = mr_1 + mr_2$.

(3) Unitary law: $m1 = m$.

(In the above m, m_1, m_2 are arbitrary elements from M and r, r_1, r_2 are arbitrary elements from R).

We point out explicitly that according to this definition all modules in the following are unitary. If M is a right R-module, then we write also M_R or $M = M_R$ in order to indicate the ring which is involved. An analogous definition holds for left modules. If S and R are two rings then M is an S-R-*bimodule* if M is a left S-module and a right R-module (with the same additive abelian group) and if, additionally, the following associative law holds:

$$s(mr) = (sm)r \quad \text{for arbitrary} \quad s \in S, m \in M, r \in R.$$

We write also ${}_SM_R$ for the S-R-bimodule.

If we speak of a module, respectively of an R-module, then we mean a one-sided R-module, in which however the side is not fixed. Statements on R-modules hold correspondingly both for right R-modules and for left R-modules.

It is well known that an R-module is called a linear vector space over R if R is a field (or skew field). Further the modules over the ring $\mathbb{Z}$ of natural numbers are the abelian groups (written additively).

If M is a right R-module we denote the neutral element of the additive group of M by 0_M and that of the additive group of R by 0_R, as with linear vector spaces it then follows that

$$0_M r = 0_M, \qquad m0_R = 0_M,$$

and also

$$-(mr) = (-m)r = m(-r) \quad \text{for arbitrary} \quad m \in M, r \in R.$$

In the following we write 0, as is usual, both for 0_M and for 0_R.

2.2 SUBMODULES AND IDEALS

In regard to mathematical structures, the substructures, subgroups, subfields and subspaces of topological spaces, generally play an important role.

In the investigation of modules, the submodules, which are about to be defined, are correspondingly important.

2.2.1 *Definition.* Let M be a right R-module. A subset A of M is called a *submodule* of M, notationally $A \hookrightarrow M$ (or also $A_R \hookrightarrow M_R$) if A is a right R-module with respect to the restriction of the addition and module multiplication of M to A.

We use the notation $A \hookrightarrow M$ for the submodule relationship, in order to have available $A \subset M$ for set-theoretic inclusion. Further we denote

$$A \hookrightarrow\!\!\!\!\!\neq\; M :\Leftrightarrow A \text{ is a proper submodule of } M$$

$$A \not\hookrightarrow M :\Leftrightarrow A \text{ is not a submodule of } M$$

We remark that from $A \not\hookrightarrow M$ it does not necessarily follow that $A \not\subset M$.

2.2.2 LEMMA. *Let M be a right R-module. If A is a subset of M and $A \neq \emptyset$ then the following are equivalent:*

(1) $A \hookrightarrow M$.

(2) *A is a subgroup of the additive group of M and for all $a \in A$ and all $r \in R$ we have $ar \in A$ (where ar is the module multiplication in M).*

(3) *For all $a_1, a_2 \in A$, $a_1 + a_2 \in A$ (with respect to addition in M) and for all $a \in A$ and all $r \in R$, we have $ar \in A$.*

Proof. This follows exactly as for linear subspaces of linear vector spaces. It is left to the reader as an exercise. □

Analogous assertions hold for submodules of left modules and bimodules.

We observe that we can think of a ring R as a right R-module R_R, as a left R-module $_RR$ and as an R-R bimodule $_RR_R$ respectively. A *right ideal, left ideal* or *two-sided ideal* of R is then a submodule of R_R, of $_RR$ or of $_RR_R$ respectively. If R is commutative then we need not distinguish between right, left and two-sided ideals and we speak then only of ideals.

Examples and remarks

(1) Every module M possesses the trivial submodules 0 and M, where 0 is the submodule which contains only the zero element of M.

(2) Let M be arbitrary and let $m_0 \in M$. Then, as we see immediately from 2.2.2,

$$m_0R := \{m_0r \mid r \in R\}$$

is a submodule of M which is called the *cyclic submodule* of M generated by m_0.

(3) If M_K is a vector space over the field K then the submodules are called (linear) subspaces.

(4) In the ring $\mathbb{Z}$ of natural numbers every ideal is cyclic.

(5) Cyclic ideals of a ring are called *principal ideals* and a commutative ring is called a *principal ideal ring* if every ideal is a principal ideal.

(6) A field K has only the trivial ideals 0 and K.

2.2.3 *Definition*

(1) A module $M = M_R$ is called *cyclic* $:\Leftrightarrow$

$$\exists m_0 \in M[M = m_0 R].$$

(2) A module $M = M_R$ is called *simple* $:\Leftrightarrow$

$$M \neq 0 \wedge \forall A \hookrightarrow M[A = 0 \vee A = M],$$

i.e. $M \neq 0$ and 0 and M are the only submodules of M.

(3) A ring R is called *simple* $:\Leftrightarrow$

$$R \neq 0 \wedge \forall A \hookrightarrow {}_R R_R[A = 0 \vee A = R],$$

i.e. $R \neq 0$ and 0 and R are the only two-sided ideals of R.

(4) A submodule $A \hookrightarrow M$ is called a *minimal*, respectively a *maximal*, submodule of M $:\Leftrightarrow$

$$0 \underset{\neq}{\hookrightarrow} A \wedge \forall B \hookrightarrow M[B \underset{\neq}{\hookrightarrow} A \Rightarrow B = 0]$$

$$\text{resp. } A \underset{\neq}{\hookrightarrow} M \wedge \forall B \hookrightarrow M[A \underset{\neq}{\hookrightarrow} B \Rightarrow B = M].$$

In the same way we speak of *simple, minimal* and *maximal ideals.* As already mentioned, cyclic ideals are called *principal ideals.*

We emphasize in addition that the minimal submodules are previously the simple submodules. The minimal (=simple), respectively maximal, submodules of a module are, if they exist, evidently minimal, respectively maximal, elements in the ordered set of non-zero, respectively proper, submodules under the ordering by inclusion.

2.2.4 LEMMA. *M is simple* $\Leftrightarrow$

$$M \neq 0 \wedge \forall m \in M[m \neq 0 \Rightarrow mR = M].$$

Proof. "$\Rightarrow$": Let $m \neq 0$, then $m = m1 \in mR$, so $mR \neq 0$, and hence $mR = M$.

"$\Leftarrow$": Let $0 \underset{\neq}{\hookrightarrow} A \hookrightarrow M$ and $0 \neq a \in A$, then $aR = M$, but $aR \hookrightarrow A$, so $A = M$. □

Examples

(1) $\mathbb{Z}$ contains no minimal (=simple) ideal, for if $n\mathbb{Z} \neq 0$ then, for example, $2n\mathbb{Z}$ is a non-zero ideal property contained within $n\mathbb{Z}$. The maximal ideals of $\mathbb{Z}$ are exactly the prime ideals $p\mathbb{Z}$, p = prime number. The proof of this follows from the fact that

$$m\mathbb{Z} \hookrightarrow n\mathbb{Z} \Leftrightarrow n \mid m.$$

(2) $\mathbb{Q}_{\mathbb{Z}}$ has no minimal and no maximal submodules.

Let

$$0 \subsetneq A \hookrightarrow \mathbb{Q}_{\mathbb{Z}}$$

and let

$$a \in A, \qquad a \neq 0,$$

then

$$0 \subsetneq 2a\mathbb{Z} \subsetneq a\mathbb{Z} \hookrightarrow A \hookrightarrow \mathbb{Q}.$$

Thus A cannot be minimal. Reference to 2.3.7 shows that there are no maximal submodules.
(3) In a vector space $V = V_K$ the minimal (=simple) subspaces are just the one-dimensional subspaces and these are given precisely in the form vK by the elements $v \in V$, $v \neq 0$. If V is n-dimensional, then the maximal subspaces are exactly the $(n-1)$-dimensional subspaces. If V is not finite-dimensional, then there are likewise maximal subspaces (a fact which is well known from linear algebra and which will here be shown later).
(4) If K is a skew field, then K_K is simple as also is K as a ring (i.e. ${}_KK_K$ is simple).

This follows immediately from the fact that every element $\neq 0$ of K possesses an inverse.
(5) Let $R := K_n$ be the ring of $n \times n$ square matrices with coefficients in a skew field. Without proof we mention (proof follows later) that although R is simple (as a ring) nevertheless R_R is not for $n > 1$.

We take this opportunity to recall the definition of an algebra.

2.2.5 *Definition.* An *algebra* is a pair (R, K), where
(I) R is a ring.
(II) K is a commutative ring.
(III) R is a right K-module for which we have

$$\forall r_1, r_2 \in R \, \forall k \in K[(r_1 r_2)k = r_1(r_2 k) = (r_1 k) r_2].$$

Our assumptions on rings and modules presuppose that R has a unit element and that K operates unitarily on R. The algebra (R, K) will also be called a K-algebra R or an algebra over K.

There is no significance in our defining R as a "right K-algebra". Since K is commutative we can from the definition

$$kr := rk, \qquad r \in R, k \in K$$

pass over immediately to a "left K-algebra".

If 1 is the unit element of R then $1K := \{1k \mid k \in K\}$ is a subring of the centre of R. Conversely every ring is an algebra over every subring of its centre. At the same time the *centre* of a ring is, as we know, the set of those elements $a \in R$, such that for every $r \in R$ we have: $ar = ra$. The centre of R is a commutative subring of R, which contains the unit element of R.

2.3 INTERSECTION AND SUM OF SUBMODULES

2.3.1 LEMMA. *Let Γ be a set of submodules of a module M, then*

$$\bigcap_{A \in \Gamma} A := \{m \in M | \forall A \in \Gamma [m \in A]\}$$

is a submodule of M.

Proof. This follows with the help of 2.2.2 as in the case of subspaces of linear vector spaces. □

Remark. We note that when $\Gamma = \varnothing$ this definition yields

$$\bigcap_{A \in \varnothing} A := M.$$

From 2.3.1 there follows immediately the corollary.

COROLLARY $\bigcap_{A \in \Gamma} A$ *is the biggest submodule of M which is contained in all $A \in \Gamma$.*

Examples

$$2\mathbb{Z} \cap 3\mathbb{Z} = 6\mathbb{Z}, \qquad \bigcap_{p = \text{prime}} p\mathbb{Z} = 0.$$

2.3.2 LEMMA. *Let X be a subset of the module M_R. Then*

$$A := \begin{cases} \left\{ \sum_{j=1}^{n} x_j r_j | x_j \in X \wedge r_j \in R \wedge n \in \mathbb{N} \right\}, & \text{if } X \neq \varnothing \\ 0 & \text{if } X = \varnothing \end{cases}$$

is a submodule of M.

Proof. For $X = \varnothing$ the assertion is clear. Let now $X \neq \varnothing$. The proof now follows with the help of 2.2.2:

$$\sum_{i=1}^{m} x_i r_i, \quad \sum_{j=1}^{n} x'_j r'_j \in A \Rightarrow \sum_{i=1}^{m} x_i r_i + \sum_{j=1}^{n} x'_j r'_j \in A,$$

$$\sum_{j=1}^{n} x_j r_j \in A, \quad r \in R \Rightarrow \sum_{j=1}^{n} x_j r_j r \in A. \qquad □$$

2.3.3 *Definition.* The module defined in 2.3.2 is called the *submodule of M generated by X* and is denoted by $|X)$.

It is important that this submodule, which, if $X \neq \varnothing$, is the set of all finite linear combinations $\Sigma x_j r_j$ with $x_j \in X$, can also be characterized by the following property.

2.3.4 LEMMA. $|X) =$ smallest submodule of M that contains X

$$= \bigcap_{C \hookrightarrow M \wedge X \subset C} C.$$

Proof. If $X = \varnothing$ and consequently $|X) = 0$ then the assertion is trivially satisfied.

If $X \neq \varnothing$ and C is a submodule which contains X, then along with $x_j \in X$, $x_j r_j$ and all finite sums of such elements lie in C, and it follows that $|X) \hookrightarrow C$. Because X is also a subset of $|X)$ (since $x = x1 \in |X)$), $|X)$ is in fact the smallest submodule of M containing X.

Let

$$D := \bigcap_{C \hookrightarrow M \wedge X \subset C} C.$$

Since by definition X is a subset of D and D is a submodule it follows that $|X) \hookrightarrow D$. But on the other hand $|X)$ occurs as a C in the intersection and it follows that $D \hookrightarrow |X)$ thus $|X) = D$. □

In the case of an S-R-bimodule M the submodule generated by a subset of M is given by

$$(X) := \begin{cases} \left\{ \sum_{j=1}^{n} s_j x_j r_j \,\middle|\, x_j \in X \wedge s_j \in S \wedge r_j \in R \wedge n \in \mathbb{N} \right\}, & \text{if } X \neq \varnothing \\ 0, & \text{if } X = \varnothing. \end{cases}$$

As before it then follows: $(X) =$ smallest submodule of ${}_S M_R$ which contains X

$$= \bigcap_{C \hookrightarrow M \wedge X \subset C} C.$$

A corresponding notation is used for ideals.

2.3.5 *Definition.* Let again $M = M_R$.

(1) A subset X of a module M is called a *generating set of M*: $\Leftrightarrow |X) = M$.

(2) A module (or right ideal) is called *finitely generated* : $\Leftrightarrow$ there exists a finite generating set.

(3) A module (respectively right ideal) is called *cyclic* (respectively *principal right ideal*) : $\Leftrightarrow$ there exists a generating element (see 2.2.3).

(4) A subset X of a module M is called *free* : $\Leftrightarrow$ for every finite subset $\{x_1, \ldots, x_m\} \subset X$ (with $x_i \neq x_j$ for $i \neq j$ $(i, j = 1, \ldots, m)$) it follows from

$$\sum_{i=1}^{m} x_i r_i = 0 \quad \text{with} \quad r_i \in R$$

that $r_i = 0$ $(i = 1, \ldots, m)$.

(5) A subset X of a module M is called a *basis* of M : $\Leftrightarrow X$ is a generating set of $M \wedge X$ is free.

If $X \neq \varnothing$ is a generating set of M then this means that every element $m \in M$ may be written as a finite linear combination

$$m = \sum_{j=1}^{n} x_j r_j, \qquad x_j \in X, r_j \in R.$$

It is here obvious that $n \in \mathbb{N}$ is not fixed in general but depends on m. Further the coefficients r_j and, in fact, also the $x_j \in X$ that occur are not uniquely determined by m. Of course if $X = \{x_1, \ldots, x_t\}$ is finite then every element $m \in M$ can be written in the form

$$m = \sum_{j=1}^{t} x_j r_j$$

since the missing summands $x_j r_j$ can be added as $x_j 0$. Furthermore the coefficients r_j may not be uniquely determined. The coefficients are however uniquely determined if a basis is being considered.

2.3.6 LEMMA. *Let $X \neq \varnothing$ be a generating set of $M = M_R$. Then we have: X is a basis $\Leftrightarrow$ for every $m \in M$ the representation*

$$m = \sum_{j=1}^{n} x_j r_j \quad \textit{with } x_j \in X, r_j \in R$$

is unique in the following sense: If

$$m = \sum_{j=1}^{n} x_j r_j = \sum_{j=1}^{n} x_j r'_j \wedge x_i \neq x_j \qquad \text{for } i \neq j, (i, j = 1, \ldots, n),$$

then necessarily

$$r_j = r'_j \ (j = 1, \ldots, n).$$

Proof. "$\Rightarrow$": If we have

$$m = \sum_{j=1}^{n} x_j r_j = \sum_{j=1}^{n} x_j r'_j \wedge x_i \neq x_j \qquad \text{for } i \neq j \ (i, j = 1, \ldots, n),$$

then it follows that

$$0 = \sum_{j=1}^{n} x_j(r_j - r'_j)$$

and since X is free, it is immediate that $r_j - r'_j = 0$, thus $r_j = r'_j$ $(j = 1, \ldots, n)$.

"$\Leftarrow$": Let

$$\sum_{j=1}^{n} x_j r_j = 0 \wedge x_i \neq x_j \qquad \text{for } i \neq j \ (i, j = 1, \ldots, n).$$

Since also we have $0 = \sum_{j=1}^{n} x_j 0$, it follows that $r_j = 0$ $(j = 1, \ldots, n)$ i.e. X is free.

Remark. If $X = \{x_1, \cdots, x_t\}$ is a finite generating set (with $x_i \neq x_j$ for $i \neq j$) then we have: X is a basis $\Leftrightarrow$ for every $m \in M$ the coefficients $r_j \in R$ in the representation

$$m = \sum_{j=1}^{t} x_j r_j$$

are uniquely determined.

We point out that these statements on uniqueness do not make sense in the case of an infinite basis X. For an infinite X we cannot replace the missing indices by means of summands of the form $x_j 0 = 0$, since infinite sums—even of zero—are not defined! Statements of uniqueness must be formulated in the sense of 2.3.6.

Examples

(1) Every module M has trivially M itself as a generating set (for every $m \in M$ is a finite linear combination of the form $m = m1$, $1 \in R$).

(2) If R is a ring, then $\{1\}$ is a basis of R_R (and of ${}_RR$).

(3) We now consider properties of $\mathbb{Q}_{\mathbb{Z}}$.

2.3.7 PROPOSITION. *If finitely many arbitrary elements are omitted from an arbitrary generating set X of $\mathbb{Q}_{\mathbb{Z}}$, then the set with these elements omitted is again a generating set of $\mathbb{Q}_{\mathbb{Z}}$.*

Proof. It suffices to show that an arbitrary element x_0 can be omitted from X, since the proposition then follows by induction for finitely many.

Since X is a generating set $x_0/2$ can be represented in the form

$$\frac{x_0}{2} = x_0 z_0 + \sum_{x_i \neq x_0} x_i z_i, \qquad x_i \in X,\ z_i \in \mathbb{Z}.$$

Then it follows that

$$x_0 = x_0 2z_0 + \sum_{x_i \neq x_0} x_i 2z_i \Rightarrow x_0 n = \sum_{x_i \neq x_0} x_i 2z_i,$$

where $n = 1 - 2z_0 \in \mathbb{Z} \wedge n \neq 0$. Let now

$$\frac{x_0}{n} = x_0 z'_0 + \sum_{x_j \neq x_0} x_j z'_j, \qquad x_j \in X,\ z'_j \in \mathbb{Z},$$

hence

$$x_0 = x_0 n z'_0 + \sum_{x_j \neq x_0} x_j n z'_j = \sum_{x_i \neq x_0} x_i 2z_i z'_0 + \sum_{x_j \neq x_0} x_j n z'_j$$

$$= \sum_{x_k \neq x_0} x_k z''_k, \qquad x_k \in X,\ z''_k \in \mathbb{Z}.$$

Thus x_0 lies in the submodule generated by $X \backslash \{x_0\}$, and since X is a generating set of $\mathbb{Q}_{\mathbb{Z}}$, then so also is $X \backslash \{x_0\}$.

From this result it further follows: There is no finite set of generators of $\mathbb{Q}_{\mathbb{Z}}$, since otherwise $\mathbb{Q}_{\mathbb{Z}}$ would be generated by the empty set and it would follow that $\mathbb{Q}_{\mathbb{Z}} = 0$ ↯.

There is no maximal submodule of $\mathbb{Q}_{\mathbb{Z}}$. Suppose that A were to be one such and that $q \in \mathbb{Q}$, $q \notin A$, then from 2.2.2

$$q\mathbb{Z} + A := \{qz + a \mid z \in \mathbb{Z} \wedge a \in A\}$$

is a submodule of $\mathbb{Q}_{\mathbb{Z}}$. Since this contains A properly it follows that

$$q\mathbb{Z} + A = \mathbb{Q}.$$

Thus $A \cup \{q\}$, and then also A by itself, would be a generating set of $\mathbb{Q}_{\mathbb{Z}}$ from which it would follow that $A = \mathbb{Q}$ ↯.

It has already been established previously that $\mathbb{Q}_{\mathbb{Z}}$ does not also have a simple (=minimal) submodule. Obviously $\mathbb{Q}_{\mathbb{Z}}$ does not have a basis for if we omit an element from a basis then the remaining set of elements is no longer a generating set (since the omitted element is not linearly representable by the remaining elements).

(4) As the next example we prove the theorem that every vector space over a skew field has a basis. For this we make our first application of Zorn's lemma, which is needed again later in other proofs. We shall therefore formulate it here.

ZORN'S LEMMA. *Let A be an ordered set. If every totally ordered subset of A has an upper bound in A then A has a maximal element.*

We take this opportunity to remark upon the known fact that Zorn's lemma is equivalent to each of the following assertions:

1. Axiom of choice.
2. Principle of Well-ordering.

In this book we shall make use of Zorn's lemma and of the Axiom of Choice.

2.3.8 AXIOM. *Every vector space over a skew field has a basis*

Proof. Let K be a skew field and let V_K be a vector space over K. Let Φ denote the set of all free subsets of V. Since the empty set is free, Φ is non-empty. Φ is an ordered set under inclusion of subsets as order relation. In order to apply Zorn's Lemma, we must show that every totally ordered subset Γ of Φ has an upper bound in Φ. If $\Gamma = \varnothing$ then every element from Φ is an upper bound of Γ. Let now $\Gamma = \{X_j \mid j \in J\} \neq \varnothing$, then we show that

$$X := \bigcup_{j \in J} X_j$$

is free and hence represents an upper bound of Γ in Φ. Let $x_1, \ldots, x_n$ be distinct elements from X. Since Γ is totally ordered, there is an $X_j \in \Gamma$ with $x_1, \ldots, x_n \in X_j$. Since X_j is free, $\{x_1, \ldots, x_n\}$ is free and consequently X is free.

By Zorn's Lemma there exists then a maximal element Y in Φ. We show that Y is a basis of V over K. Since Y is free we only need to show that $|Y) = V$. If $V = 0$ then it follows that $Y = \varnothing$ and from the definition of $|Y)$ it follows that $|Y) = V$. If $V \neq 0$ then it follows that $Y \neq \varnothing$. Let now $v \in V$ with $v \notin Y$, then by virtue of the maximality of Y $Y \cup \{v\}$ cannot be free. Thus there exist distinct $y_1, \ldots, y_n \in Y$ together with $k, k_1, \ldots, k_n \in K$ with

$$vk + \sum_{j=1}^{n} y_j k_j = 0,$$

in which not all $k, k_1, \ldots, k_n$, are equal to 0. $k = 0$ is not possible since then (because Y is free) it would follow that $k_j = 0$ $(j = 1, \ldots, n)$. From $k \neq 0$ it follows that

$$v = vkk^{-1} = \sum_{j=1}^{n} y_j(-k_j k^{-1}) \in |Y),$$

thus $V = |Y)$. □

After examining these examples we continue with our general considerations.

PROPOSITION. *Let* $\Lambda = \{A_i \mid i \in I\}$ *be a set of submodules* $A_i \hookrightarrow M_R$. *Then*

$$\left|\bigcup_{i\in I} A_i\right) = \begin{cases} \left\{\sum_{i\in I'} a_i \mid a_i \in A_i \wedge I' \subset I \wedge I' \text{ is finite}\right\}, & \text{if } \Lambda \neq \emptyset, \\ 0, & \text{if } \Lambda = \emptyset, \end{cases}$$

i.e., in the case $\Lambda \neq \emptyset$, $\left|\bigcup_{i\in I} A_i\right)$ *is the set of all finite sums* $\sum a_i$ *with* $a_i \in A_i$.

Proof. In the case $I \neq \emptyset$, $\left|\bigcup_{i\in I} A_i\right)$ is by definition the set of all finite sums

$$\sum_{j=1}^{n} a_j r_j \quad \text{with} \quad a_j \in \bigcup_{i\in I} A_i.$$

If we bring together all summands $a_j r_j$ which lie in a fixed A_i to form a sum a_i' and if we treat with the remaining summands similarly then it follows that

$$\sum_{j=1}^{n} a_j r_j = \sum_{i\in I'} a_i',$$

thus we have

$$\left|\bigcup_{i\in I} A_i\right) \hookrightarrow \left\{\sum_{i\in I'} a_i \mid a_i \in A_i \wedge I' \subset I \wedge I' \text{ is finite}\right\}$$

The converse inclusion is clear. □

2.3.9 *Definition.* Let $\Lambda = \{A_i \mid i \in I\}$ be a set of submodules $A_i \hookrightarrow M$, then

$$\sum_{i\in I} A_i := \left|\bigcup_{i\in I} A_i\right)$$

is called the *sum of the submodules* $\{A_i \mid i \in I\}$.

If $\Lambda = \{A_1, \ldots, A_n\}$ then every element from $\sum_{j=1}^{n} A_j$ can be written in the form

$$\sum_{j=1}^{n} a_j \quad \text{with} \quad a_j \in A_j,$$

the missing summands a_j can be added as $a_j = 0$. Generally it should be emphasized that the representation $\sum_{i\in I} a_i$ of the elements of the sum need not be unique. If it is unique then a particular case occurs with which we have to concern ourselves in the next section.

We are now able to characterize the maximal submodules of a module.

2.3.10 LEMMA. *Let $A \hookrightarrow\neq M$. Then the following are equivalent:*

(1) *A is a maximal submodule of M.*

(2) $\forall m \in M \quad [m \notin A \Rightarrow M = mR + A]$.

Proof. "(1)⇒(2)": Let $m \notin A$. Then $A \hookrightarrow\neq mR + A$ and hence (2) holds.

"(2)⇒(1)": Let $A \hookrightarrow\neq B \hookrightarrow M$ and let $m \in B$, $m \notin A$. Then $M = mR + A \hookrightarrow B + A \hookrightarrow B \hookrightarrow M$ and thus $B = M$. Hence (1) holds. □

As we have seen, $\mathbb{Q}_{\mathbb{Z}}$ does not have a maximal submodule. In this connection the following theorem is of interest.

2.3.11 THEOREM. *If the module M_R is finitely generated then every proper submodule of M is contained in a maximal submodule of M.*

Proof. Let $\{m_1, \ldots, m_t\}$ be a system of generators of M. Let $A \hookrightarrow\neq M$, then the set

$$\Phi := \{B \mid A \hookrightarrow B \hookrightarrow\neq M\}$$

is non-empty since $A \in \Phi$. Moreover it is also ordered under inclusion. In order to be able to apply Zorn's lemma, we must show that every totally ordered subset $\Gamma \subset \Phi$ possesses an upper bound in Φ. To this end let

$$C := \bigcup_{B \in \Gamma} B,$$

then it follows that $A \hookrightarrow C$. Suppose $C = M$, then we should have $\{m_1, \ldots, m_t\} \subset C$ and it would follow that there must be a $B \in \Gamma$ with $\{m_1, \ldots, m_t\} \subset B$, giving therefore $B = M$ ↯. Thus we have established that $C \in \Phi$. According to Zorn there exists then a maximal element D in Φ. In order to show that D is a maximal submodule of M_R, let $D \hookrightarrow L \hookrightarrow\neq M_R$. Then it follows that $L \in \Phi$ and since D is maximal in Φ it follows that $D = L$. □

If $M \neq 0$ and if M is finitely generated then it follows with $A = 0$ that M has a maximal submodule.

2.3.12 COROLLARY. *Every finitely generated module $M \neq 0$ has a maximal submodule.*

In order to be able to "dualize" the notion of finite generation, we must first of all state an equivalent reformulation.

2.3.13 THEOREM. *The module M_R is finitely generated if and only if there is in every set $\{A_i | i \in I\}$ of submodules $A_i \hookrightarrow M$ with*

$$\sum_{i \in I} A_i = M$$

a finite subset $\{A_i | i \in I_0\}$ (i.e. $I_0 \subset I$ and I_0 is finite) such that

$$\sum_{i \in I_0} A_i = M.$$

Proof. Let M be finitely generated, i.e. $M = m_1 R + \ldots + m_t R$.

Since $\sum_{i \in I} A_i = M$ every m_j is a finite sum of elements from the A_i. Clearly there is a finite subset $I_0 \subset I$ such that

$$m_1, \ldots, m_t \in \sum_{i \in I_0} A_i.$$

Then it follows that

$$M = m_1 R + \ldots + m_t R \hookrightarrow \sum_{i \in I_0} A_i \hookrightarrow M,$$

thus the assertion holds.

To prove the converse we consider the set of submodules $\{mR | m \in M\}$. Then there is a finite subset $\{m_1 R, \ldots, m_t R\}$ with

$$m_1 R + \ldots + m_t R = M,$$

thus M is finitely generated.

We can now formulate the dual notion.

2.3.14 *Definition.* The module M_R is said to be *finitely cogenerated* $:\Leftrightarrow$ for every set $\{A_i | i \in I\}$ of submodules $A_i \hookrightarrow M$ with $\bigcap_{i \in I} A_i = 0$ there is a finite subset $\{A_i | i \in I_0\}$ (i.e. $I_0 \subset I$ and I_0 is finite) with $\bigcap_{i \in I_0} A_i = 0$.

We shall return later to this concept. For the present we may point out two examples.

(1) $\mathbb{Z}_{\mathbb{Z}}$ is not finitely cogenerated since

$$\bigcap_{\text{prime } p} p\mathbb{Z} = 0,$$

but for finitely many primes $p_1, \ldots, p_n$ we have

$$\bigcap_{i=1}^{n} p_i\mathbb{Z} = p_1 \ldots p_n\mathbb{Z} \neq 0$$

(2) A vector space V over a field K is finitely cogenerated if and only if it has finite dimension. The proof is left to the reader as an exercise.

As in the case of vector spaces the *modular law* holds also for modules over an arbitrary ring.

2.3.15 LEMMA (MODULAR LAW). *From A, B, $C \hookrightarrow M$ and $B \hookrightarrow C$ it follows that*

$$(A+B)\cap C = (A\cap C)+(B\cap C) = (A\cap C)+B.$$

Proof. Let $a+b=c\in(A+B)\cap C$ where $a\in A$, $b\in B$, $c\in C$. It then follows from $B\hookrightarrow C$ that $a=c-b\in A\cap C$, thus $a+b=c\in(A\cap C)+B$ and hence $(A+B)\cap C\hookrightarrow(A\cap C)+B$.

Let now $d\in A\cap C$, $b\in B$. Then since $B\hookrightarrow C$ it follows that $d+b\in(A+B)\cap C$ and thus also that $(A\cap C)+B\hookrightarrow(A+B)\cap C$. □

We observe that for A, B, $C\hookrightarrow M$ and without the assumption $B\hookrightarrow C$ we already have

$$(A\cap C)+(B\cap C)\hookrightarrow(A+B)\cap C.$$

However the reverse inclusion does not necessarily hold.

2.4 INTERNAL DIRECT SUMS

2.4.1 M is called the *internal direct sum* of the set $\{B_i \mid i\in I\}$ of submodules $B_i\hookrightarrow M$, in symbols:

$$M=\bigoplus_{i\in I} B_i :\Leftrightarrow \begin{cases} (1)\ M=\sum_{i\in I} B_i \wedge \\ (2)\ \forall j\in I\left[B_j\cap \sum_{\substack{i\in I\\ i\neq j}} B_i=0\right]. \end{cases}$$

$M=\bigoplus_{i\in I} B_i$ is also said to be a *direct decomposition* of M into the sum of the submodules $\{B_i \mid i\in I\}$.

In the case of a finite index set, say $I=\{1,\ldots,n\}$ M is also written as $M=B_1\oplus\ldots\oplus B_n$.

2.4.2 LEMMA. *Let* $\{B_i | i \in I\}$ *be a set of submodules* $B_i \in M$ *and let* $M = \sum_i B_i$. *Then* (2) *of the previous definition is equivalent to*:

For every $x \in M$ *the representation* $x = \sum_{i \in I'} b_i$ *with* $b_i \in B_i$, $I' \subset I$, I' *finite, is unique in the following sense*:

If

$$x = \sum_{i \in I'} b_i = \sum_{i \in I'} c_i \quad \textit{with} \quad b_i, c_i \in B_i,$$

then it follows that

$$\forall i \in I'[b_i = c_i].$$

Proof "⇒": Let (2) hold and let $x = \sum_{i \in I'} b_i = \sum_{i \in I'} c_i$ then it follows that

$$\forall j \in I' \left[b_j - c_j = \sum_{\substack{i \in I' \\ i \neq j}} c_i - b_i \in B_j \cap \sum_{\substack{i \in I' \\ i \neq j}} B_i \right].$$

Since

$$B_j \cap \sum_{\substack{i \in I' \\ i \neq j}} B_i \hookrightarrow B_j \cap \sum_{\substack{i \in I \\ i \neq j}} B_i = 0$$

it follows that $b_j = c_j$ for all $j \in I'$.

"⇐": Let

$$b \in B_j \cap \sum_{\substack{i \in I \\ i \neq j}} B_i,$$

then $b = b_j \in B_j$ and there is a finite subset $I' \subset I$ with $j \notin I'$ so that

$$b = b_j = \sum_{i \in I'} b_i, \qquad b_i \in B_i.$$

If we add to the left-hand side the summands $0 \in B_i$, $i \in I'$ and to the right-hand side the summand $0 \in B_j$, then the same finite index set $I' \cup \{j\}$ appears on both sides and from uniqueness it follows that $b = b_j = 0$, i.e. (2) holds. □

2.4.3 *Definitions*

(1) A submodule $B \hookrightarrow M$ is called a *direct summand* of $M :\Leftrightarrow \exists C \hookrightarrow M[M = B \oplus C]$.

(2) A module $M \neq 0$ is called *directly indecomposable*: $\Leftrightarrow$ 0 and M are the only direct summands of M.

Examples and Remarks

(1) Let $V = V_K$ be a vector space and let $\{x_i \mid i \in I\}$ be a basis of V. Then clearly we have

$$V = \bigoplus_{i \in I} x_i K.$$

Further every subspace of V is a direct summand, as we show later in a more general context.

(2) In $\mathbb{Z}_{\mathbb{Z}}$ the ideal $n\mathbb{Z}$ with $n \neq 0$, $n \neq \pm 1$ is not a direct summand. Suppose $\mathbb{Z} = n\mathbb{Z} \oplus m\mathbb{Z}$. Then $nm \in n\mathbb{Z} \cap m\mathbb{Z}$. Hence $m = 0$ and so $\mathbb{Z} = n\mathbb{Z}$, i.e. $n = \pm 1$ ↯. From this it follows that $\mathbb{Z}_{\mathbb{Z}}$ is directly indecomposable.

(3) Every simple module M is directly indecomposable for it has only 0 and M as submodules.

(4) Every module M which has a largest proper submodule or, in the set of non-zero submodules, a smallest submodule, is directly indecomposable. The proof may be left to the reader.

2.5 FACTOR MODULES AND FACTOR RINGS

The definition of factor modules holds as in the case of factor spaces of linear vector spaces since only properties of linearity are employed in the definition.

Let $C \hookrightarrow M_R$. Then, in particular, C is a subgroup of the additive group of M. Clearly the *factor group* $M/C = \{m + C \mid m \in M\}$ exists under the addition

$$(m_1 + C) + (m_2 + C) := (m_1 + m_2) + C.$$

A module multiplication can now be defined on M/C so that M/C becomes a right module termed a *factor module* or a *residue class module of M modulo C or also of M by C.*

2.5.1 *Definition*

$$(m + C)r := mr + C, \qquad m \in M, r \in R.$$

In order to show that M/C is indeed a right R-module, it is sufficient to show that

$$M/C \times R \to M/C \quad \text{with} \quad (m + C, r) \mapsto mr + C$$

is a mapping, since the other module properties follow directly from those of M.

Let $m_1+C=m_2+C$. Then $m_1=m_2+c$, $c\in C$. Hence $m_1r+C=(m_2+c)r+C=m_2r+cr+C=m_2r+C$.

Factor modules of left modules and of bimodules are defined correspondingly. Let now R be a ring and C a two-sided ideal of R. The factor group of the additive group of R modulo C, R/C, can again be made into a ring which is then called the *factor ring or residue class ring of R modulo C (or by C)*.

2.5.2 *Definition*

$$(r_1+C)(r_2+C) := r_1r_2+C, \qquad r_1, r_2\in R.$$

As before, we see easily that this multiplication is independent of the representatives of the residue classes, i.e. in fact it represents an operation. The other ring properties of R/C again follow immediately from those of R.

If R is a ring with a unit element 1—as is always assumed here—then $1+C$ is the unit element of R/C. We have now to examine some relations between the properties of two-sided ideals and properties of the associated factor rings. For this we need some concepts and simple facts.

2.5.3 *Definition*. Let A, B be two-sided ideals of R. We put

$$AB := (\{ab \mid a\in A \wedge b\in B\}),$$

i.e. AB is the additive group generated by all products ab with $a\in A$, $b\in B$; AB is easily seen to be an ideal and is called the *product of the ideals A and B*.

We then deduce immediately the following.

Remark.

$$AB=\left\{\sum_{j=1}^{n} a_jb_j \mid a_j\in A \wedge b_j\in B \wedge n\in\mathbb{N}\right\}.$$

2.5.4 *Definitions*. Let C be a two-sided ideal of R.

(1) Let C be called a *strongly prime ideal* of R :$\Leftrightarrow$

$$C\neq R \wedge \forall r_1, r_2\in R[r_1r_2\in C \Rightarrow (r_1\in C \vee r_2\in C)].$$

(2) Let C be called a *prime ideal* of R :$\Leftrightarrow$

$$C\neq R \wedge \forall A, B \hookrightarrow {}_RR_R[AB\hookrightarrow C \Rightarrow (A\hookrightarrow C \vee B\hookrightarrow C)],$$

i.e. if the product AB of two two-sided ideals A, B lies in C then at least one of these ideals lies in C.

(3) $r \in R$ is called a *left zero divisor* $:\Leftrightarrow r \neq 0$ and there exists $s \in R$, $s \neq 0$ and $rs = 0$; analogously for a right *zero divisor.*

(4) R is said to have *no zero divisors* $\Leftrightarrow$ there exists no right or left zero divisor in R.

(5) Let $r \in R$; $r' \in R$ is called a *right inverse*, respectively a *left inverse*, respectively an *inverse element* of $r :\Leftrightarrow$

$$rr' = 1, \text{ resp. } r'r = 1 \text{ resp. } rr' = r'r = 1.$$

We remark that from the existence of a right zero divisor it follows that there is a left zero divisor (and conversely). If r' is a right inverse and r'' is a left inverse element of r, then it follows that

$$r' = 1r'' = (r''r)r' = r''(rr') = r''1 = r''.$$

It is also immediate from this that an inverse element (if it exists) is uniquely determined. It is denoted by r^{-1}.

2.5.5 LEMMA

(1) *C is a strongly prime ideal of $R \Rightarrow C$ is a prime ideal of R.*

(2) *If R is commutative then the converse of* (1) *also holds.*

Proof. (1) Let $A, B \hookrightarrow {}_RR_R$ and let $AB \hookrightarrow C$. Suppose $A \not\hookrightarrow C$. Then

$$\exists a_0 \in A[a_0 \notin C].$$

Since $a_0b \in C \wedge a_0 \notin C \Rightarrow b \in C$ for all $b \in B$ it follows that $B \hookrightarrow C$.

(2) Let $r_1, r_2 \in C$. Since R is commutative, r_1R and r_2R are two-sided ideals. Since $r_1r_2 \in C$ it follows that

$$r_1Rr_2R = r_1r_2R \hookrightarrow C.$$

Since C is a prime ideal it follows that

$$r_1R \hookrightarrow C \vee r_2R \hookrightarrow C \quad \text{and so} \quad r_1 \in C \vee r_2 \in C. \qquad \square$$

2.5.6 THEOREM. *Let C be a two-sided ideal of R. Then the following hold:*

(1) *C is a strongly prime ideal in $R \Leftrightarrow R/C$ has no zero divisors.*

(2) *C is a prime ideal in $R \Leftrightarrow$ the zero ideal is a prime ideal in R/C.*

(3) *C is a maximal two-sided ideal in $R \Leftrightarrow R/C$ is simple.*

(4) *C is a maximal right ideal in $R \Leftrightarrow R/C$ is a skew field.*

Proof. (1) "$\Rightarrow$": For brevity put $\bar{R} := R/C$ and $\bar{r} := r + C$. Let $\bar{r}_1, \bar{r}_2 \in \bar{R}$ and suppose $\bar{r}_1\bar{r}_2 = \bar{0}$. Then $r_1r_2 + C = (r_1 + C)(r_2 + C) = C$ and so $r_1r_2 \in C$. Hence $r_1 \in C$ or $r_2 \in C$, i.e. $\bar{r}_1 = \bar{0}$ or $\bar{r}_2 = \bar{0}$.

(1) "$\Leftarrow$": Let $r_1, r_2 \in R$ and suppose $r_1 r_2 \in C$. Then $\bar{r}_1 \bar{r}_2 = (r_1 + C)(r_2 + C) = r_1 r_2 + C = C$. Thus $\bar{r}_1 = \bar{0}$ or $\bar{r}_2 = \bar{0}$, i.e. $r_1 \in C$ or $r_2 \in C$.

(4) "$\Rightarrow$": Let $0 \neq \bar{r} \in \bar{R}$. Then $r \notin C$ and so $R = rR + C$, since, from $r \notin C$, $rR + C$ is a right ideal properly containing C and, from the maximality of C, must be equal to R. Consequently there is $r' \in R$ and $c \in C$ with

$$1 = rr' + c \Rightarrow \bar{1} = rr' + C = (r + C)(r' + C) = \bar{r}\bar{r}',$$

i.e. every element $\neq 0$ of $\bar{R}$ has a right inverse.

From $\bar{1} = \bar{r}\bar{r}' \neq 0$ it follows that $\bar{r}' \neq 0$, thus there exists $\overline{r''} \in \bar{R}$ with $\overline{r' r''} = \bar{1}$. Hence $\bar{r} = \overline{r''}$ and so $\overline{r'}$ is an inverse of $\bar{r}$ and $\bar{R}$ is a skew field.

(4) "$\Leftarrow$": Let $r \in R$ and $r \notin C$. Then $\bar{r} \neq 0$ and so $\exists \bar{r}' \in \bar{R}[\bar{r}\bar{r}' = \bar{r}'\bar{r} = 1]$. Then $rr' + C = 1 + C$ and so, for some $c \in C$, $rr' + c = 1$. Hence $R = rR + C$ which implies that C is a maximal right ideal in R (from 2.3.10). (2) and (3) are proved similarly to (1) and (4). The proof is remitted to the reader as an exercise. Furthermore we shall later get to know of a precise relationship between the lattice of ideals of R and of R/C, from which all assertions of this theorem follow directly.

Examples

(1) Factor spaces of vector spaces are well-known.

(2)

$$\mathbb{Z}/n\mathbb{Z} = \begin{cases} \text{field of } p \text{ elements}, & \text{if } n = p \text{ prime} \\ \text{ring with zero divisors}, & \text{if } n \neq p \wedge n \neq 0 \\ & \qquad\qquad \wedge\, n \neq \pm 1 \\ 0 & \text{if } n = \pm 1 \\ \mathbb{Z} \text{ (up to isomorphism)}, & \text{if } n = 0. \end{cases}$$

(3) Let $K[x]$ be the polynomial ring in the indeterminate x with coefficients in a field K. Let $f(x) \in K[x]$ and let $f(x)$ be irreducible, then $K[x]/f(x)K[x]$ is a finite dimensional extension field of K (more precisely, an extension field of an isomorphic copy of K).

EXERCISES

(1)

Show that in the definition of a module the commutativity of the addition follows from the other assumptions.

(2)

Exhibit a module M without a finite set of generators in which every proper submodule is contained in a maximal submodule.

(3)

(a) Let $A, B, C \hookrightarrow M = M_R$. Show that from $A \subset B \cup C$ it follows that $A \hookrightarrow B \vee A \hookrightarrow C$.

(b) Give an example of a module M and submodules $A, B, C, D \hookrightarrow M_R$ such that

$$A \subset B \cup C \cup D \wedge A \not\hookrightarrow B \wedge A \not\hookrightarrow C \wedge A \underset{\neq}{\hookrightarrow} D.$$

(4)

Let A be a two sided ideal of a ring R. Prove: A is a maximal right ideal $\Leftrightarrow A$ is a maximal left ideal.

(5)

Let

$$M = M_R \wedge x \in M \wedge x \neq 0 \wedge \Lambda := \{A \mid A \hookrightarrow M \wedge x \notin A\}.$$

Prove:

(a) Λ is non-empty and Λ has a maximal element (with respect to inclusion as ordering).

(b) If $R = K$ is a field, then every maximal element from Λ is a maximal submodule of M.

(6)

Exhibit in the set $\Lambda := \{A \mid A \hookrightarrow \mathbb{Q}_{\mathbb{Z}} \wedge 1 \notin A\}$ a maximal element B and a submodule $C \hookrightarrow \mathbb{Q}_{\mathbb{Z}}$ with

$$B \underset{\neq}{\hookrightarrow} C \underset{\neq}{\hookrightarrow} \mathbb{Q}_{\mathbb{Z}}.$$

(7)

Let $\{B_i \mid i = 1, 2, 3, \ldots\}$ be a set of submodules of $M = M_R$ with

$$M = \sum_{i=1}^{\infty} B_i.$$

Prove that the following are equivalent.

(1) $\forall j = 1, 2, \ldots \left[B_j \cap \sum_{i=j+1}^{\infty} B_i = 0 \right].$

(2) $M = \bigoplus_{i=1}^{\infty} B_i.$

(8)

(a) Give an example of a module M with a maximal free subset which is not a set of generators.

(b) Give an example of a module $\neq 0$ which is not a vector space and in which every maximal free subset is a basis. (Hint: use a suitable $\mathbb{Z}$-module.)

(9)

Let $V = V_K$ be a vector space, let X be a free subset of V and let Y be a set of generators of V with $X \subset Y$. Show: there exists a basis Z of V with $X \subset Z \subset Y$.

(10)

(a) Exhibit a module M and a submodule $A \hookrightarrow M$ such that there exist different submodules $B_1 \hookrightarrow M$, $B_2 \hookrightarrow M$ with

$$M = A \oplus B_1 = A \oplus B_2.$$

(b) Obtain an example of a module M which is not simple and in which for every submodule $A \hookrightarrow M$ there exists exactly one $B \hookrightarrow M$ with $M = A \oplus B$.

(11)

Let X be a finite set, $X = \{x_1, \ldots, x_n\}$, and let $R := \mathbb{R}^X$ be the set of all mappings $f: X \to \mathbb{R}$ (where $\mathbb{R}$ is the field of real numbers). Prove the following.

(a) R is a commutative ring under the following definitions:

$$\begin{aligned}(f+g)(x_i) &= f(x_i)+g(x_i)\\ (f\circ g)(x_i) &= f(x_i)\cdot g(x_i)\end{aligned} \qquad (f, g \in R,\ i = 1, \ldots, n).$$

(b) R is a principal ideal ring.

(c) Every ideal is an intersection of maximal ideals and the intersection of all maximal ideals is 0.

(d) Every ideal is a direct summand.

(e) R is a direct sum of simple ideals.

(12)

Let $\{A_i \mid i \in I\}$ be a set of submodules of a module M and let $B \hookrightarrow M$.

(a) Prove $\sum_{i\in I} (A_i \cap B) \hookrightarrow \left(\sum_{i\in I} A_i\right) \cap B$.

(b) Prove $\left(\bigcap_{i\in I} A_i\right) + B \hookrightarrow \bigcap_{i\in I} (A_i + B)$.

(c) Give an example for which there holds:

$$\sum_{i\in I}(A_i \cap B) \neq \left(\sum_{i\in I} A_i\right) \cap B.$$

(d) Give an example for which there holds:

$$\left(\bigcap_{i\in I} A_i\right) + B \neq \bigcap_{i\in I}(A_i + B).$$

(13)

Definition. A ring R is called *regular* (in the sense of von Neumann):

$$:\Leftrightarrow \forall r \in R\, \exists r' \in R\,[rr'r = r].$$

Prove: the following conditions are equivalent.

(1) R is regular.
(2) Every cyclic right ideal of R is a direct summand of R_R.
(3) Every cyclic left ideal of R is a direct summand of $_RR$.
(4) Every finitely generated right ideal of R is a direct summand of R_R.
(5) Every finitely generated left ideal of R is a direct summand of $_RR$.

Chapter 3

Homomorphisms of Modules and Rings

3.1 DEFINITIONS AND SIMPLE PROPERTIES

The structure-preserving mappings of modules are called homomorphisms. These are defined in the same way as are linear mappings of linear vector spaces.

3.1.1 *Definition.* Let A and B be both right R-modules or left S-modules or S-R-bimodules respectively. A *homomorphism* α *of* A *into* B is a mapping

$$\alpha: A \to B$$

which satisfies

(1) $\forall a_1, a_2 \in A \forall r_1, r_2 \in R[\alpha(a_1 r_1 + a_2 r_2) = \alpha(a_1) r_1 + \alpha(a_2) r_2]$ or

(2) $\forall a_1, a_2 \in A \forall s_1, s_2 \in S[\alpha(s_1 a_1 + s_2 a_2) = s_1 \alpha(a_1) + s_2 \alpha(a_2)]$ or

(3) $\forall a_1, a_2 \in A \forall s_1, s_2 \in S \forall r_1, r_2 \in R[\alpha(s_1 a_1 r_1 + s_2 a_2 r_2) = s_1 \alpha(a_1) r_1 + s_2 \alpha(a_2) r_2]$.

respectively.

The notation

$$\alpha: A_R \to B_R$$

indicates that A and B are right R-modules and that α is a homomorphism. Analogously for the other cases. To emphasize the ring and also the side involved in a homomorphism $\alpha: A_R \to B_R$ we shall also speak of α as an *R-module homomorphism* or a *right module homomorphism.* Instead of the notation $\alpha(a)$ for the image of $a \in A$ by α we shall also write merely αa. In the case of $\alpha: {}_S A \to {}_S B$ let $a\alpha$ denote the image of a by α; then the equation in (2) assumes the following form:

$$(s_1 a_1 + s_2 a_2)\alpha = s_1(a_1 \alpha) + s_2(a_2 \alpha).$$

A homomorphism is thus written on the side opposite to the operation of the ring. If there is to be any deviation from this notational rule we shall especially indicate it. Generally for a mapping $\alpha: A \to B$ we use the symbol $a \mapsto \alpha(a)$ for the elements in correspondence; we combine $\alpha: A \to B$ and $a \mapsto \alpha(a)$ in the following notation:

$$\alpha: A \ni a \mapsto \alpha(a) \in B,$$

which we have already used in Chapter 1. The following notions, which are customary, are also used for homomorphisms:

$$\textit{Domain of } \alpha = \mathrm{Dom}(\alpha) := A.$$
$$\textit{Codomain of } \alpha = \mathrm{Cod}(\alpha) := B.$$

$$\textit{Image of } \alpha = \mathrm{Im}(\alpha) := \{\alpha(a) \mid a \in A\}.$$

α is an *injection* $:\Leftrightarrow \forall a_1, a_2 \in A[\alpha(a_1) = \alpha(a_2) \Rightarrow a_1 = a_2]$

(i.e. α is one-one).

α is a *surjection* $:\Leftrightarrow \mathrm{Im}\,\alpha = \mathrm{Cod}(\alpha)$

(i.e. α is a mapping "onto").

α is a *bijection* $:\Leftrightarrow \alpha$ is an injection $\wedge$ α is a surjection

In the following, if we speak of homomorphisms of modules without indicating the side then the concepts and considerations are to be regarded as holding for a one-sided module. All is exemplified only for right modules, where it is clear that everything holding for right modules holds, as appropriate, for left modules. In the main everything remains valid for bimodules, but there is no need to pursue this in detail.

Examples of homomorphisms

(1) The 0-homomorphism of A into B:

$$0: A \ni a \mapsto 0 \in B.$$

(2) The identity injection = inclusion of a submodule $A \hookrightarrow B$

$$\iota: A \ni a \mapsto a \in B.$$

(3) The natural (canonical) homomorphism of a module A onto the factor module A/C, where $C \hookrightarrow A$:

$$\nu: A \ni a \mapsto a + C \in A/C.$$

It is immediately clear in cases 1 and 2 that we are in fact considering homomorphisms; for ν it follows directly from the definition of the module A/C:

$$\nu(a_1r_1+a_2r_2)=(a_1r_1+a_2r_2)+C=(a_1r_1+C)+(a_2r_2+C)$$
$$=(a_1+C)r_1+(a_2+C)r_2=\nu(a_1)r_1+\nu(a_2)r_2.$$

The homomorphisms 0, ι, ν are used always in the following with the same meaning but with changing notations for domain and codomain. For the identity mapping of a module A, which is a special case of inclusion, we write 1_A.

Let α and β be homomorphisms with $\mathrm{Cod}(\alpha)=\mathrm{Dom}(\beta)$. Suppose

$$\alpha: A\to B, \qquad \beta: B\to C,$$

then the composition of the mappings α, β, denoted by $\beta\alpha$, is obviously again a homomorphism in fact of A into C. For $a\in A$ we then have $(\beta\alpha)a=\beta(\alpha a)$.

As is easily seen, a mapping $\alpha: A\to B$ is a bijection precisely if there exists a (uniquely determined) inverse mapping $\alpha^{-1}: B\to A$ with $\alpha^{-1}\alpha=1_A$, $\alpha\alpha^{-1}=1_B$. If α is a bijective homomorphism then α^{-1} is also a homomorphism: let $b_1=\alpha(a_1)$, $b_2=\alpha(a_2)$ be arbitrary elements from B and let r_1, $r_2\in R$, then we have

$$\alpha^{-1}(b_1r_1+b_2r_2)=\alpha^{-1}(\alpha(a_1)r_1+\alpha(a_2)r_2)$$
$$=\alpha^{-1}(\alpha(a_1r_1+a_2r_2))=a_1r_1+a_2r_2$$
$$=\alpha^{-1}(b_1)r_1+\alpha^{-1}(b_2)r_2.$$

In the following let $\alpha: A\to B$ always denote a homomorphism.

For $U\subset A$, $V\subset B$, there is defined:

$$\alpha(U):=\{\alpha(u)\mid u\in U\}$$
$$\alpha^{-1}(V):=\{a\mid a\in A\wedge\alpha(a)\in V\}.$$

We remark that α^{-1} is itself in general not defined, if it is, then α is bijective.

3.1.2 LEMMA

(1) $U\hookrightarrow A\Rightarrow\alpha(U)\hookrightarrow B$.

(2) $V\hookrightarrow B\Rightarrow\alpha^{-1}(V)\hookrightarrow A$.

Proof. (1) Let u_1, $u_2\in U$, thus

$$\alpha(u_1),\alpha(u_2)\in\alpha(U)\wedge r_1, r_2\in R\Rightarrow\alpha(u_1)r_1+\alpha(u_2)r_2=\alpha(u_1r_1+u_2r_2)\in\alpha(U),$$

since $u_1r_1+u_2r_2\in U$.

(2) Let $a_1, a_2 \in \alpha^{-1}(V)$, thus

$$\alpha(a_1), \alpha(a_2) \in V \wedge r_1, r_2 \in R \Rightarrow \alpha(a_1 r_1 + a_2 r_2) = \alpha(a_1) r_1 + \alpha(a_2) r_2 \in V$$
$$\Rightarrow a_1 r_1 + a_2 r_2 \in \alpha^{-1}(V). \qquad \square$$

3.1.3 *Definition*

Kernel of $\alpha = \mathrm{Ker}(\alpha) := \alpha^{-1}(0)$.

Image of $\alpha = \mathrm{Im}(\alpha) := \alpha(A)$.

Cokernel of $\alpha = \mathrm{Coker}(\alpha) := \mathrm{Cod}(\alpha)/\mathrm{Im}(\alpha) = B/\alpha(A)$.

Coimage of $\alpha = \mathrm{Coim}(\alpha) := \mathrm{Dom}(\alpha)/\mathrm{Ker}(\alpha) = A/\alpha^{-1}(0)$.

We had previously introduced $\mathrm{Im}(\alpha)$. By virtue of 3.1.2 we know that $\mathrm{Ker}(\alpha)$ and $\mathrm{Im}(\alpha)$ are submodules so that the definitions of Cokernel and Coimage are meaningful.

For the category $\boldsymbol{M}_R$ of right R-modules, which was introduced in 1.2.5 (recall, that all modules are now unitary), we make use of all of the notation from Chapter 1. In particular, by employing the concepts from 1.1.3, we now wish to characterize injective, surjective and bijective homomorphisms. First of all we repeat these concepts for the category $\boldsymbol{M}_R$.

3.1.4 *Definition.* A homomorphism $\alpha: A_R \to B_R$ is called a *monomorphism* $\Leftrightarrow$

$$\forall C \in \boldsymbol{M}_R \forall \gamma_1, \gamma_2 \in \mathrm{Hom}_R(C, A)[\alpha\gamma_1 = \alpha\gamma_2 \Rightarrow \gamma_1 = \gamma_2];$$

an *epimorphism* $\Leftrightarrow$

$$\forall C \in \boldsymbol{M}_R \forall \beta_1, \beta_2 \in \mathrm{Hom}_R(B, C)[\beta_1\alpha = \beta_2\alpha \Rightarrow \beta_1 = \beta_2];$$

a *bimorphism* $\Leftrightarrow$

$$\alpha \text{ is an epimorphism} \wedge \alpha \text{ is a monomorphism};$$

an *isomorphism* $\Leftrightarrow$

$$\exists \alpha' \in \mathrm{Hom}_R(B, A)[\alpha'\alpha = 1_A \wedge \alpha\alpha' = 1_B].$$

3.1.5 THEOREM. *Let $\alpha: A \to B$ be a homomorphism, then we have:*
(1) *α is an injection $\Leftrightarrow$ α is a monomorphism.*
(2) *α is a surjection $\Leftrightarrow$ α is an epimorphism.*
(3) *α is a bijection $\Leftrightarrow$ α is a bimorphism*
$\Leftrightarrow$ α is an isomorphism.

Proof. (1) "$\Rightarrow$": Let $\alpha\gamma_1 = \alpha\gamma_2$ with $\gamma_1, \gamma_2 \in \mathrm{Hom}_R(C, A)$. Suppose $\gamma_1 \neq \gamma_2$ Then

$$\exists c \in C[\gamma_1(c) \neq \gamma_2(c)].$$

Hence

$$\alpha(\gamma_1(c)) \neq \alpha(\gamma_2(c))$$

and so

$$\alpha\gamma_1 \neq \alpha\gamma_2 \quad ↯.$$

Thus $\gamma_1 = \gamma_2$ must hold.

(1) "$\Leftarrow$": Let $\alpha(a_1) = \alpha(a_2)$. Then $\alpha(a_1) - \alpha(a_2) = \alpha(a_1 - a_2) = 0$. Let

$$\gamma_1 = \iota\colon (a_1 - a_2)R \ni (a_1 - a_2)r \mapsto (a_1 - a_2)r \in A$$

$$\gamma_2 = 0\colon (a_1 - a_2)R \ni (a_1 - a_2)r \mapsto 0 \in A,$$

Then

$$\gamma_1, \gamma_2 \in \mathrm{Hom}_R((a_1 - a_2)R, A)$$

and we have

$$\alpha(\gamma_1((a_1 - a_2)r)) = \alpha((a_1 - a_2)r) = \alpha(a_1 - a_2)r = 0$$

$$\alpha(\gamma_2((a_1 - a_2)r)) = \alpha(0) = 0$$

i.e. $\alpha\gamma_1 = \alpha\gamma_2$. By assumption it follows that

$$\gamma_1 = \gamma_2 \Rightarrow \gamma_1(a_1 - a_2) = a_1 - a_2 = \gamma_2(a_1 - a_2) = 0 \Rightarrow a_1 = a_2.$$

(2) "$\Rightarrow$": Let $\beta_1\alpha = \beta_2\alpha$ with $\beta_1, \beta_2 \in \mathrm{Hom}_R(B, C)$. Suppose

$$\beta_1 \neq \beta_2 \Rightarrow \exists b \in B[\beta_1(b) \neq \beta_2(b)].$$

Since α is surjective, there exists $a \in A$ such that $\alpha(a) = b$. Hence

$$\beta_1\alpha(a) = \beta_1(b) \neq \beta_2(b) = \beta_2\alpha(a) \Rightarrow \beta_1\alpha \neq \beta_2\alpha \quad ↯.$$

Thus $\beta_1 = \beta_2$ must hold.

(2) "$\Leftarrow$": Let

$$\beta_1 = \nu\colon B \to B/\mathrm{Im}(\alpha)$$

$$\beta_2 = 0\colon B \to B/\mathrm{Im}(\alpha).$$

Then $\beta_1, \beta_2 \in \mathrm{Hom}_R(B, B/\mathrm{Im}(\alpha))$ and $\beta_1\alpha = \beta_2\alpha = 0$. By assumption it follows then that $\beta_1 = \beta_2$ i.e. $B = \mathrm{Im}(\alpha)$ and consequently α is surjective.

(3) "bijection$\Leftrightarrow$bimorphism" follows from (1) and (2). Further it is clear that every bijection is an isomorphism since, as we have previously shown,

if α is a bijection then α^{-1} is a homomorphism. Conversely let α be an isomorphism. Then it follows from $\alpha'\alpha = 1_A$ that α is injective and from $\alpha'\alpha = 1_B$ that α is surjective. (Obviously, then, $\alpha^{-1} = \alpha'$.) □

3.1.6 LEMMA. *Let $\alpha: A \to B$ and $\beta: B \to C$ be homomorphisms. Then we have:*

α, β are monomorphisms $\Rightarrow \beta\alpha$ is a monomorphism.
α, β are epimorphisms $\Rightarrow \beta\alpha$ is an epimorphism.
$\beta\alpha$ is a monomorphism $\Rightarrow \alpha$ is a monomorphism.
$\beta\alpha$ is an epimorphism $\Rightarrow \beta$ is an epimorphism.

Proof. (1) Let $\gamma_1, \gamma_2 \in \mathrm{Hom}_R(M, A)$. Since β and α are monomorphisms, then we have: $\beta\alpha\gamma_1 = \beta\alpha\gamma_2 \Rightarrow \alpha\gamma_1 = \alpha\gamma_2 \Rightarrow \gamma_1 = \gamma_2$; thus $\beta\alpha$ is a monomorphism. Analogously for epimorphisms.

(2) Let again $\gamma_1, \gamma_2 \in \mathrm{Hom}_R(M, A)$. Since $\beta\alpha$ is a monomorphism, then we have: $\alpha\gamma_1 = \alpha\gamma_2 \Rightarrow \beta\alpha\gamma_1 = \beta\alpha\gamma_2 \Rightarrow \gamma_1 = \gamma_2$; thus α is a monomorphism. Analogously for epimorphisms. □

3.1.7 *Definition.* Two modules A, B are called *isomorphic*, notationally $A \cong B :\Leftrightarrow$ there exists an isomorphism $\alpha: A \to B$.

REMARK. *$\cong$ is an equivalence relation of the class of all right R-*modules.

Proof

(1) $A \cong A$, since 1_A is an isomorphism.

(2) Let $\alpha: A \to B$ be an isomorphism. Then $\alpha^{-1}: B \to A$ is an isomorphism, i.e., from $A \cong B$ it follows that $B \cong A$.

(3) Let $\alpha: A \to B$, $\beta: B \to C$ be isomorphisms. Then so is $\beta\alpha$ since $\alpha^{-1}\beta^{-1}\beta\alpha = 1_A$ and $\beta\alpha\alpha^{-1}\beta^{-1} = 1_C$, i.e. from $A \cong B$ and $B \cong C$ it follows that $A \cong C$. □

3.1.8. LEMMA. *Let $\alpha: A \to B$ be a homomorphism. Then we have:*

(1) *α is a monomorphism* $\Leftrightarrow \mathrm{Ker}(\alpha) = 0$.

(2) $U \hookrightarrow A \Rightarrow \alpha^{-1}(\alpha(U)) = U + \mathrm{Ker}(\alpha)$.

(3) $V \hookrightarrow B \Rightarrow \alpha(\alpha^{-1}(V)) = V \cap \mathrm{Im}(\alpha)$.

(4) *Let also $\beta: B \to C$ be a homomorphism. Then*

$$\mathrm{Ker}(\beta\alpha) = \alpha^{-1}(\mathrm{Ker}(\beta)) \wedge \mathrm{Im}(\beta\alpha) = \beta(\mathrm{Im}(\alpha)).$$

Proof. (1) "$\Rightarrow$": α is a monomorphism $\Rightarrow \alpha$ is an injection (from 3.1.5)$\Rightarrow$ $\mathrm{Ker}(\alpha) = 0$ (for $\alpha(0) = 0$).

(1) "$\Leftarrow$": Let $\alpha(a_1) = \alpha(a_2)$.

Then $\alpha(a_1 - a_2) = 0 \Rightarrow a_1 - a_2 \in \text{Ker}(\alpha) = 0 \Rightarrow a_1 = a_2$. Hence α is an injection $\Rightarrow$ α is a monomorphism (from 3.1.5).

(2) "$\alpha^{-1}(\alpha(U)) \hookrightarrow U + \text{Ker}(\alpha)$": Let $a \in \alpha^{-1}(\alpha(U))$. Then

$$\alpha(a) \in \alpha(U) \quad \text{and so} \quad \exists u \in U[\alpha(a) = \alpha(u)].$$

Then

$$\alpha(a-u) = 0 \Rightarrow a - u \in \text{Ker}(\alpha) \Rightarrow a \in U + \text{Ker}(\alpha).$$

(2) "$U + \text{Ker}(\alpha) \hookrightarrow \alpha^{-1}(\alpha(U))$": Let $u \in U$ and $k \in \text{Ker}(\alpha)$. Then

$$\alpha(u+k) = \alpha(u) + \alpha(k) = \alpha(u) + 0 = \alpha(u) \in \alpha(U).$$

Hence $u + k \in \alpha^{-1}(\alpha(U))$.

(3) Exercise for the reader.

(4) $a \in \text{Ker}(\beta\alpha) \Leftrightarrow \beta\alpha(a) = 0 \Leftrightarrow \alpha(a) \in \text{Ker}(\beta) \Leftrightarrow a \in \alpha^{-1}(\text{Ker}(\beta))$.

$$\text{Im}(\beta\alpha) = \beta\alpha(A) = \beta(\alpha(A)) = \beta(\text{Im}(\alpha)).$$ □

From the lemma there follows directly:

Let $U \hookrightarrow A$ and let α be a monomorphism $\alpha: A \to B$. Thus $U = \alpha^{-1}(\alpha(U))$, i.e. we obtain every submodule U of A in the form $\alpha^{-1}(V)$ with $V \hookrightarrow B$ (substitute $V = \alpha(U)$); let $V \hookrightarrow B$ and let α be an epimorphism $\alpha: A \to B$. Thus $V = \alpha(\alpha^{-1}(V))$, i.e. we obtain every submodule V of B in the form $\alpha(U)$ with $U \hookrightarrow A$ (substitute $U = \alpha^{-1}(V)$).

In the following use is made as need arises of both of these facts without specific mention.

3.1.9 COROLLARY. *If*

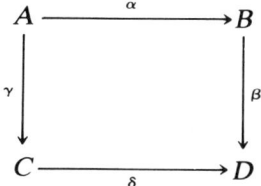

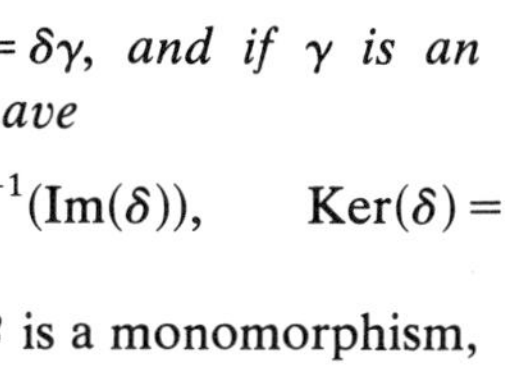

is commutative, i.e. $\beta\alpha = \delta\gamma$, *and if* γ *is an epimorphism and* β *is a monomorphism, then we have*

$$\text{Im}(\alpha) = \beta^{-1}(\text{Im}(\delta)), \qquad \text{Ker}(\delta) = \gamma(\text{Ker}(\alpha)).$$

Proof. From 3.1.8, since β is a monomorphism,

$$\begin{aligned}\text{Im}(\alpha) = \beta^{-1}(\beta(\text{Im}(\alpha)) \Rightarrow \text{Im}(\alpha) &= \beta^{-1}(\text{Im}(\beta\alpha)) \\ &= \beta^{-1}(\text{Im}(\delta\gamma)) = \beta^{-1}(\text{Im}(\delta))\end{aligned}$$

since γ is an epimorphism. Further from 3.1.8, since γ is an epimorphism $\operatorname{Ker}(\delta)=\gamma(\gamma^{-1}(\operatorname{Ker}(\delta)))$. Thus from 3.1.8 $\operatorname{Ker}(\delta)=\gamma(\operatorname{Ker}(\delta\gamma))$ and so $\operatorname{Ker}(\delta)=\gamma(\operatorname{Ker}(\beta\alpha))=\gamma(\operatorname{Ker}(\alpha))$, since β is a monomorphism. □

We apply ourselves now to the question of the behaviour of sums and intersections of submodules with respect to homomorphisms and inverse mappings (for this see also Exercise 1).

3.1.10 LEMMA. *Let a homomorphism $\alpha: A \to B$ be given together with a set $\{A_i \mid i \in I\}$ of $A_i \hookrightarrow A$ and a set $\{B_i \mid i \in I\}$ of $B_i \hookrightarrow B$. Then we have*

(a) $$\alpha\Big(\sum_{i\in I} A_i\Big)=\sum_{i\in I}\alpha(A_i), \qquad \alpha^{-1}\Big(\bigcap_{i\in I} B_i\Big)=\bigcap_{i\in I}\alpha^{-1}(B_i).$$

(b) $$\alpha^{-1}\Big(\sum_{i\in I} B_i\Big)\hookleftarrow\sum_{i\in I}\alpha^{-1}(B_i), \qquad \alpha\Big(\bigcap_{i\in I} A_i\Big)\hookrightarrow\bigcap_{i\in I}\alpha(A_i).$$

(c) *Let now $B_i \hookrightarrow \operatorname{Im}(\alpha)$ for all $i \in I$, then we have*

$$\alpha^{-1}\Big(\sum_{i\in I} B_i\Big)=\sum_{i\in I}\alpha^{-1}(B_i).$$

Let now $\operatorname{Ker}(\alpha) \hookrightarrow A_i$ for all $i \in I$, then we have

$$\alpha\Big(\bigcap_{i\in I} A_i\Big)=\bigcap_{i\in I}\alpha(A_i).$$

Proof. The assertions in (a) and (b) are easy to verify and are left to the reader as an exercise. It remains to prove (c). From consideration of (a) and 3.1.8 it follows that:

$$\begin{aligned}\alpha^{-1}\Big(\sum_{i\in I} B_i\Big)&=\alpha^{-1}\Big(\sum_{i\in I}(B_i\cap\operatorname{Im}(\alpha))\Big)=\alpha^{-1}\Big(\sum_{i\in I}\alpha\alpha^{-1}(B_i)\Big)\\&=\alpha^{-1}\alpha\Big(\sum_{i\in I}\alpha^{-1}(B_i)\Big)=\Big(\sum_{i\in I}\alpha^{-1}(B_i)\Big)+\operatorname{Ker}(\alpha)\\&=\sum_{i\in I}\alpha^{-1}(B_i)\end{aligned}$$

and also

$$\begin{aligned}\alpha\Big(\bigcap_{i\in I} A_i\Big)&=\alpha\Big(\bigcap_{i\in I}(A_i+\operatorname{Ker}(\alpha))\Big)=\alpha\Big(\bigcap_{i\in I}\alpha^{-1}\alpha(A_i)\Big)\\&=\alpha\alpha^{-1}\Big(\bigcap_{i\in I}\alpha(A_i)\Big)=\Big(\bigcap_{i\in I}\alpha(A_i)\Big)\cap\operatorname{Im}(\alpha)=\bigcap_{i\in I}\alpha(A_i).\end{aligned}$$

□

3.1.11 COROLLARY. *Let $U_R \hookrightarrow M_R$, then we have: M/U is finitely cogenerated (2.3.14)$\Leftrightarrow$in every set $\{A_i | i \in I\}$ of submodules $A_i \hookrightarrow M$ with*

$$\bigcap_{i \in I} A_i = U$$

there is a finite subset $\{A_i | i \in I_0\}$ (i.e. I_0 finite) with

$$\bigcap_{i \in I_0} A_i = U.$$

Proof. "$\Rightarrow$": Let $\nu: M \to M/U$ denote the natural epimorphism. $\bigcap_{i \in I} A_i = U$ implies that $U = \mathrm{Ker}(\nu) \hookrightarrow A_i$ so that 3.1.10(c) can be applied. Therefore it follows that

$$\bigcap_{i \in I} \nu(A_i) = \nu\Big(\bigcap_{i \in I} A_i\Big) = \nu(U) = 0 \hookrightarrow N/U.$$

By assumption there is then a finite subset $I_0 \subset I$ with

$$\bigcap_{i \in I_0} \nu(A_i) = 0.$$

Then it follows from 3.1.10(a) that

$$\nu^{-1}(0) = U = \nu^{-1}\Big(\bigcap_{i \in I_0} \nu(A_i)\Big) = \bigcap_{i \in I_0} \nu^{-1}\nu(A_i) = \bigcap_{i \in I_0} (A_i + U) = \bigcap_{i \in I_0} A_i.$$

"$\Leftarrow$": Let now $\{\Lambda_i | i \in I\}$ be a set of submodules $\Lambda_i \hookrightarrow M/U$ with

$$\bigcap_{i \in I} \Lambda_i = 0.$$

Then it follows from 3.1.10(a) that

$$\nu^{-1}(0) = U = \nu^{-1}\Big(\bigcap_{i \in I} \Lambda_i\Big) = \bigcap_{i \in I} \nu^{-1}(\Lambda_i).$$

By assumption there is a finite subset $I_0 \subset I$ with

$$\bigcap_{i \in I_0} \nu^{-1}(\Lambda_i) = U.$$

From $U = \mathrm{Ker}(\nu) \hookrightarrow \nu^{-1}(\Lambda_i)$ it follows from 3.1.10(c) that

$$\nu\Big(\bigcap_{i \in I_0} \nu^{-1}(\Lambda_i)\Big) = \bigcap_{i \in I_0} \nu\nu^{-1}(\Lambda_i) = \bigcap_{i \in I_0} (\Lambda_i \cap \mathrm{Im}(\nu)) = \bigcap_{i \in I_0} \Lambda_i = \nu(U) = 0.$$

□

A *lattice*, respectively a *complete lattice*, is an ordered set, in which every two-element subset, respectively subset, has an infimum and a supremum.

The set of all submodules of a module is, under $\hookrightarrow$ as order-relation, a complete lattice, in which the infimum is the intersection and the supremum is the sum of the submodules. Let now A_R be given, then denote the lattice of submodules of A by Lat(A). Let $\alpha: A \to L$ be a homomorphism, and let C denote Ker(α), N denote Im(α). Then we consider the sublattice

$$\text{Lat}(A, \bar{C}) := \{U \mid C \hookrightarrow U \hookrightarrow A\}$$

of Lat(A) and the sublattice

$$\text{Lat}(L, N) := \{V \mid V \hookrightarrow N\} (= \text{Lat}(N))$$

of Lat(L). With these notations the following relationship holds.

3.1.12 LEMMA. *A bijection $\hat{\alpha}$ is defined by*

$$\hat{\alpha}: \text{Lat}(A, \bar{C}) \ni U \mapsto \alpha(U) \in \text{Lat}(L, N)$$

with respect to which there holds:
(1) $\hat{\alpha}(U_1 + U_2) = \hat{\alpha}(U_1) + \hat{\alpha}(U_2)$
(2) $\hat{\alpha}(U_1 \cap U_2) = \hat{\alpha}(U_1) \cap \hat{\alpha}(U_2)$,

which means that $\hat{\alpha}$ is a lattice isomorphism between

$$\text{Lat}(\text{Dom}(\alpha), \overline{\text{Ker}(\alpha)}) \quad \text{and} \quad \text{Lat}(\text{Cod}(\alpha), \text{Im}(\alpha)) = \text{Lat}(\text{Im}(\alpha)).$$

Proof. For this proof we use 3.1.8.

"$\hat{\alpha}$ Injective": Let $\alpha(U_1) = \alpha(U_2)$ hold for

$$U_1, U_2 \in \text{Lat}(A, \bar{C}).$$

Then

$$\alpha^{-1}(\alpha(U_1)) = U_1 + \text{Ker}(\alpha) = \alpha^{-1}(\alpha(U_2)) = U_2 + \text{Ker}(\alpha).$$

From

$$\text{Ker}(\alpha) = C \hookrightarrow U_i \qquad (i = 1, 2)$$

it follows that $U_1 = U_2$.

"$\hat{\alpha}$ Surjective": Let $V \hookrightarrow N = \text{Im}(\alpha)$.

Then

$$\alpha^{-1}(0) = C \hookrightarrow \alpha^{-1}(V) \hookrightarrow A \wedge \alpha(\alpha^{-1}(V)) = V \cap N = V,$$

i.e. $\hat{\alpha}(\alpha^{-1}(V)) = V$.

(1) $\hat{\alpha}(U_1 + U_2) = \alpha(U_1 + U_2) = \alpha(U_1) + \alpha(U_2) = \hat{\alpha}(U_1) + \hat{\alpha}(U_2)$.

(2) Trivially we have

$$\hat{\alpha}(U_1 \cap U_2) \hookrightarrow \hat{\alpha}(U_1) \cap \hat{\alpha}(U_2).$$

Let now $x \in \hat{\alpha}(U_1) \cap \hat{\alpha}(U_2)$, i.e. $x = \alpha(u_1) = \alpha(u_2)$ with

$$u_1 \in U_1,\ u_2 \in U_2.$$

Then

$$\alpha(u_1 - u_2) = 0 \Rightarrow u_1 - u_2 = c \in \operatorname{Ker}(\alpha) = C \Rightarrow u_1 = u_2 + c.$$

From $C \hookrightarrow U_2$ it follows that

$$u_1 = u_2 + c \in U_1 \cap U_2$$

and so

$$x = \alpha(u_1) \in \alpha(U_1 \cap U_2) \Rightarrow \hat{\alpha}(U_1) \cap \hat{\alpha}(U_2) \hookrightarrow \hat{\alpha}(U_1 \cap U_2). \qquad \square$$

3.1.13 COROLLARY. *Let $C \hookrightarrow A$ and let $\nu: A \to A/C$. Then*

$$\hat{\nu}: \operatorname{Lat}(A, \bar{C}) \ni U \mapsto \nu(U) \in \operatorname{Lat}(A/C)$$

is a lattice isomorphism.

3.1.14 COROLLARY. *Maximal $C \hookrightarrow A \Leftrightarrow A/C$* is simple.

As an exercise the reader may give a new and complete proof of 2.5.6.

3.2 RING HOMOMORPHISMS

We now make some remarks on ring homomorphisms.

3.2.1 *Definition.* Let R and S be rings. Then a *ring homomorphism*

$$\rho: R \to S$$

is a mapping, for which for all $r_1, r_2 \in R$ we have:

$$\rho(r_1 + r_2) = \rho(r_1) + \rho(r_2),$$

$$\rho(r_1 r_2) = \rho(r_1)\rho(r_2).$$

ρ is called *unitary*, if—as is here always assumed—R and S are rings with unit elements and ρ maps the unit element of R onto that of S.

For the category of rings we also use the concepts introduced in 1.1.3.

3.2.2 LEMMA. *Let $\rho: R \to S$ be a ring homomorphism. Then there holds:*

(1) *ρ is an injection $\Rightarrow \rho$ is a monomorphism.*
(2) *ρ is a surjection $\Rightarrow \rho$ is an epimorphism.*
(3) *ρ is a bijection $\Leftrightarrow \rho$ is an isomorphism.*
$\Rightarrow \rho$ is a bimorphism.

Proof. As in the proof of 3.1.5. It should be stressed that the converse of (1) does indeed hold but not the converse of (2) and (3) (see exercises). In this respect the category of rings differs from that of modules.

We call two rings R and S *isomorphic*, notationally $R \cong S$, if there exists an isomorphism of R with S. Obviously $\cong$ is an equivalence relation in the class of all rings. An isomorphism of R with R is called an *automorphism*.

As for modules there exist ring homomorphisms ι and ν as well as 0, in the case that the zero ring is admitted. Let C be a two-sided ideal in a ring R, then ν is defined by

$$\nu: R \ni r \mapsto r + C \in R/C$$

where R/C is the residue class ring (2.5.2). Further it is clear that the image of a (unitary) subring with respect to a (unitary) ring homomorphism ρ is again a (unitary) subring of $\mathrm{Cod}(\rho)$. In particular $\mathrm{Im}(\rho)$ is a (unitary) subring of $\mathrm{Cod}(\rho)$.

For the most part the ideals of a ring are more important than the subrings. Consequently we establish

3.2.3 LEMMA. *Let $\rho: R \to S$ be a ring homomorphism and let V be a two-sided ideal in S, then $\overset{-1}{\rho}(V)$ is a two-sided ideal in R.*

Proof. Let $u_1, u_2 \in \overset{-1}{\rho}(V)$ and $r \in R$, then we have

$$\rho(u_1 + u_2) = \rho(u_1) + \rho(u_2) \in V \Rightarrow u_1 + u_2 \in \overset{-1}{\rho}(V),$$

$$\rho(u_1 r) = \rho(u_1)\rho(r) \in V \Rightarrow u_1 r \in \overset{-1}{\rho}(V)$$

and analogously

$$r u_1 \in \overset{-1}{\rho}(V) \Rightarrow \overset{-1}{\rho}(V)$$

is a two-sided ideal in R. □

It follows from the lemma that $\mathrm{Ker}(\rho)$ is a two-sided ideal in R for which the residue class ring $R/\mathrm{Ker}(\rho)$ exists. As a special case $\mathrm{Ker}(\nu) = C$ for $\nu: R \to R/C$. It is now to be shown that to every unitary ring homomorphism

$$\rho: R \to S$$

there exists a functor (see 1.3)

$$F_\rho: \mathbf{M}_S \to \mathbf{M}_R.$$

For this purpose to every module M_S there is associated a module M_R in the following manner: Let the additive group M^+ of M_R be equal to that of M_S, the structure of an R-module is defined by

$$mr := m\rho(r), \qquad m \in M^+, r \in R.$$

Direct verification establishes that M_R is a unitary R-module. Let now

$$\varphi: M_S \to N_S$$

be given. Then evidently we have

$$\varphi(mr) = \varphi(m\rho(r)) = \varphi(m)\rho(r) = \varphi(m)r.$$

Thus every S-homomorphism is also an R-homomorphism. In order to show that F_ρ with $F_\rho(M_S) = M_R$, $F_\rho(\varphi) = \varphi$ is a functor it remains only to observe that

$$F_\rho(1_{M_S}) = 1_{M_R}, \qquad F_\rho(\psi\varphi) = \psi\varphi = F_\rho(\psi)F_\rho(\varphi).$$

Such a functor F_ρ is usually known as a "*change of rings*".

Since every S-homomorphism is an R-homomorphism it follows, as a consequence, that $\mathrm{Hom}_S(M, N) \hookrightarrow \mathrm{Hom}_R(M, N)$. If ρ is surjective then we have, in fact, $\mathrm{Hom}_S(M, N) = \mathrm{Hom}_R(M, N)$. The S-submodules of M_S are evidently also R-submodules and in the case of a surjective ρ the S-submodules coincide with the R-submodules.

Examples of ring homomorphisms

(1) Let R be a unitary subring of S and let $\rho = \iota$ be the inclusion mapping.

(2) To every ring S with unit element 1 there is a ring homomorphism

$$\rho: \mathbb{Z} \ni z \mapsto z1 \in S,$$

and the corresponding functor F_ρ is the forgetful functor of M_S in the category of abelian groups.

(3) Let C be a two-sided ideal of R and let

$$\nu: R \to R/C$$

be the natural epimorphism. Then every R/C-module is also an R-module and for $M_{R/C}$, $N_{R/C}$ we have

$$\mathrm{Hom}_{R/C}(M, N) = \mathrm{Hom}_R(M, N).$$

3.3 GENERATORS AND COGENERATORS

Generators and cogenerators are categorical concepts, which play an important role in the modern development of the theory of modules and also in other categories. We present here the definitions and some simple consequences. We shall later return to these concepts several times.

3.3.1 *Definition*

(a) The module B_R is called a *generator* (of M_R) $:\Leftrightarrow$

$$\forall M \in \mathsf{M}_R\left[M = \sum_{\varphi \in \mathrm{Hom}_R(B,M)} \mathrm{Im}(\varphi)\right].$$

(b) The module C_R is called a *cogenerator* (of M_R) $:\Leftrightarrow$

$$\forall M \in \mathsf{M}_R\left[0 = \bigcap_{\varphi \in \mathrm{Hom}_R(M,C)} \mathrm{Ker}(\varphi)\right].$$

For arbitrary modules B, M

$$\mathrm{Im}(B, M) := \sum_{\varphi \in \mathrm{Hom}_R(B,M)} \mathrm{Im}(\varphi)$$

is itself, as a sum of submodules of M, a submodule of M. The property that B is a generator means that $\mathrm{Im}(B, M)$ is as large as possible for every M and so equals M.

For arbitrary modules C, M

$$\mathrm{Ker}(M, C) := \bigcap_{\varphi \in \mathrm{Hom}_R(M,C)} \mathrm{Ker}(\varphi)$$

is itself, as an intersection of submodules of M, a submodule of M. The property that C is a generator means that $\mathrm{Ker}(M, C)$ is as small as possible for every M and so equals 0.

An example of a generator of M_R is immediately available: R_R is a generator. Namely let $m \in M$, then the homomorphism

$$\varphi_m : R \ni r \mapsto mr \in M$$

exists with $\varphi_m(1) = m1 = m$. From this it follows that

$$M = \sum_{m \in M} \mathrm{Im}(\varphi_m) \hookrightarrow \mathrm{Im}(R, M) \hookrightarrow M,$$

thus we have $\mathrm{Im}(R, M) = M$.

Cogenerators of M_R also exist; however, examples can best be presented later when we have injective modules at our disposal.

3.3.2 COROLLARY

(a) *If B is a generator and if A is a module with* $\mathrm{Im}(A, B) = B$ *then A is also a generator.*

(b) *Every module which can be mapped epimorphically onto R_R is a generator.*

(c) *If C is a cogenerator and if D is a module with* $\mathrm{Ker}(C, D) = 0$ *then D is also a cogenerator.*

Proof. (a) Evidently we have:

$$\sum_{\substack{\psi \in \mathrm{Hom}_R(A,B)\\ \varphi \in \mathrm{Hom}_R(B,M)}} \mathrm{Im}(\varphi\psi) = \sum_{\varphi,\psi} \varphi(\mathrm{Im}(\psi)) = \sum_{\varphi} \varphi\Big(\sum_{\psi} \mathrm{Im}(\psi)\Big)$$

$$= \sum_{\varphi} \varphi(B) = \sum_{\varphi} \mathrm{Im}(\varphi) = M.$$

(b) It follows from (a) that R_R is a generator.

(c) Evidently we have:

$$\bigcap_{\substack{\varphi \in \mathrm{Hom}_R(M,C)\\ \psi \in \mathrm{Hom}_R(C,D)}} \mathrm{Ker}(\psi\varphi) = \bigcap_{\varphi,\psi} \varphi^{-1}\,\mathrm{Ker}(\psi)) = \bigcap_{\varphi} \varphi^{-1}\Big(\bigcap_{\psi} \mathrm{Ker}(\psi)\Big)$$

$$= \bigcap_{\varphi} \varphi^{-1}(0) = \bigcap_{\varphi} \mathrm{Ker}(\varphi) = 0. \qquad \square$$

Generators and cogenerators can be characterized in the following manner by properties of homomorphisms.

3.3.3 THEOREM

(a) *B is a generator* $\Leftrightarrow$

$$\forall \mu \in \mathrm{Hom}_R(M, N),\ \mu \neq 0 \qquad \exists \varphi \in \mathrm{Hom}_R(B, M)[\mu\varphi \neq 0].$$

(b) *C is a cogenerator C* $\Leftrightarrow$

$$\forall \lambda \in \mathrm{Hom}_R(L, M),\ \lambda \neq 0 \qquad \exists \varphi \in \mathrm{Hom}_R(M, C)[\varphi\lambda \neq 0].$$

Proof. (a) "$\Rightarrow$": Since $\mu \neq 0$ there is an $m \in M$ with $\mu(m) \neq 0$. As B is a generator, there is a representation

$$m = \sum_{i=1}^{k} \varphi_i(b_i), \qquad \varphi_i \in \mathrm{Hom}_R(B, M),\ b_i \in B,$$

hence we have

$$0 \neq \mu(m) = \sum_{i=1}^{k} \mu\varphi_i(b_i),$$

and consequently there is a φ_i with $\mu\varphi_i \neq 0$.

(a) "$\Leftarrow$": Suppose $\mathrm{Im}(B, M) \neq M$, then let

$$\nu : M \to M \;\mathrm{Im}(B, M)$$

be the natural epimorphism. Since $\nu \neq 0$ there is a $\varphi \in \mathrm{Hom}_R(B, M)$ with $\nu\varphi \neq 0$, consequently we have $\mathrm{Im}(\varphi) \not\hookrightarrow \mathrm{Im}(B, M)$ in contradiction to the definition of $\mathrm{Im}(B, M)$.

(b) "$\Rightarrow$": Since $\lambda \neq 0$ there is an $l \in L$ with $\lambda(l) \neq 0$. As C is a cogenerator, there is a $\varphi \in \mathrm{Hom}_R(M, C)$ with $\lambda(l) \notin \mathrm{Ker}(\varphi)$. Hence we have $\varphi\lambda(l) \neq 0$, thus $\varphi\lambda \neq 0$.

(b) "$\Leftarrow$": Suppose $\mathrm{Ker}(M, C) \neq 0$, then let

$$\iota : \mathrm{Ker}(M, C) \to M$$

be the inclusion mapping. Since $\iota \neq 0$ there is a $\varphi \in \mathrm{Hom}_R(M, C)$ with $\varphi\iota \neq 0$. Consequently we have $\mathrm{Ker}(M, C) \not\hookrightarrow \mathrm{Ker}(\varphi)$ in contradiction to the definition of $\mathrm{Ker}(M, C)$. □

3.4 FACTORIZATION OF HOMOMORPHISMS

It is often expedient to factorize a given homomorphism into a product of two homomorphisms where at least one, or even both, factors are to possess certain "pleasant" properties. The homomorphism theorem is the first and particularly important example of such a factorization.

3.4.1 HOMOMORPHISM THEOREM

(a) *Every module homomorphism*

$$\alpha : A \to B$$

has a factorization $\alpha = \alpha'\nu$ *where*

$$\nu : A \to A/\mathrm{Ker}(\alpha)$$

is the natural epimorphism (*see* 3.1) *and* α' *is the monomorphism defined by*

$$\alpha' : A/\mathrm{Ker}(\alpha) \ni a + \mathrm{Ker}(\alpha) \mapsto \alpha(a) \in B,$$

α' *is an isomorphism if and only if* α *is an epimorphism.*

(b) *Every ring homomorphism*

$$\rho : R \to S$$

has a factorization $\rho = \rho'\nu$ *where*

$$\nu : R \to R/\mathrm{Ker}(\rho)$$

is the natural epimorphism and ρ' *is the monomorphism defined by*

$$\rho': R/\mathrm{Ker}(\rho) \ni r + \mathrm{Ker}(\rho) \mapsto \rho(r) \in S,$$

ρ' *is an isomorphism if and only if* ρ *is surjective.*

Remark. The equation $\alpha = \alpha'\nu$ is exactly equivalent to the commutativity of the diagram

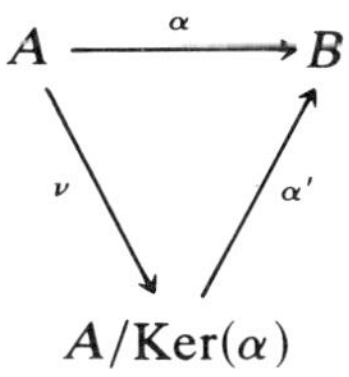

(analogously for the equation $\rho = \rho'\nu$).

Proof. It suffices to go through the proof of (a) since that of (b) proceeds entirely analogously.

It is first of all to be established that α' is a mapping: Let $a + \mathrm{Ker}(\alpha) = a_1 + \mathrm{Ker}(\alpha)$. Then $a_1 = a + u$, $u \in \mathrm{Ker}(\alpha)$. Hence

$$\alpha'(a_1 + \mathrm{Ker}(\alpha)) = \alpha(a_1) = \alpha(a+u) = \alpha(a) + \alpha(u) = \alpha(a) = \alpha'(a + \mathrm{Ker}(\alpha)),$$

then α' is obviously a homomorphism. In order to see that α' is a monomorphism, let (as in 3.1.8)

$$\alpha'(a_1 + \mathrm{Ker}(\alpha)) = \alpha(a_1) = 0.$$

Thus $a_1 \in \mathrm{Ker}(\alpha)$ and so

$$a_1 + \mathrm{Ker}(\alpha) = 0 + \mathrm{Ker}(\alpha).$$

Hence $\mathrm{Ker}(\alpha') = 0$.

Let now $a \in A$ be arbitrary, then we have:

$$\alpha'(\nu(a)) = \alpha'(a + \mathrm{Ker}(\alpha)) = \alpha(a).$$

Thus

$$\alpha = \alpha'\nu.$$

Since α' is a monomorphism and, as $\mathrm{Im}(\alpha') = \mathrm{Im}(\alpha)$, α' is then precisely an isomorphism if α is an epimorphism. □

3.4.2 COROLLARY

(a) *If* $\alpha: A \to B$ *is a module homomorphism then*

$$\hat{\alpha}: A/\mathrm{Ker}(\alpha) \ni a + \mathrm{Ker}(\alpha) \mapsto \alpha(a) \in \mathrm{Im}(\alpha)$$

is an isomorphism, thus we have

$$A/\mathrm{Ker}(\alpha) \cong \mathrm{Im}(\alpha).$$

(b) *If $\rho: R \to S$ is a ring homomorphism then*

$$\hat{\rho}: R/\mathrm{Ker}(\rho) \ni r + \mathrm{Ker}(\rho) \mapsto \rho(r) \in \mathrm{Im}(\rho)$$

is an isomorphism, thus we have

$$R/\mathrm{Ker}(\rho) \cong \mathrm{Im}(\rho) \qquad \text{(as rings)}.$$

Proof. (a) We obtain $\hat{\alpha}$ from α' by means of the restriction of $\mathrm{Cod}(\alpha') = \mathrm{Cod}(\alpha)$ to $\mathrm{Im}(\alpha)$.

(b) Analogously. □

Since the results on ring homomorphisms, which have so far appeared, suffice for later considerations, we confine ourselves from now on to module homomorphisms. Thus let A, B, C, as well as all homomorphisms, be from a module category, in which right, left or bi-modules may be considered.

3.4.3 FIRST ISOMORPHISM THEOREM. *Let $B \hookrightarrow A \wedge C \hookrightarrow A$, then we have*

$$(B+C)/C \cong B/(B \cap C).$$

Proof. For the proof we consider the homomorphisms

$$\nu: B + C \to (B+C)/C$$

with $\mathrm{Ker}(\nu) = C$ and

$$\alpha := \nu|B: B \to (B+C)/C$$

with $\mathrm{Ker}(\alpha) = B \cap C$. We now apply 3.4.2:

$$(B+C)/C \cong \mathrm{Im}(\nu) = \nu(B+C) = \nu(B) + \nu(C) = \nu(B),$$

$$B/(B \cap C) \cong \mathrm{Im}(\alpha) = \alpha(B) = \nu(B)$$

$$\Rightarrow \quad (B+C)/C \cong B/(B \cap C). \qquad \square$$

We can also prove this theorem without invoking 3.4.2 by verifying that

$$B/(B \cap C) \ni b + (B \cap C) \mapsto b + C \in (B+C)/C$$

is an isomorphism. This may be left to the reader as an exercise.

3.4.4 COROLLARY. $A = B \oplus C \Rightarrow A/C \cong B.$

Proof.

$$A/C = (B+C)/C \cong B/(B \cap C) = B/0 \cong B. \qquad \square$$

As a further deduction we give *Zassenhaus's Lemma* which is used in an essential way in the next chapter. It indicates that perhaps a modification must first be achieved in order to be able to apply the first Isomorphism Theorem.

3.4.5 LEMMA. *Let* $U' \hookrightarrow U \hookrightarrow A \wedge V' \hookrightarrow V \hookrightarrow A$ *then we have*

$$(U'+(U\cap V))/(U'+(U\cap V'))\cong(V'+(U\cap V))/(V'+(U'\cap V)).$$

Proof. We show that the left-hand side is isomorphic to

$$(U\cap V)/((U'\cap V)+(V'\cap U)).$$

Since this expression is symmetric in U and V, the right-hand side is then also isomorphic to it, from which the assertion follows.

As $U\cap V' \hookrightarrow U\cap V$ we have

$$U'+(U\cap V)=(U\cap V)+(U'+(U\cap V')),$$

and further according to the modular law (2.3.15)

$$\begin{aligned}(U\cap V)\cap(U'+(U\cap V'))&=(U\cap V\cap U')+(U\cap V')\\ &=(U'\cap V)+(U\cap V').\end{aligned}$$

From the First Isomorphism Theorem it follows therefore that

$$\begin{aligned}&(U'+(U\cap V))/(U'+(U\cap V'))\\ &\quad=((U\cap V)+(U'+(U\cap V')))/(U'+(U\cap V'))\\ &\quad\cong(U\cap V)/((U\cap V)\cap(U'+(U\cap V')))\\ &\quad=(U\cap V)/((U'\cap V)+(U\cap V')).\end{aligned}$$

□

3.4.6 SECOND ISOMORPHISM THEOREM. *Let* $C \hookrightarrow B \hookrightarrow A$, *then we have*

$$A/B\cong(A/C)/(B/C).$$

Proof. Let

$$\nu_1: A\to A/C$$

$$\nu_2: A/C\to(A/C)/(B/C),$$

where ν_2 is well-defined, since, from $C \hookrightarrow B \hookrightarrow A$, B/C is also a submodule of A/C.

Since ν_1 and ν_2 are epimorphisms, $\nu_2\nu_1$ is an epimorphism (3.1.6) and consequently 3.4.2 implies that

$$A/\mathrm{Ker}(\nu_2\nu_1)\cong(A/C)/(B/C).$$

But according to 3.1.8 we have

$$\begin{aligned}\operatorname{Ker}(\nu_2\nu_1) &= \overset{-1}{\nu_1}(\operatorname{Ker}(\nu_2)) = \overset{-1}{\nu_1}(B/C) = \overset{-1}{\nu_1}(\nu_1(B)) \\ &= B + \operatorname{Ker}(\nu_1) = B + C = B,\end{aligned}$$

from which the assertion follows. □

Example. $\mathbb{Z}/3\mathbb{Z} \cong (\mathbb{Z}/6\mathbb{Z})/(3\mathbb{Z}/6\mathbb{Z})$.

Finally a result is to be presented which can be considered as the generalization of the Homomorphism Theorem 3.4.1.

3.4.7 THEOREM. *Let* $\alpha: A \to B$ *be a homomorphism and let* $\varphi: A \to C$ *be an epimorphism with* $\operatorname{Ker}(\varphi) \hookrightarrow \operatorname{Ker}(\alpha)$. *Then there exists a homomorphism* $\lambda: C \to B$ *with*

(1) $\alpha = \lambda\varphi$.
(2) $\operatorname{Im}(\lambda) = \operatorname{Im}(\alpha)$.
(3) λ *is a monomorphism* $\Leftrightarrow \operatorname{Ker}(\varphi) = \operatorname{Ker}(\alpha)$

Remark. (1) means that the diagram

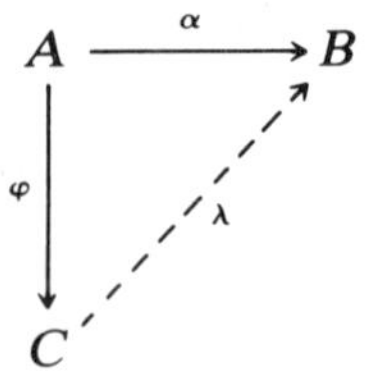

is commutative.

Proof. Since φ is an epimorphism, for an arbitrary $c \in C$ there is an $a \in A$ with $\varphi(a) = c$. To every $c \in C$ let there be chosen a fixed $a_c \in A$ with $\varphi(a_c) = c$ (Axiom of Choice). Then a mapping is defined by

$$\lambda: C \to B \quad \text{with} \quad \lambda(c) := \alpha(a_c).$$

In order to show that λ is indeed a homomorphism it must first of all be established that λ is independent of the choice of the a_c with $\varphi(a_c) = c$.

Let $c = \varphi(a) = \varphi(a_c)$ with $a, a_c \in A$.

Then

$$\varphi(a - a_c) = 0$$

and so

$$a - a_c \in \operatorname{Ker}(\varphi) \hookrightarrow \operatorname{Ker}(\alpha) \qquad \text{(by assumption)}.$$

Hence

$$\alpha(a-a_c)=0 \Rightarrow \alpha(a)=\alpha(a_c)=\lambda(c).$$

It now follows immediately that λ is a homomorphism: Let $c_1=\varphi(a_1)$, $c_2=\varphi(a_2)$ with $a_1, a_2 \in A$ and let $r_1, r_2 \in R$. Then

$$\varphi(a_1r_1+a_2r_2)=\varphi(a_1)r_1+\varphi(a_2)r_2=c_1r_1+c_2r_2$$

$$\Rightarrow \quad \lambda(c_1r_1+c_2r_2)=\alpha(a_1r_1+a_2r_2)=\alpha(a_1)r_1+\alpha(a_2)r_2$$

$$=\lambda(c_1)r_1+\lambda(c_2)r_2.$$

(1) and (2) follow directly from the definition of λ. For the proof of (3) first let λ be a monomorphism. By assumption we have $\mathrm{Ker}(\varphi) \hookrightarrow \mathrm{Ker}(\alpha)$.

To prove that $\mathrm{Ker}(\alpha) \hookrightarrow \mathrm{Ker}(\varphi)$ let $a \in \mathrm{Ker}(\alpha)$, since $0=\alpha(a)=\lambda(\varphi(a))$ it then follows that $\varphi(a)=0$, thus $a \in \mathrm{Ker}(\varphi)$ holds. Suppose now that $\mathrm{Ker}(\varphi)=\mathrm{Ker}(\alpha)$, then it follows from $\lambda(c)=0$ and $c=\varphi(a)$ that $\alpha(a)=0$ holds, thus $a \in \mathrm{Ker}(\alpha)=\mathrm{Ker}(\varphi)$ and hence $c=\varphi(a)=0$. □

We draw attention to two special cases of 3.4.7:

(1) Let $\alpha: A \to B$, $A' \hookrightarrow \mathrm{Ker}(\alpha)$, $C=A/A'$, $\varphi=\nu: A \to A/A'$ then the diagram

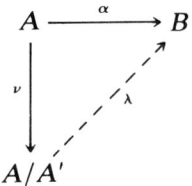

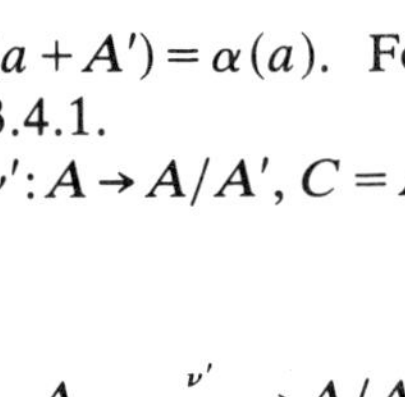

is commutative where $\lambda(a+A')=\alpha(a)$. For $A'=\mathrm{Ker}(\alpha)$ this is the Homomorphism Theorem 3.4.1.

(2) Let $A'' \hookrightarrow A' \hookrightarrow A$, $\alpha=\nu': A \to A/A'$, $C=A/A''$, $\varphi=\nu'': A \to A/A''$ then the diagram

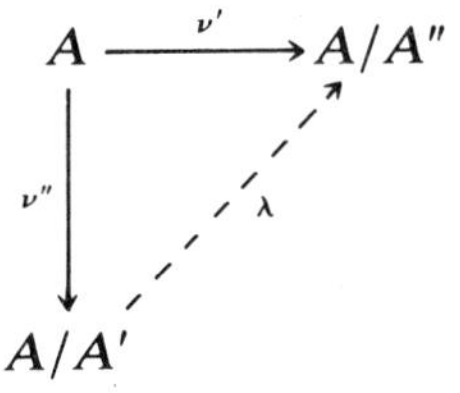

is commutative, where $\lambda(a+A'')=a+A'$.

Let now $\lambda = \beta\alpha$ be a given factorization of a given homomorphism λ.

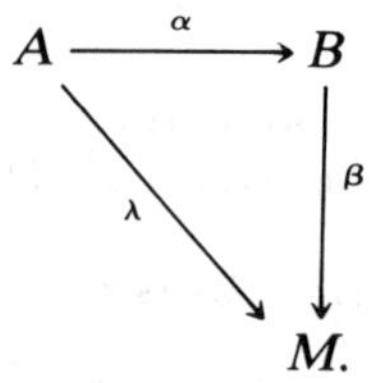

We inquire into the relationship between the properties of λ and the "decomposition properties" of B. Before we begin with this, we recall the definition of the (internal) direct sum (2.4), which is now needed for two summands only. In this case we have:

$$B = B_0 \oplus B_1 \Leftrightarrow B = B_0 + B_1 \wedge B_0 \cap B_1 = 0.$$

3.4.8 *Definition*

(1) The submodule $B_0 \hookrightarrow B$ is called a *direct summand* of $B :\Leftrightarrow$ there exists a submodule $B_1 \hookrightarrow B$ with $B = B_0 \oplus B_1$.

(2) A monomorphism $\alpha: A \to B$ is said to *split* $:\Leftrightarrow \mathrm{Im}(\alpha)$ is a direct summand in B.

(3) An epimorphism $\beta: B \to C$ is said to *split* $:\Leftrightarrow \mathrm{Ker}(\beta)$ is a direct summand in B.

3.4.9 LEMMA. *Let the diagram*

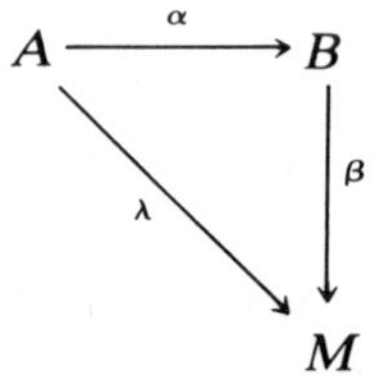

be commutative, i.e. $\lambda = \beta\alpha$. *Then*

(1) $\mathrm{Im}(\alpha) + \mathrm{Ker}(\beta) = \beta^{-1}(\mathrm{Im}(\lambda))$,

(2) $\mathrm{Im}(\alpha) \cap \mathrm{Ker}(\beta) = \alpha(\mathrm{Ker}(\lambda))$.

Proof. (1) $\lambda = \beta\alpha \Rightarrow \mathrm{Im}(\lambda) = \mathrm{Im}(\beta\alpha) = \beta(\mathrm{Im}(\alpha)) \Rightarrow \beta^{-1}(\mathrm{Im}(\lambda)) = \beta^{-1}(\beta(\mathrm{Im}(\alpha))) = \mathrm{Im}(\alpha) + \mathrm{Ker}(\beta)$ by 3.1.8.

(2) $\mathrm{Ker}(\lambda) = \mathrm{Ker}(\beta\alpha) = \alpha^{-1}(\mathrm{Ker}(\beta))$ by 3.1.8

$$\Rightarrow \alpha(\mathrm{Ker}(\lambda)) = \alpha(\alpha^{-1}(\mathrm{Ker}(\beta))) = \mathrm{Im}(\alpha) \cap \mathrm{Ker}(\beta)$$

by 3.1.8. □

3.4.10 COROLLARY
(a) λ *is an epimorphism* $\Rightarrow \operatorname{Im}(\alpha)+\operatorname{Ker}(\beta)=\beta^{-1}(M)=B$.
(b) λ *is a monomorphism* $\Rightarrow \operatorname{Im}(\alpha)\cap\operatorname{Ker}(\beta)=\alpha(0)=0$.
(c) λ *is an isomorphism* $\Rightarrow \operatorname{Im}(\alpha)\oplus\operatorname{Ker}(\beta)=B$.

Proof. Direct consequence from 3.4.9. □

3.4.11 COROLLARY
(1) *For* $\alpha: A\to B$ *the following are equivalent*:
  (a) α *is a split monomorphism.*
  (b) *There exists a homomorphism* $\beta: B\to A$ *with* $\beta\alpha=1_A$.
(2) *For* $\beta: B\to C$ *the following are equivalent*:
  (a) β *is a split epimorphism.*
  (b) *There exists a homomorphism* $\gamma: C\to B$ *with* $\beta\gamma=1_C$.

Proof. (1) "(a)$\Rightarrow$(b)": Let $B=\operatorname{Im}(\alpha)\oplus B_1$ and let $\pi: B\to\operatorname{Im}(\alpha)$ be the projection of B onto $\operatorname{Im}(\alpha)$ defined by

$$\pi(\alpha(a)+b_1) := \alpha(a), \qquad \alpha(a)\in\operatorname{Im}(\alpha),\ b_1\in B_1.$$

Further call $\alpha_0: A\ni a\mapsto\alpha(a)\in\operatorname{Im}(\alpha)$, i.e. let α_0 be the isomorphism defined by the restriction of the domain B of α to $\operatorname{Im}(\alpha)$.

For $\beta := \alpha_0^{-1}\pi$ we then have

$$\beta\alpha(a)=\alpha_0^{-1}\pi\alpha(a)=\alpha_0^{-1}(\alpha(a))=a, \qquad a\in A,$$

thus $\beta\alpha=1_A$.

(1) "(a)$\Leftarrow$(b)": Since $\beta\alpha=1_A$ α is a monomorphism which splits by 3.4.10 (c).

(2) "(a)$\Rightarrow$(b)": Let $B=\operatorname{Ker}(\beta)\oplus B_1$, and let $\iota: B_1\ni b\mapsto b\in B$ be the inclusion mapping of B_1 into B. Further let β_1 denote the restriction of β onto B_1, then β_1 is an isomorphism (since β is an epimorphism and $\operatorname{Ker}(\beta)\cap B_1=0$). For $\gamma := \iota\beta_1^{-1}$ we then have

$$\beta\gamma(c)=\beta\iota\beta_1^{-1}(c)=\beta(\beta_1^{-1}(c))=c, \qquad c\in C,$$

thus $\beta\gamma=1_C$.

(2) "$(a)\Leftarrow(b)$": Since $\beta\gamma=1_C$ β is an epimorphism, which splits by 3.4.10(c). □

We point out, in particular, the special case, in which α is the inclusion mapping of a submodule $A\hookrightarrow B$ and $\beta: B\to B/A$ is the natural epimorphism.

3.5 THE THEOREM OF JORDAN–HÖLDER–SCHREIER

We now consider finite chains of submodules of a module A. Let

$$0 = B_0 \hookrightarrow B_1 \hookrightarrow B_2 \hookrightarrow \ldots \hookrightarrow B_{k-1} \hookrightarrow B_k = A,$$

$$0 = C_0 \hookrightarrow C_1 \hookrightarrow C_2 \hookrightarrow \ldots \hookrightarrow C_{l-1} \hookrightarrow C_l = A.$$

We denote the first of these two chains by B and the second by C. Then we have the following.

3.5.1 *Definitions*

(1) *Length of the chain* $\mathsf{B} := k$.

(2) *The factors of the chain* B are the factor modules B_i/B_{i-1}, $i = 1, \ldots, k$. The ith factor of B is B_i/B_{i-1}.

(3) The chains B and C are said to be *isomorphic*, $\mathsf{B} \cong \mathsf{C} :\Leftrightarrow$ there exists a bijection δ between the index set I of B and the index set J of C such that we have:

$$B_i/B_{i-1} \cong C_{\delta(i)}/C_{\delta(i)-1}, \qquad i = 1, \ldots, k.$$

(4) C is called a *refinement* of B and B a *subchain* of $\mathsf{C} :\Leftrightarrow$ either $\mathsf{B} = \mathsf{C}$ (trivial refinement) or B is obtained from C by omitting certain of the C_j from C.

(5) The chain B of A is called a *composition series* $:\Leftrightarrow \forall i = 1, \ldots, k$ $[B_{i-1}$ maximal in $B_i]$ ($\Leftrightarrow \forall i = 1, \ldots, k$ $[B_i/B_{i-1}$ simple$]$ by 3.1.14).

(6) The module A is said to be of *finite length* $:\Leftrightarrow A = 0 \vee A$ has a composition series.

Remark. If $\mathsf{B} \cong \mathsf{C}$ *holds and if* $B_i = B_{i-1}$ *for a fixed i, then there is a j so that, if* B_i *in* B *and* C_j *in* C *are omitted, the chains, resulting in this way, are again isomorphic.*

Proof. The proof follows directly from the fact that $B_i = B_{i-1}$ has the consequence that first of all $B_i/B_{i-1} = 0$ and thereby $C_{\delta(i)}/C_{\delta(i)-1} = 0$ thus $C_{\delta(i)} = C_{\delta(i)-1}$. From the omission of B_i resp. $B_{\delta(i)}$ precisely the factor $B_i/B_{i-1} = 0 = C_{\delta(i)}/C_{\delta(i)-1}$ is thus omitted whereas the other factors are unchanged. □

We shall make use of this remark in the following without especial mention. It is further clear that the isomorphism defined in (3) is an equivalence relation in the set of all chains of A of the form B.

Examples

(1) Let $V = V_K$ be a vector space and let $\{x_1, \ldots, x_n\}$ be a basis of V. Then

$$0 \hookrightarrow x_1K \hookrightarrow x_1K + x_2K \hookrightarrow \ldots \hookrightarrow \sum_{i=1}^{n-1} x_iK \hookrightarrow \sum_{i=1}^{n} x_iK = V$$

is a composition series of V.

(2) Every chain of $\mathbb{Z}_{\mathbb{Z}}$ can be properly refined. If

$$0 \hookrightarrow B_1 \hookrightarrow \ldots \hookrightarrow \mathbb{Z}$$

is such a chain with $B_1 \neq 0$ (which does not entail a restriction) then, since $\mathbb{Z}$ does not contain a simple ideal, B_1 cannot be simple. Thus between 0 and B_1 an ideal different from both can be inserted. Consequently $\mathbb{Z}_{\mathbb{Z}}$ does not have a composition series.

(3) In $\mathbb{Q}_{\mathbb{Z}}$ every chain

$$0 \hookrightarrow B_1 \hookrightarrow B_2 \hookrightarrow \ldots \hookrightarrow B_k = \mathbb{Q}_{\mathbb{Z}}$$

with $0 \neq B_1$ and $B_{k-1} \neq Q$ can be properly refined both between 0 and B_1 and also between B_{k-1} and $\mathbb{Q}$, since $\mathbb{Q}_{\mathbb{Z}}$ contains neither a minimal (=simple) nor a maximal submodule. Accordingly $\mathbb{Q}_{\mathbb{Z}}$ does not have a composition series.

We prove now the Jordan–Hölder–Schreier Theorem, from which we then obtain as a most important corollary that, if a module has a composition series, the series is uniquely determined up to isomorphism.

3.5.2 Jordan–Hölder–Schreier Theorem

Any two (finite!) chains of a module have isomorphic refinements.

Proof. Let B and C be given finite chains of the module A. The modules

$$B_{i,j} = B_i + (B_{i+1} \cap C_j), \qquad j = 0, \ldots, l$$

are inserted between B_i and B_{i+1} $(i = 0, \ldots, k-1)$, and so we obviously have

$$B_i = B_{i,0} \hookrightarrow B_{i,1} \hookrightarrow \ldots \hookrightarrow B_{i,l} = B_{i+1}.$$

Analogously the modules

$$C_{i,j} = C_j + (C_{j+1} \cap B_i), \qquad i = 0, \ldots, k$$

are inserted between C_j and C_{j+1} $(j = 0, \ldots, l-1)$ and we have

$$C_j = C_{0,j} \hookrightarrow C_{1,j} \hookrightarrow \ldots \hookrightarrow C_{k,j} = C_{j+1}.$$

The refined chains are then denoted by B^* and C^*; they both have the same length kl. From 3.4.5 it follows that

$$B_{i,j+1}/B_{i,j} \cong C_{i+1,j}/C_{i,j} \qquad \begin{cases} i=0,\ldots,k-1 \\ j=0,\ldots,l-1 \end{cases}.$$

Since in these kl isomorphisms precisely all of the kl factors of B^* and precisely all of the kl factors of C^* appear, it follows that $B^* \cong C^*$. □

3.5.3 COROLLARY. *Let A be a module of finite length. Then we have:*
(1) *Every chain B of the form*

$$0 = B_0 \subsetneq B_1 \subsetneq \ldots \subsetneq B_k = A$$

can be refined to a composition series.
(2) *Any two composition series of A are isomorphic.*

Proof. (1) By assumption there is a composition series C of A. According to the Jordan–Hölder–Schreier Theorem B and C have isomorphic refinements B^* and C^*. Since C, as a composition series, can only be trivially refined, there is (from the remark following 3.5.1) a refinement B° of B with $B^\circ \cong C$. Since all the factors in C are simple, so also are the factors of B°, consequently B° is a composition series.
(2) Let now B and C be composition series and let in the terminology of (1): $B^\circ \cong C$. Since B° is a refinement of B and both are composition series, it follows that $B = B^\circ$ and therefore $B \cong C$. □

3.5.4 *Definition.* Let A be a module of finite length. Then let the *length* of $A = \mathrm{Le}(A) :=$ length of one (and therefore of any) composition series of A.

3.5.5 COROLLARY. *Let $A \hookrightarrow M$. Then we have: M is a module of finite length if and only if A and M/A are modules of finite length. If the length is finite then we have*

$$\mathrm{Le}(M) = \mathrm{Le}(A) + \mathrm{Le}(M/A).$$

Proof. If $0 = A$ or $A = M$ then the assertion is clear. Let now $0 \subsetneq A \subsetneq M$ and let M be of finite length. Then the chain

$$0 \hookrightarrow A \hookrightarrow M$$

can be refined to a composition series:

$$0 \hookrightarrow A_1 \hookrightarrow \ldots \hookrightarrow A_k = A \hookrightarrow \ldots \hookrightarrow A_n = M.$$

The initial part of the chain up to $A_k = A$ is a composition series of A. We claim that

$$0 = A/A \hookrightarrow A_{k+1}/A \hookrightarrow \ldots \hookrightarrow A_n/A = M/A$$

is a composition series of M/A. This holds, since according to the Second Isomorphism Theorem

$$(A_{k+i+1}/A)/(A_{k+i}/A) \cong A_{k+i+1}/A_{k+i}$$

is simple. From the preceding it follows that $\mathrm{Le}(M) = \mathrm{Lc}(A) + \mathrm{Le}(M/A)$. Let now A and M/A be of finite length and let

$$0 \hookrightarrow A_1 \hookrightarrow \ldots \hookrightarrow A_k = A, \qquad 0 \hookrightarrow \bar{B}_1 \hookrightarrow \ldots \hookrightarrow \bar{B}_l = M/A$$

be composition series of A and M/A respectively. Let $\nu: M \to M/A$ and $B_i := \nu^{-1}(\bar{B}_i)$. Then we have $A \hookrightarrow B_i$ and $\nu(B_i) = B_i/A = \bar{B}_i$. Since $\bar{B}_{i+1}/\bar{B}_i$ is simple and as

$$(B_{i+1}/A)/(B_i/A) \cong B_{i+1}/B_i,$$

B_{i+1}/B_i is also simple. Consequently

$$0 \hookrightarrow A_1 \hookrightarrow \ldots \hookrightarrow A_k = A \hookrightarrow B_1 \hookrightarrow \ldots \hookrightarrow B_l = M$$

is a composition series of M, i.e. M is of finite length. □

In particular the proof has shown how from composition series for A and M/A such a series for M can be manufactured.

Example. The $\mathbb{Z}$-module $\mathbb{Z}/6\mathbb{Z}$ has two composition series

$$0 \hookrightarrow 2\mathbb{Z}/6\mathbb{Z} \hookrightarrow \mathbb{Z}/6\mathbb{Z}, \qquad 0 \hookrightarrow 3\mathbb{Z}/6\mathbb{Z} \hookrightarrow \mathbb{Z}/6\mathbb{Z}.$$

The factors of the first are

$$2\mathbb{Z}/6\mathbb{Z} \cong \mathbb{Z}/3\mathbb{Z}, \qquad (\mathbb{Z}/6\mathbb{Z})/(2\mathbb{Z}/6\mathbb{Z}) \cong \mathbb{Z}/2\mathbb{Z},$$

those of the second are

$$3\mathbb{Z}/6\mathbb{Z} \cong \mathbb{Z}/2\mathbb{Z}, \qquad (\mathbb{Z}/6\mathbb{Z})/(3\mathbb{Z}/6\mathbb{Z}) \cong \mathbb{Z}/3\mathbb{Z},$$

from which the isomorphism of the two chains follows immediately. The significance of the Jordan–Hölder–Schreier Theorem for modules of finite length becomes clear from the following consideration. Let A be a module of finite length, let B be an arbitrary submodule of A, let C be a maximal submodule of B, then B/C is a composition-factor (=factor of a composition series) of A. Thus let us consider the chain

$$0 \hookrightarrow C \hookrightarrow B \hookrightarrow A$$

(correspondingly the shorter chain in case that $C=0$ or $B=A$ resp.). This can be refined to a composition series, in which no module is inserted between C and B since C is maximal in B. Consequently B/C is in fact a composition factor of A, i.e. up to isomorphism one of the uniquely determined finitely many composition factors of A.

3.6 FUNCTORIAL PROPERTIES OF Hom

As we have already observed in Chapter 1, Hom_R is a functor of the category $\mathbf{M}_R$ (or ${}_S\mathbf{M}$ or ${}_S\mathbf{M}_R$), contravariant in the first argument and covariant in the second, into the category $\mathbf{S}$ of sets:

$$\mathrm{Hom}_R\colon \mathrm{Obj}(\mathbf{M}_R)\times \mathrm{Obj}(\mathbf{M}_R)\ni (A,B)\mapsto \mathrm{Hom}_R(A,B)\in \mathrm{Obj}(\mathbf{S})$$

$$\mathrm{Hom}_R\colon \mathrm{Mor}(\mathbf{M}_R)\times \mathrm{Mor}(\mathbf{M}_R)\ni (\alpha,\gamma)\mapsto \mathrm{Hom}_R(\alpha,\gamma)\in \mathrm{Mor}(\mathbf{S}),$$

where $\mathrm{Hom}_R(A,B)$ is the set of homomorphisms of A into B and $\mathrm{Hom}_R(\alpha,\gamma)$ is defined in the following manner: For

$$\alpha: A\to B, \qquad \gamma: C\to D$$

let

$$\mathrm{Hom}_R(\alpha,\gamma)\colon \mathrm{Hom}_R(B,C)\ni \beta\mapsto \gamma\beta\alpha\in \mathrm{Hom}_R(A,D).$$

If $R=K$ is a field, i.e. $\mathbf{M}_K$ is the category of K-vector spaces, then $\mathrm{Hom}_K(A,B)$ becomes again a vector space over K in a well known manner by means of the definition

$$(\alpha_1+\alpha_2)(a) := \alpha_1(a)+\alpha_2(a)$$

$$(\alpha k)(a) := \alpha(ak),$$

(with $\alpha_1,\alpha_2\in \mathrm{Hom}_K(A,B)$, $a\in A$, $k\in K$) a vector space over K, and Hom_K can now be considered as a functor in the category $\mathbf{M}_K$ itself (and not only in $\mathbf{S}$). This property is now to be generalized. Let now R be once more an arbitrary ring with a unit element. By the following definition $\mathrm{Hom}_R(A,B)$ becomes an abelian group. For $\alpha_1,\alpha_2\in \mathrm{Hom}_R(A,B)$, $\alpha_1+\alpha_2\in \mathrm{Hom}_R(A,B)$ is defined by

$$(\alpha_1+\alpha_2)(a)=\alpha_1(a)+\alpha_2(a), \qquad a\in A.$$

The group-theoretic properties of $\mathrm{Hom}_R(A,B)$, which follow from those of B, are then easy to verify: in particular the zero mapping of A into B is the zero element of $\mathrm{Hom}_R(A,B)$ and the mapping $-\alpha$ with

$$(-\alpha)(a) := -\alpha(a)$$

is the homomorphism inverse to $\alpha\in \mathrm{Hom}_R(A,B)$.

With this interpretation of $\mathrm{Hom}_R(A, B)$, Hom_R becomes a functor in the category A of abelian groups. For this purpose we further establish that $\mathrm{Hom}_R(\alpha, \gamma)$ is now a group homomorphism of $\mathrm{Hom}_R(B, C)$ into $\mathrm{Hom}_R(A, D)$:

$$\begin{aligned}\mathrm{Hom}_R(\alpha, \gamma)(\beta_1+\beta_2) &= \gamma(\beta_1+\beta_2)\alpha \\ &= \gamma\beta_1\alpha + \gamma\beta_2\alpha \\ &= \mathrm{Hom}_R(\alpha, \gamma)(\beta_1)+\mathrm{Hom}_R(\alpha, \gamma)(\beta_2),\end{aligned}$$

since

$$\begin{aligned}(\gamma(\beta_1+\beta_2)\alpha)(a) &= \gamma((\beta_1+\beta_2)(\alpha(a))) \\ &= \gamma(\beta_1(\alpha(a))+\beta_2(\alpha(a))) = \gamma(\beta_1(\alpha(a)))+\gamma(\beta_2(\alpha(a))) \\ &= (\gamma\beta_1\alpha)(a)+(\gamma\beta_2\alpha)(a) = (\gamma\beta_1\alpha+\gamma\beta_2\alpha)(a).\end{aligned}$$

Let now S be also a ring with a unit element, let $A = {}_SA_R$ and as before let $B = B_R$. Then $\mathrm{Hom}_R(A, B)$ becomes by the definition

$$(\alpha s)(a) := \alpha(sa), \qquad \alpha \in \mathrm{Hom}_R(A, B),\ a \in A,\ s \in S,$$

a right S-module, as is immediately verifiable.

Further let T be a ring with a unit element and let $A = A_R$ and also $B = {}_TB_R$. Then by the definition

$$(t\alpha)(a) := t\alpha(a), \qquad \alpha \in \mathrm{Hom}_R(A, B),\ a \in A,\ t \in T,$$

$\mathrm{Hom}_R(A, B)$ becomes a left T-module. If we have simultaneously $A = {}_SA_R$, $B = {}_TB_R$ then it follows that

$$\mathrm{Hom}_R(A, B) = {}_T\mathrm{Hom}_R(A, B)_S,$$

i.e. $\mathrm{Hom}_R(A, B)$ becomes a T-S-bimodule.

3.6.1 *Definition.* The *centre* of the ring R is

$$Z(R) := \{s \mid s \in R \wedge \forall r \in R[sr = rs]\}.$$

Remark. $Z(R)$ is a commutative subring of R, which contains the unit element of R.

If we put $S := Z(R)$ and let $A = A_R$, then, by the following definition, A becomes an S-R-bimodule,

$$sa := as, \qquad s \in S = Z(R),\ a \in A,$$

as is easily verified.

Since this holds for every R-module, it follows that $\mathrm{Hom}_R(A, B)$ can be considered as an $S = Z(R)$-module, right, left or two-sided. As we realize easily, Hom_R can then also be understood to be a functor in the category M_S, ${}_S\mathsf{M}$, ${}_S\mathsf{M}_S$ respectively. If R is commutative, i.e. $S = Z(R) = R$, then Hom_R is a functor in M_R, as in the case of a vector space over a field.

In order to avoid confusion in complicated cases, we write, for example, in the situation ${}_SA_R$, ${}_TB_R$ also

$$\mathrm{Hom}_R({}_SA_R, {}_TB_R),$$

where the index R of the Hom_R indicates that an R-homomorphism is involved, and the indices S and T imply that $\mathrm{Hom}_R({}_SA_R, {}_TB_R)$ is to be considered in the previously employed sense as a T-S bimodule. In the situation ${}_RA_S$, ${}_RB_T$ then $\mathrm{Hom}_R({}_RA_S, {}_RB_T)$ is an S-T bimodule, and from our convention at the beginning of 3.1

$$a(s\alpha t) = (as)(\alpha t) = (as\alpha)t = as\alpha t,$$

indicates that $a \in A$ is first of all multiplied by $s \in S$; then $\alpha \in \mathrm{Hom}_R(A, B)$ is applied to as and the image multiplies $t \in T$.

If we consider Hom_R with respect to a fixed second argument M_R as a functor of the first argument, then the following notational conventions are used:

$$\mathrm{Hom}_R(-, M)\colon \mathrm{Obj}(\mathsf{M}_R) \ni A \mapsto \mathrm{Hom}_R(A, M) \in \mathrm{Obj}(\mathsf{S})$$

$$\mathrm{Hom}_R(-, M)\colon \mathrm{Mor}(\mathsf{M}_R) \ni \alpha \mapsto \mathrm{Hom}_R(\alpha, M) \ni \mathrm{Mor}(\mathsf{S}),$$

in which we are to have

$$\mathrm{Hom}_R(\alpha, M) := \mathrm{Hom}_R(\alpha, 1_M).$$

Analogously for the second argument.

3.7 THE ENDOMORPHISM RING OF A MODULE

As mentioned in the previous section, for every module A $\mathrm{Hom}_R(A, A)$ is an additive abelian group. In addition we know that the composition $\beta\alpha$ of two homomorphisms

$$\alpha\colon A \to B, \qquad \beta\colon B \to C$$

is again a homomorphism. Consequently in $\mathrm{Hom}_R(A, A)$ the product of any two elements is defined by composition and this product is associative (being the composition of mappings).

3.7.1 THEOREM. $\operatorname{Hom}_R(A, A)$ *is a ring with a unit element if addition and multiplication are defined as*:

$$(\alpha_1+\alpha_2)(a)=\alpha_1(a)+\alpha_2(a)$$

$$(\alpha_1\alpha_2)=\alpha_1(\alpha_2(a)).$$

Proof. By virtue of the preceding explanation it remains to show that the distributive law holds:

$$((\alpha_1+\alpha_2)\alpha_3)(a)=(\alpha_1+\alpha_2)(\alpha_3(a))=\alpha_1(\alpha_3(a))+\alpha_2(\alpha_3(a))$$

$$=(\alpha_1\alpha_3)(a)+(\alpha_2\alpha_3)(a)=(\alpha_1\alpha_3+\alpha_2\alpha_3)(a)$$

$$\Rightarrow \quad (\alpha_1+\alpha_2)\alpha_3=\alpha_1\alpha_3+\alpha_2\alpha_3.$$

$$(\alpha_3(\alpha_1+\alpha_2))(a)=\alpha_3((\alpha_1+\alpha_2)(a))=\alpha_3(\alpha_1(a)+\alpha_2(a))$$

$$=\alpha_3(\alpha_1(a))+\alpha_3(\alpha_2(a))=(\alpha_3\alpha_1)(a)+(\alpha_3\alpha_2)(a)$$

$$=(\alpha_3\alpha_1+\alpha_3\alpha_2)(a)$$

$$\Rightarrow \quad \alpha_3(\alpha_1+\alpha_2)=\alpha_3\alpha_1+\alpha_3\alpha_2.$$

The unit element of $\operatorname{Hom}_R(A, A)$ is the identity mapping on A. □

3.7.2 *Definition.* The ring given in 3.7.1 is called the *endomorphism ring* of A (also called the R-endomorphism ring of A), and is denoted by $\operatorname{End}(A_R)$.

Example. If $V=V_K$ is a vector space then $\operatorname{End}(V_K)$ is the ring of linear mappings of V into itself.

Remark. If V_K is a vector space of dimension n with $0<n<\infty$ then $\operatorname{End}(V_K)$ is isomorphic as a ring to the ring of all $n\times n$ square matrices with coefficients in K. The proof of this fact is given later in a more general context.

We wish now to determine $\operatorname{End}(R_R)$ for an arbitrary ring R. To this end we consider for a fixed $r_0\in R$ the mapping

$$r_0^{(l)}: R\ni x\mapsto r_0x\in R.$$

From the distributive and associative laws we have $r_0^{(l)}\in\operatorname{Hom}_T(R_R, R_R)$; $r_0^{(l)}$ is said to be the *left multiplication* induced by r_0. Let now $\varphi\in\operatorname{End}(R_R)$, then for an arbitrary $x\in R$ and the unit element $1\in R$ we have:

$$\varphi(x)=\varphi(1x)=\varphi(1)x=\varphi(1)^{(l)}(x),$$

i.e. $\varphi = \varphi(1)^{(l)}$. Evidently $\mathrm{End}(R_R)$ consists precisely of all left multiplications, as a result of which we then write

$$R^{(l)} = \mathrm{End}(R_R).$$

3.7.3 LEMMA. *The mapping*

$$\rho: R \ni r \mapsto r^{(l)} \in R^{(l)}$$

is a ring isomorphism.

Proof. For $r_1, r_2, x \in R$ we have

$$\begin{aligned} (r_1+r_2)^{(l)}(x) &= (r_1+r_2)x = r_1x + r_2x \\ &= r_1^{(l)}(x) + r_2^{(l)}(x) = (r_1^{(l)} + r_2^{(l)})(x) \\ \Rightarrow \quad (r_1+r_2)^{(l)} &= r_1^{(l)} + r_2^{(l)}. \\ (r_1r_2)^{(l)}(x) &= (r_1r_2)x = r_1(r_2x) = r_1^{(l)}(r_2^{(l)}(x)) \\ &= (r_1^{(l)}r_2^{(l)})(x) \\ \Rightarrow \quad (r_1r_2)^{(l)} &= r_1^{(l)}r_2^{(l)}. \end{aligned}$$

Thus ρ is a ring homomorphism.

Let now $r_1x = r_2x$. Then for $x = 1: r_1 = r_11 = r_21 = r_2$. Thus we have $r_1^{(l)} = r_2^{(l)}$ and so $r_1 = r_2$, i.e. ρ is injective. It is clear that ρ is surjective. □

Analogously we can consider the ring $R^{(r)}$ of right multiplications of R, and we have analogously

$$R \cong R^{(r)} = \mathrm{End}({}_RR).$$

There follows now an important result on the endomorphism ring of a simple module. First of all we prove something more general.

3.7.4 LEMMA. *Let A and B be two simple R-modules. Then every homomorphism of A into B is either* 0 *or an isomorphism.*

Proof. Let $\alpha: A \to B$ be a homomorphism. From $\mathrm{Ker}(\alpha) \hookrightarrow A$ we have either $\mathrm{Ker}(\alpha) = A$, thus $\alpha = 0$ or $\mathrm{Ker}(\alpha) = 0$, i.e. α is a monomorphism. From $\mathrm{Im}(\alpha) \hookrightarrow B$ we have either $\mathrm{Im}(\alpha) = 0$, thus $\alpha = 0$ or $\mathrm{Im}(\alpha) = B$, i.e. α is an epimorphism. From both assertions: $\alpha \neq 0 \Rightarrow \alpha$ is an isomorphism. □

3.7.5 LEMMA (SCHUR). *The endomorphism ring of a simple module is a skew field.*

Proof. From 3.7.4 every non-zero endomorphism is an automorphism and thus has an inverse element in the endomorphism ring. Consequently the endomorphism ring is a skew field. □

We return once more to the general situation in which an arbitrary module A_R is given, and let $S := \operatorname{End}(A_R)$. In our notation the endomorphisms operate on the left of A. If we write for $\alpha \in S$, $a \in A$ instead of $\alpha(a)$ merely αa, we may verify easily that A is a left S-module. From

$$\alpha(ar) = \alpha(a)r = (\alpha a)r, \qquad \alpha \in S, a \in A, r \in R$$

A is in fact an S-R-bimodule. We shall come back later many times to this bimodule structure, the relationship between the structure of A_R, ${}_SA$ and ${}_SA_R$ will indeed play a role in certain considerations.

3.8 DUAL MODULES

As in the special case of vector spaces the concept of the dual module and the consequential relationships play an important role in the theory of modules. The main result of the following considerations consists of showing that (as with vector spaces, see 1.4.4) the passage to the bidual is a functor Δ, and that a functorial morphism exists between the identity functor and Δ.

We prove at once the following more general theorem:

3.8.1 THEOREM. *Let ${}_TL_R$ be given. Then*

(1) $$\operatorname{Hom}_R(-, {}_TL_R): \mathsf{M}_R \to {}_T\mathsf{M}$$

with

$$\operatorname{Hom}_R(-, {}_TL_R): \operatorname{Obj}(\mathsf{M}_R) \ni A \mapsto \operatorname{Hom}_R(A, {}_TL_R) \in \operatorname{Obj}({}_T\mathsf{M})$$

$$\operatorname{Hom}_R(-, {}_TL_R): \operatorname{Mor}(\mathsf{M}_R) \ni \alpha \mapsto \operatorname{Hom}_R(\alpha, 1_L) \in \operatorname{Mor}({}_T\mathsf{M})$$

is a contravariant functor.

(2) *Let*

$$\Delta_L := \operatorname{Hom}_T({}_T\operatorname{Hom}_R(-, {}_TL_R), {}_TL_R),$$

then

$$\Delta_L: \mathsf{M}_R \to \mathsf{M}_R$$

is a covariant functor.

(3) *For $A \in \mathsf{M}_R$ let*

$$\Phi_A: A \to \Delta_L(A)$$

with

$$\Phi_A(a)\colon \mathrm{Hom}_R(A, L) \ni \varphi \mapsto \varphi(a) \in L,$$

then

$$\Phi = (\Phi_A | A \in M_R)$$

is a functorial morphism between the identity functor 1_{M_R} and Δ_L.

Proof. (1) As already established earlier, $\mathrm{Hom}_R(A_R, {}_TL_R)$ is a left T-module, and for $\alpha \in \mathrm{Hom}_R(A, B)$ we have

$$\mathrm{Hom}_R(\alpha, 1_L)\colon \mathrm{Hom}_R(B, L) \ni \psi \mapsto \psi\alpha \in \mathrm{Hom}_R(A, L).$$

It remains to be established that $\mathrm{Hom}_R(\alpha, 1_L)$ is a left T-homomorphism; this follows immediately from

$$t\psi \cdot \alpha = t \cdot \psi\alpha, \qquad t \in T.$$

Finally we have

$$\mathrm{Hom}_R(1_A, 1_L) = 1_{\mathrm{Hom}_R(A,L)},$$

$$\mathrm{Hom}_R(\beta\alpha, 1_L) = \mathrm{Hom}_R(\alpha, 1_L)\mathrm{Hom}_R(\beta, 1_L),$$

and so everything is proved.

(2) The functor Δ_L is the composition of the functors

$$\mathrm{Hom}_R(-, {}_TL_R)\colon M_R \to {}_TM$$

and (of the analogously defined functor)

$$\mathrm{Hom}_T(-, {}_TL_R)\colon {}_TM \to M_R.$$

(3) Φ_A is an R-homomorphism. Let

$$a_1, a_2 \in A, \qquad r_1, r_2 \in E, \qquad \varphi \in \mathrm{Hom}_R(A, L),$$

then we have

$$\begin{aligned}\varphi\Phi(a_1r_1 + a_2r_2) &= \varphi(a_1r_1 + a_2r_2)\\ &= \varphi(a_1)r_1 + \varphi(a_2)r_2\\ &= \varphi\Phi(a_1)r_1 + \varphi\Phi(a_2)r_2\\ &= \varphi(\Phi(a_1)r_1 + \Phi(a_2)r_2),\end{aligned}$$

which was to be shown.

It remains to be proved that the diagram

$$\begin{array}{ccc} A & \xrightarrow{\Phi_A} & \Delta_L(A) \\ {\scriptstyle\alpha}\downarrow & & \downarrow{\scriptstyle\Delta_L(\alpha)} \\ B & \xrightarrow{\Phi_B} & \Delta_L(B) \end{array}$$

is commutative. For $a \in A$, $\psi \in \operatorname{Hom}_R(B, L)$ we have on the one hand

$$\psi\Phi_B(\alpha a) = \psi(\alpha a),$$

and on the other hand

$$\psi\Delta_L(\alpha)\Phi_A(a) = \psi\alpha\Phi_A(a) = \psi\alpha(a) = \psi(\alpha a),$$

which was to be shown. □

Of particular interest is the special case $T = R$ and ${}_TL_R = {}_RR_R$. We assume this in the following definition.

3.8.2 *Definition*

(1) For A_R

$$A^* := \operatorname{Hom}_R(A, R)$$

is called the *dual* and

$$A^{**} := \Delta(A) := \Delta_R(A) = \operatorname{Hom}_R({}_R\operatorname{Hom}_R(A_R, R_R), {}_RR)$$

the *bidual module* to A_R.

(2) For $\alpha: A_R \to B_R$

$$\alpha^* := \operatorname{Hom}_R(\alpha, R) = \operatorname{Hom}_R(\alpha, 1_R)$$

is called the *dual* and

$$\alpha^{**} := \operatorname{Hom}_R(\operatorname{Hom}_R(\alpha, R), R)$$

is called the *bidual homomorphism* to α.

(3) For $a \in A$ $a^{**} := \Phi_A(a)$ is called the *bidual element* to a.

For many considerations it is of interest to know which properties are possessed by the homomorphism

$$\Phi_A: A \ni a \mapsto a^{**} \in A^{**}.$$

If A_R is a finite-dimensional vector space then it is well known that Φ_A is an isomorphism. In general this is not the case. Different possibilities are characterized by particular denotations:

3.8.3 *Definition*. Let $\Phi_A: A \ni a \mapsto a^{**} \in A^{**}$.
(1) A_R is called *torsionless* $:\Leftrightarrow \Phi_A$ is a monomorphism.
(2) A_R is called *reflexive* $:\Leftrightarrow \Phi_A$ is an isomorphism.
Since later we have to consider numerous applications of these ideas, we here omit examples.

3.9 EXACT SEQUENCES

In homological algebra, complexes and exact sequences play an important role. They are a part of the fundamental concepts and are used, in particular, in the definition of the functors Ext and Tor. Although in this book we do not go further into homological concepts, nevertheless at least complexes and exact sequences are to be presented. Their usefulness appears subsequently in an application in Chapter 12 of this book.

Let R be a ring and let

$$\boldsymbol{A} := \ldots \xrightarrow{\alpha_{i-2}} A_{i-1} \xrightarrow{\alpha_{i-1}} A_i \xrightarrow{\alpha_i} A_{i+1} \xrightarrow{\alpha_{i+1}} \ldots$$

be a sequence of homomorphisms of right R-modules $A_i \xrightarrow{\alpha_i} A_{i+1}$, finite or infinite on one or other or both sides. For example $\boldsymbol{A}$ can have the form

$$\boldsymbol{A} = 0 \to A_1 \xrightarrow{\alpha_1} A_2 \xrightarrow{\alpha_2} A_3 \xrightarrow{\alpha_3} \ldots$$

or

$$\boldsymbol{A} = \ldots \xrightarrow{\alpha_{-4}} A_{-3} \xrightarrow{\alpha_{-3}} A_{-2} \xrightarrow{\alpha_{-2}} A_{-1} \to 0$$

or

$$\boldsymbol{A} = 0 \to A \xrightarrow{f} M \xrightarrow{g} W \to 0$$

where

$$0 \to A \qquad \text{resp. } W \to 0$$

is, as appropriate to the case, an unambiguously determined R-homomorphism. Finally the enumeration can also be inverted as for example in

$$\boldsymbol{A} = \ldots \xrightarrow{\alpha_3} A_3 \xrightarrow{\alpha_2} A_2 \xrightarrow{\alpha_1} A_1 \to 0.$$

3.9.1 *Definition* (a) A sequence $\boldsymbol{A}$ is called a *complex* $:\Leftrightarrow$ for every subsequence of the form

$$A_{i-1} \xrightarrow{\alpha_{i-1}} A_i \xrightarrow{\alpha_i} A_{i+1},$$

$$\operatorname{Im}(\alpha_{i-1}) \hookrightarrow \operatorname{Ker}(\alpha_i)$$

holds.

(b) A sequence (or complex) $\mathbf{A}$ is called *exact* : $\Leftrightarrow$ for every subsequence of the form

$$A_{i-1} \xrightarrow{\alpha_{i-1}} A_i \xrightarrow{\alpha_i} A_{i+1},$$

$$\operatorname{Im}(\alpha_{i-1}) = \operatorname{Ker}(\alpha_i)$$

holds.

(c) An exact sequence $\mathbf{A}$ is called a *split exact sequence* : $\Leftrightarrow$ for every subsequence of the form

$$A_{i-1} \xrightarrow{\alpha_{i-1}} A_i \xrightarrow{\alpha_i} A_{i+1},$$

$$\operatorname{Im}(\alpha_{i-1}) = \operatorname{Ker}(\alpha_i)$$

is a direct summand of A_i.

(d) If $\mathbf{A}$ is a complex then the sequence

$$\ldots, \operatorname{Ker}(\alpha_i)/\operatorname{Im}(\alpha_{i-1}), \operatorname{Ker}(\alpha_{i+1})/\operatorname{Im}(\alpha_i), \ldots$$

is called the *homology* of $\mathbf{A}$ and $\operatorname{Ker}(\alpha_i)/\operatorname{Im}(\alpha_{i-1})$ is called the ith *homology module* of $\mathbf{A}$.

(e) An exact sequence of the form

$$0 \to A \xrightarrow{f} M \xrightarrow{g} W \to 0$$

is called a *short exact sequence*.

We point out that a sequence $\mathbf{A}$ is a complex if and only if (for all occurring index pairs i, $i-1$)

$$\alpha_i \alpha_{i-1} = 0$$

holds (for $\alpha_i \alpha_{i-1} = 0 \Leftrightarrow \operatorname{Im}(\alpha_{i-1}) \hookrightarrow \operatorname{Ker}(\alpha_i)$).

All of these concepts are mentioned for the sake of completeness; in this book (in Chapter 12) we shall however only have short exact sequences to consider. We confine ourselves now to what we need there.

We begin first of all by making clear what it means for the short sequence

$$0 \to A \xrightarrow{f} M \xrightarrow{g} W \to 0$$

to be exact. Since the first mapping $0 \to A$ has image 0, the exactness of $0 \to A \xrightarrow{f} M$ indicates that f is a monomorphism. Since the last mapping $W \to 0$ is the zero mapping with Kernel $= W$, the exactness of $M \xrightarrow{g} W \to 0$ indicates that g is an epimorphism. From $\operatorname{Im}(f) = \operatorname{Ker}(g)$ it then follows that $M/\operatorname{Im}(f) \cong W$.

If $A \hookrightarrow M$ then we obtain in particular the short exact sequence

$$0 \to A \xrightarrow{\iota} M \xrightarrow{\nu} M/A \to 0,$$

where ι is the inclusion mapping and ν is the natural epimorphism.

The following lemma is needed later.

3.9.2 LEMMA. *Let all modules be right R-modules and let all homomorphisms be R-module homomorphisms. Let*

$$\begin{array}{ccccccccc} 0 & \longrightarrow & A & \xrightarrow{f} & M & \xrightarrow{g} & W & \longrightarrow & 0 \\ & & \Big\downarrow{\scriptstyle\alpha} & & \Big\downarrow{\scriptstyle\mu} & & \Big\downarrow{\scriptstyle\omega} & & \\ 0 & \longrightarrow & B & \xrightarrow{h} & N & \xrightarrow{k} & X & \longrightarrow & 0 \end{array}$$

be a commutative diagram (i.e. $\mu f = h\alpha$ and $\omega g = k\mu$ are to hold) with exact rows and let α, μ, ω be monomorphisms. Then μ is an isomorphism if and only if α and ω are both isomorphisms.

Proof. First let μ be an isomorphism. Let $b \in B$. Then $h(b) \in N$ and so there is $m \in M$ with $\mu(m) = h(b)$, thus $\omega g(m) = k\mu(m) = kh(b) = 0$. Since ω is a monomorphism it follows that $g(m) = 0$, thus $m \in \text{Ker}(g) = \text{Im}(f) \Rightarrow$ there is an $a \in A$ with $f(a) = m$. Hence $h\alpha(a) = \mu f(a) = \mu(m) = h(b)$ and so $h(\alpha(a) - b) = 0$, and since h is a monomorphism, it follows that $\alpha(a) = b$, i.e. α is also surjective and, in consequence, an isomorphism.

Let now $x \in X$ be given. Then there is $n \in N$ with $k(n) = x$ and so there is $m \in M$ with $\mu(m) = n$, thus $\omega g(m) = k\mu(m) = k(n) = x \Rightarrow \omega$ is likewise surjective, thus an isomorphism.

Conversely let now α and ω be two isomorphisms and let $n \in N$ be given. Then there is $w \in W$ with $\omega(w) = k(n)$. Consequently an $m \in M$ exists with $g(m) = w \Rightarrow k\mu(m) = \omega g(m) = \omega(w) = k(n) \Rightarrow k(n - \mu(m)) = 0$

$\Rightarrow$ there is $b \in B$ with $h(b) = n - \mu(m)$;

$\Rightarrow$ there exists an $a \in A$ with $\alpha(a) = b$;

$\Rightarrow \mu f(a) = h\alpha(a) = h(b) = n - \mu(m)$;

$\Rightarrow \mu(f(a) + m) = n$, thus μ is surjective and consequently is an isomorphism. □

This proof is a typical example of so-called diagram-chasing. It is clear that, without the notation of diagrams, this proof would be very obscure.

We direct our attention now to split short exact sequences. Let

$$0 \to A \xrightarrow{f} M \xrightarrow{g} W \to 0$$

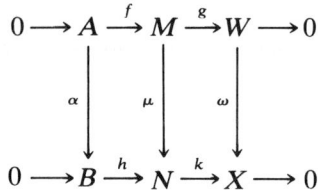

be an exact sequence. Obviously the splitting of the subsequences

$$0 \to A \xrightarrow{f} M \quad \text{and} \quad M \xrightarrow{g} W \to 0$$

is already given, so that the splitting of the given short exact sequence depends only on the splitting of

$$A \xrightarrow{f} M \xrightarrow{g} W$$

i.e., on whether $\operatorname{Im}(f) = \operatorname{Ker}(g)$ is a direct summand in M.

3.9.3 LEMMA. *Let* $\mathbf{A} = 0 \to A \xrightarrow{f} M \xrightarrow{g} W \to 0$ *be a short exact sequence.*

(a) *The following are equivalent*:
  (1) $\mathbf{A}$ *splits.*
  (2) *There exists a homomorphism* $f_0: M \to A$ *with* $f_0 f = 1_A$.
  (3) *There exists a homomorphism* $g_0: W \to M$ *with* $g g_0 = 1_W$.

(b) *If* $\mathbf{A}$ *splits, then* f_0 *and* g_0 *exist as in* (2), (3) *resp. so that*

$$0 \leftarrow A \xleftarrow{f_0} M \xleftarrow{g_0} W \leftarrow 0$$

is exact and splits.

Proof. (a) "(1)⇔(2)": 3.4.11 (1).
"(1)⇔(3)": 3.4.11 (2).

(b) Let $f_0: M \to A$ with $f_0 f = 1_A$ chosen arbitrarily. From 3.4.10 it follows that

$$M = \operatorname{Im}(f) \oplus \operatorname{Ker}(f_0) = \operatorname{Ker}(g) \oplus \operatorname{Ker}(f_0).$$

From this $g|\operatorname{Ker}(f_0)$ is an isomorphism.

Let now $h: W \to \operatorname{Ker}(f_0)$ be the inverse isomorphism and let $\iota: \operatorname{Ker}(f_0) \to M$ be the inclusion mapping, then let $g_0 := \iota h$. As $M = \operatorname{Ker}(g) \oplus \operatorname{Ker}(f_0)$ and since g is an epimorphism every element from W may be written in the form $g(x)$ with $x \in \operatorname{Ker}(f_0)$. It then follows that

$$g g_0(g(x)) = g\iota(hg(x)) = g(x),$$

thus $g g_0 = 1_W$ and also $g_0 g(x) = x$, thus $\operatorname{Im}(g_0) = \operatorname{Ker}(f_0)$. Consequently

$$0 \leftarrow A \xleftarrow{f_0} M \xleftarrow{g_0} W \leftarrow 0$$

is exact and, from $g g_0 = 1_W$, $f_0 f = 1_A$, splits by (a). □

EXERCISES

(1)

Let $A, L \in M_R$, $\alpha: A \to L$, $B \hookrightarrow A$, $C \hookrightarrow A$, $M \hookrightarrow L$, $N \hookrightarrow L$.

(a) Prove: The following statements are equivalent:

(1) $\alpha(B \cap C) = \alpha(B) \cap \alpha(C)$.

(2) $(B + \mathrm{Ker}(\alpha)) \cap (C + \mathrm{Ker}(\alpha)) = B \cap C + \mathrm{Ker}(\alpha)$.

(3) $(B \cap \mathrm{Ker}(\alpha)) + (C \cap \mathrm{Ker}(\alpha)) = (B + C) \cap \mathrm{Ker}(\alpha)$.

(b) Prove: The following statements are equivalent:

(1) $\alpha^{-1}(M + N) = \alpha^{-1}(M) + \alpha^{-1}(N)$.

(2) $(M \cap \mathrm{Im}(\alpha)) + (N \cap \mathrm{Im}(\alpha)) = (M + N) \cap \mathrm{Im}(\alpha)$.

(3) $(M + \mathrm{Im}(\alpha)) \cap (N + \mathrm{Im}(\alpha)) = (M \cap N) + \mathrm{Im}(\alpha)$.

(2)

Construct an example in which the conditions in 1(a), 1(b) resp. are not satisfied.

(3)

(a) Let a module homomorphism $\varphi: M \to N$ be given and also $A \hookrightarrow M$, $V \hookrightarrow N$.

Show: $\varphi^{-1}(\varphi(A) + V) = A + \varphi^{-1}(V)$.

(b) Let a module homomorphism $\varphi: M \to N$ be given and also $B \hookrightarrow N$, $U \hookrightarrow M$.

Show: $\varphi(\varphi^{-1}(B) \cap U) = B \cap \varphi(U)$.

(4)

(a) Prove: In the category of unitary rings every monomorphism is injective. (Hint: Use $\mathbb{Z}[x]$ = polynomial ring in x with coefficients in $\mathbb{Z}$).

(b) Prove: $\iota: \mathbb{Z} \to \mathbb{Q}$ is an epimorphism in the category of unitary rings.

(5)

Determine all composition series of $\mathbb{Z}/30\mathbb{Z}$ and exhibit all isomorphisms between them.

(6)

(a) Determine the following groups:

$$\begin{array}{ll} \mathrm{Hom}_{\mathbb{Z}}(\mathbb{Q}, \mathbb{Z}), & \mathrm{Hom}_{\mathbb{Z}}(\mathbb{Q}/\mathbb{Z}, \mathbb{Z}) \\ \mathrm{Hom}_{\mathbb{Z}}(\mathbb{Q}, \mathbb{Q}), & \mathrm{Hom}_{\mathbb{Z}}(\mathbb{Z}/n\mathbb{Z}, \mathbb{Q}) \end{array} \quad \text{for } n \in \mathbb{N}.$$

(b) Show for $M \in M_R$: $\mathrm{Hom}_R(R, M) \cong M$.

(7)

Let A be an additive abelian group and let End(A) be the endomorphism ring of A, where for $\alpha \in \text{End}(A)$ and $a \in A$ αa is the image of a by α. Further let R be a ring and let

$$\rho: R \to \text{End}(A)$$

be a ring homomorphism, unitary ring homomorphism resp.

(a) Show: By the definition

$$ra := \rho(r)a, \qquad a \in A, r \in R,$$

A becomes a left R-module, unitary left R-module resp.

(b) Show: Every left R-module, unitary left R-module resp., ${}_RA$ with A as additive group is obtained in the manner outlined above.

(c) Construct an example of an additive abelian group A and a ring R so that A is a unitary left R-module in two different ways.

(d) Formulate the corresponding relations for right R-modules without altering the multiplication in End(A).

(8)

Prove: For every vector space V_K the endomorphism ring End(V_K) is regular (for the definition see Chapter 2, Exercise 13).

Chapter 4

Direct Products, Direct Sums, Free Modules

In the structure theory of modules we attempt, on the one hand, to reduce a given module to simpler modules by means of additive decomposition or residue class decomposition. On the other hand we endeavour to construct new modules from given modules. Obviously this construction is not arbitrarily undertaken; a guiding principle is the question of modules with known universal properties. We have already become acquainted with such universal properties in respect of products and coproducts in categories (1.5). Products and coproducts are now to be investigated in the category of modules.

4.1 CONSTRUCTION OF PRODUCTS AND COPRODUCTS

We begin by recalling some known set-theoretic concepts. Let $(A_i|i \in I)$ be a family of sets A_i with index set $I \neq \varnothing$. Then the product $\prod_{i\in I} A_i$ of the family $(A_i|i \in I)$ is the set of the mappings

$$\alpha: I \to \bigcup_{i\in I} A_i$$

with $\alpha(i) \in A_i$ for all $i \in I$.

Notation

(1) $a_i := \alpha(i)$ is called the ith component of α.
(2) $(a_i) := (\alpha(i)) := \alpha$.

Thus we obviously have for $(a_i), (a_i') \in \prod_{i\in I} A_i$:

$$(a_i) = (a_i') \Leftrightarrow \forall i \in I[a_i = a_i'].$$

We observe that I need not be countable. If however I is countable, say $I=\{1, 2, 3, \ldots\}$, then the notation

$$(a_1a_2a_3\ldots) := (a_i) = \alpha$$

is also used. If I is finite, say $I=\{1, 2, \ldots, n\}$, then let

$$(a_1a_2\ldots a_n) := (a_i) = \alpha.$$

If now $A_i \in \mathcal{M}_R$ holds for all $i \in I$ then, by a componentwise definition, $\prod_{i\in I} A_i$ becomes a unitary right R-module.

4.1.1 *Definition.* Let $(a_i), (b_i) \in \prod_{i\in I} A_i$, $r \in R$.

Addition: $(a_i)+(b_i) := (a_i+b_i)$.

Module multiplication: $(a_i)r := (a_ir)$.

If again we write $\alpha = (a_i)$, $\beta = (b_i)$ then instead of the above we have

$$(\alpha+\beta)(i) := \alpha(i)+\beta(i), \qquad i \in I$$

$$(\alpha r)(i) := \alpha(i)r, \qquad i \in I.$$

The proof, that with respect to this definition $\prod_{i\in I} A_i$ is an object from $\mathcal{M}_R$, is trivial. In particular the zero mapping

$$I \ni i \mapsto 0_i \in \bigcup_{i\in I} A_i,$$

where 0_i is the zero of A_i, is the zero element of $\prod A_i$ and $-\alpha := (-a_i)$ is the element inverse to $\alpha = (a_i)$ with respect to addition.

4.1.2 *Definition.* An element $(a_i) \in \prod_{i\in I} A_i$ is said to be of *finite support*: $\Leftrightarrow$ the set of the $i \in I$ with $a_i \neq 0$ is finite (where the empty set is considered also as finite).

We see then from the criterion for submodules that the set of all elements from $\prod A_i$ of finite support is a submodule of $\prod A_i$.

4.1.3 *Definition*

(1) If $(A_i \mid i \in I)$ is a family of objects from $\mathcal{M}_R$ then $\prod_{i\in I} A_i \in \mathcal{M}_R$ is called the *direct product* of the family $(A_i \mid i \in I)$.

(2) The submodule of all elements of finite support of $\prod_{i\in I} A_i$ is called the *external direct sum* of the family $(A_i \mid i \in I)$ and is denoted by $\coprod_{i\in I} A_i$.

4.1.4 *Remark.* If I is finite then we have

$$\prod_{i\in I} A_i = \coprod_{i\in I} A_i.$$

In 1.5 we had defined in an arbitrary category the product and coproduct of a family of objects. We now have to show that the direct product, direct sum resp., is, together with a certain family of homomorphisms, the product in $\boldsymbol{M}_R$, coproduct in $\boldsymbol{M}_R$ resp.

For $j \in I$ we consider the following mappings:

$$\pi_j\colon \prod_{i\in I} A_i \ni (a_i) \mapsto a_j \in A_j$$

$$\sigma\colon \coprod_{i\in I} A_i \ni (a_i) \mapsto (a_i) \in \prod_{i\in I} A_i$$

$$\eta_j\colon A_j \ni a_j \mapsto \boldsymbol{a}_j \in \coprod_{i\in I} A_i, \qquad \text{with } \boldsymbol{a}_j(i) = \begin{cases} 0 & \text{for } i \neq j \\ a_j & \text{for } i = j \end{cases}$$

We then easily verify the following properties.

4.1.5 LEMMA

(1) *π_j and $\pi_j\sigma$ are epimorphisms.*

(2) *η_j and $\sigma\eta_j$ are monomorphisms.*

(3) $\pi_k\sigma\eta_j = \begin{cases} 1_{A_j} & \textit{for } k = j \\ 0 & \textit{for } k \neq j \end{cases}$

(4) $(\sigma\eta_j\pi_j)^2 = \sigma\eta_j\pi_j$, $(\eta_j\pi_j\sigma)^2 = \eta_j\pi_j\sigma$.

(5) *If $I = \{1, 2, \ldots, n\}$ then*

$$(\eta_j\pi_j)^2 = \eta_j\pi_j \wedge 1_{\prod A_i} = \sum_{j=1}^{n} \eta_j\pi_j.$$

4.1.6 THEOREM

(1) $\left(\prod_{i\in I} A_i, (\pi_i \mid i \in I)\right)$ *is a product of the family $(A_i \mid i \in I)$ in the category $\boldsymbol{M}_R$, i.e. for every object C from $\boldsymbol{M}_R$ and every family $(\gamma_i \mid i \in I)$ of homomorphisms*

$$\gamma_i\colon C \to A_i, \qquad i \in I$$

there exists exactly one homomorphism

$$\gamma\colon C \to \prod_{i\in I} A_i$$

satisfying

$$\gamma_i = \pi_i \gamma, \qquad i \in I.$$

(2) $\left(\coprod_{i\in I} A_i, (\eta_i | i \in I)\right)$ *is a coproduct of the family* $(A_i | i \in I)$ *in the category* M_R, *i.e. for every object* B *from* M_R *and every family* $(\beta_i | i \in I)$ *of homomorphisms*

$$\beta_i : A_i \to B, \qquad i \in I$$

there exists exactly one homomorphism

$$\beta : \coprod_{i\in I} A_i \to B$$

satisfying

$$\beta_i = \beta \eta_i, \qquad i \in I.$$

Proof. (1) We exhibit the desired $\gamma : C \to \prod_{i\in I} A_i$: Let

$$\gamma(c) := (\gamma_i(c)) \in \prod_{i\in I} A_i \qquad \text{for } c \in C.$$

Then γ is a homomorphism and we have

$$(\pi_j \gamma)(c) = \pi_j(\gamma(c)) = \gamma_j(c), \qquad c \in C,$$

thus $\gamma_j = \pi_j \gamma$, $j \in I$.

Uniqueness of γ: Let also $\gamma' : C \to \prod_{i\in I} A_i$ with $\gamma_j(c) = (\pi_j \gamma')(c) = \pi_j(\gamma'(c))$, then it follows that

$$\gamma'(c) = (\gamma_i(c)) = \gamma(c),$$

thus $\gamma' = \gamma$.

(2) We can again give the desired $\beta : \coprod_{i\in I} A_i \to B$ explicitly: Let

$$\beta((a_i)) := \sum \beta_i(a_i) \in B,$$

where the sum runs over only the $i \in I$ with $a_i \neq 0$; from the definition of $\coprod A_i$ the sum is thereby meaningful (the sum over the empty index set is, as always, put equal to 0).

The β is a homomorphism, and we have

$$(\beta \eta_j)(a_j) = \beta(\boldsymbol{a}_j) = \beta_j(a_j),$$

thus

$$\beta_j = \beta\eta_j, \qquad j \in I.$$

Uniqueness of β: Let also $\beta'\colon \coprod_{i\in I} A_i \to B$ with $\beta_j = \beta'_{\eta j}$, then it follows that

$$\boldsymbol{\beta}(\boldsymbol{a}_j) = \beta_j(a_j) = \beta'\eta_j(a_j) = \beta'(\boldsymbol{a}_j)$$

and since every element from $\coprod A_i$ is a sum of finitely many $\boldsymbol{a}_j$, we deduce that $\beta = \beta'$. □

The following notational device is common and we also employ it.

4.1.7 *Notation*. Let I be a non-empty set and let $A \in M_R$. Then let

$$A^I := \prod_{i\in I} A_i \qquad \text{with } A_i = A \quad \text{for every } i \in I.$$

$$A^{(I)} := \coprod_{i\in I} A_i \qquad \text{with } A_i = A \quad \text{for every } i \in I.$$

We call A^I, $A^{(I)}$ resp., the direct product, the direct sum resp., of I copies of A.

4.2 CONNECTION BETWEEN THE INTERNAL AND EXTERNAL DIRECT SUMS

In 2.4 the internal direct sum was introduced and in the preceding chapter we have defined the (external) direct sum. We are about to show that these concepts are not essentially different from one another so that in what follows they can mostly be identified without leading to misunderstanding.

Thus we have the monomorphism

$$\eta_j\colon A_j \ni a_j \mapsto \boldsymbol{a}_j \in \coprod_{i\in I} A_i,$$

where

$$\boldsymbol{a}_j(i) = \begin{cases} 0 & \text{for } i \neq j \\ a_j & \text{for } i = j \end{cases}$$

following on from 4.1. Let $A'_j := \eta_j(A_j)$, then A'_j is a module isomorphic to A_j.

In the case that I is the set $\{1, 2, 3, \ldots, n\}$ it follows that

$$\boldsymbol{a}_j = (0 \ldots 0 a_j 0 \ldots 0),$$

↑ jth place

and also

$$A'_j = \{(0 \dots 0 a_j 0 \dots 0) | a_j \in A_j\}.$$

4.2.1 THEOREM. *Let $(A_i | i \in I)$ be a family of R-modules. Then we have:*

$$\coprod_{i \in I} A_i = \bigoplus_{i \in I} A'_i \quad \text{and} \quad A_i \cong A'_i,$$

in other words, the external direct sum of the A_i is equal to the internal direct sum of the submodules A'_i of $\coprod_{i \in I} A_i$ isomorphic to the A_i.

Proof. From the definition of the A'_i we have $\sum_{i \in I} A'_i \hookrightarrow \coprod_{i \in I} A_i$. Let now $0 \neq (a_i) \in \coprod_{i \in I} A_i$ and let $a_{i_1} \neq 0, \dots, a_{i_n} \neq 0$, whereas $a_i = 0$ for all other $i \in I$, then it follows that

$$(a_i) \in A'_{i_1} + \dots + A'_{i_n}, \quad \text{thus} \quad \sum_{i \in I} A'_i = \coprod_{i \in I} A_i.$$

Let

$$(a_i) \in A'_j \cap \sum_{\substack{i \neq j \\ i \in I}} A'_i \Rightarrow a_i = 0 \qquad \text{for } i \neq j$$

and

$$a_j = 0 \Rightarrow (a_i) = 0.$$

As asserted above, we have finally $A_i \cong A'_i$, where $a_i \mapsto \eta_i(a_i)$. □

Warning. In the following the isomorphic modules A_i and A'_i are usually identified and so A_i is written in place of A'_i. Moreover on account of Theorem 4.2.1, the distinction between internal and external direct sums is often dropped and in both cases $\bigoplus A_i$ is written and called the direct sum. In the absence of any indication it is to be determined from the context which particular direct sum is being considered.

4.3 HOMOMORPHISMS OF DIRECT PRODUCTS AND SUMS

Let $(A_i | i \in I)$, $(B_i | i \in I)$ be two families of $A_i, B_i \in \mathcal{M}_R$. Further let $(\alpha_i | i \in I)$ be a family of homomorphisms

$$\alpha_i : A_i \to B_i, \qquad i \in I.$$

Under these assumptions we have

4.3.1 LEMMA. *The mappings defined by,*

$$\prod_{i\in I}\alpha_i\colon \prod_{i\in I}A_i \ni (a_i)\mapsto(\alpha_i(a_i))\in\prod_{i\in I}B_i,$$

$$\bigoplus_{i\in I}\alpha_i\colon \bigoplus_{i\in I}A_i \ni (a_i)\mapsto(\alpha_i(a_i))\in\bigoplus_{i\in I}B_i$$

are homomorphisms with

(1) $\prod \alpha_i$ *is* mono$\Leftrightarrow\bigoplus\alpha_i$ *is* mono$\Leftrightarrow\forall i\in I$ [α_i *is* mono];
(2) $\prod \alpha_i$ *is* epi$\Leftrightarrow\bigoplus\alpha_i$ *is* epi$\Leftrightarrow\forall i\in I$ [α_i *is* epi];
(3) $\prod \alpha_i$ *is* iso$\Leftrightarrow\bigoplus\alpha_i$ *is* iso$\Leftrightarrow\forall i\in I$ [α_i *is* iso].

Proof. Exercise for the reader.

4.3.2 LEMMA. *Assumptions as above. Let further*

$$\begin{aligned}\iota_i\colon \operatorname{Ker}(\alpha_i)\ni a_i\mapsto a_i\in A_i,\\ \iota_i'\colon \operatorname{Im}(\alpha_i)\ni b_i\mapsto b_i\in B_i,\end{aligned}\qquad i\in I,$$

then the following are isomorphisms:

(1) $\prod_{i\in I}\operatorname{Ker}(\alpha_i)\ni(a_i)\mapsto(\iota_i(a_i))\in\operatorname{Ker}\left(\prod_{i\in I}\alpha_i\right).$

(2) $\bigoplus_{i\in I}\operatorname{Ker}(\alpha_i)\ni(a_i)\mapsto(\iota_i(a_i))\in\operatorname{Ker}\left(\bigoplus_{i\in I}\alpha_i\right).$

(3) $\prod_{i\in I}\operatorname{Im}(\alpha_i)\ni(b_i)\mapsto(\iota_i'(b_i))\in\operatorname{Im}\left(\prod_{i\in I}\alpha_i\right).$

(4) $\bigoplus_{i\in I}\operatorname{Im}(\alpha_i)\ni(b_i)\mapsto(\iota_i'(b_i))\in\operatorname{Im}\left(\bigoplus_{i\in I}\alpha_i\right).$

Thus

$$\prod_{i\in I}\operatorname{Ker}(\alpha_i)\cong\operatorname{Ker}\left(\prod_{i\in I}\alpha_i\right),\qquad \bigoplus_{i\in I}\operatorname{Ker}(\alpha_i)\cong\operatorname{Ker}\left(\bigoplus_{i\in I}\alpha_i\right)$$

$$\prod_{i\in I}\operatorname{Im}(\alpha_i)\cong\operatorname{Im}\left(\prod_{i\in I}\alpha_i\right),\qquad \bigoplus_{i\in I}\operatorname{Im}(\alpha_i)\cong\operatorname{Im}\left(\bigoplus_{i\in I}\alpha_i\right).$$

Proof. Exercise for the reader.

4.3.3 LEMMA. *Let the families* $(A_i \mid i\in I)$, $(B_j\mid j\in J)$ *be given, then*

$$\operatorname{Hom}_R\left(\bigoplus_{i\in I}A_i,\prod_{j\in J}B_j\right)\ni\varphi\mapsto(\pi_j\varphi\eta_i)\in\prod_{(i,j)\in I\times J}\operatorname{Hom}_R(A_i,B_j)$$

is a group isomorphism.

Proof. It is clear that a group homomorphism is involved. It remains to prove:

"Mono": Let $0 \neq \varphi \in \mathrm{Hom}_R(\bigoplus A_i, \prod B_j)$. Then there exists

$$(a_i) \in \bigoplus A_i \quad \text{with} \quad \varphi((a_i)) \neq 0.$$

Since $(a_i) = \sum_{a_i \neq 0} \boldsymbol{a}_i$ we have $\varphi((a_i)) = \varphi(\sum \boldsymbol{a}_i) = \sum \varphi(\boldsymbol{a}_i) \neq 0$

$\Rightarrow$ there exists i with $\varphi(\boldsymbol{a}_i) = \varphi\eta_i(a_i) = 0$;

$\Rightarrow$ there exists j with $\pi_j\varphi\eta_i(a_i) \neq 0 \Rightarrow \pi_j\varphi\eta_i \neq 0$.

"Epi": Let $(\alpha_{ji}) \in \prod \mathrm{Hom}_R(A_i, B_j)$. To a fixed $i \in I$ and to the family $(\alpha_{ji} | j \in J)$, where $\alpha_{ji}: A_i \to B_j$ there is then associated by 4.1.6 a homomorphism $\beta_i: A_i \to \prod_{j \in J} B_j$ so that

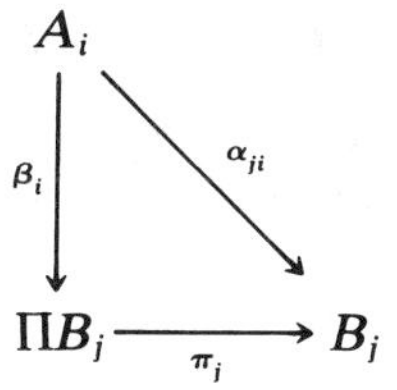

is commutative.

To the family $(\beta_i | i \in I)$ there then corresponds by 4.1.6 a $\varphi: \bigoplus A_i \to \prod B_j$, so that

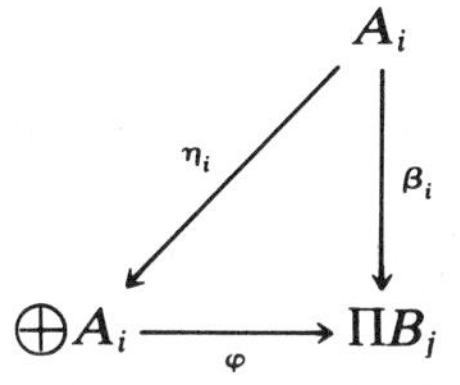

is commutative. Then follows $\alpha_{ji} = \pi_j\beta_i = \pi_j\varphi\eta_i$ from which the assertion follows. □

Special cases

$$\mathrm{Hom}_R\Big(\bigoplus_{i \in I} A_i, B\Big) \cong \prod_{i \in I} \mathrm{Hom}_R(A_i, B), \qquad \text{where } \varphi \mapsto (\varphi\eta_i).$$

$$\mathrm{Hom}_R\Big(A, \prod_{j \in J} B_j\Big) \cong \prod_{j \in J} \mathrm{Hom}_R(A, B_j), \qquad \text{where } \varphi \mapsto (\pi_j\varphi).$$

4.4 FREE MODULES

In 2.3.5 a basis of a module was defined as a free generating set. Modules that possess a basis can be characterized in the following manner.

4.4.1 LEMMA. *Let $F = F_R$. Then the following conditions are equivalent*:
(1) *F has a basis.*
(2) $F = \bigoplus_{i \in I} A_i \wedge \forall i \in I[R_R \cong A_i]$.

Proof. We remark first of all that (1) and (2) are satisfied for $F = 0$, in fact with $\varnothing$ as basis and $I = \varnothing$. This follows by convention that the sum over the empty set is equal to 0. We can therefore assume that $F \neq 0$.

"(1)⇒(2)": Let X be a basis of F and let $a \in X$. Then

$$\varphi_a : R_R \ni r \mapsto ar \in R_R$$

is evidently an epimorphism. Further, from the property of a basis, it follows from $ar = 0 = a0$ that $r = 0$ and so we obtain an isomorphism. We claim that

$$F = \bigoplus_{a \in X} aR.$$

Since X is a basis, X is also a generating set and so we have $F = \sum_{a \in X} aR$. For $a_0 \in X$ let

$$c \in a_0 R \cap \sum_{\substack{a \in X \\ a \neq a_0}} aR,$$

then there exist distinct $a_1, \ldots, a_n \in X$, $a_i \neq a_0$ and $r_0, r_1, \ldots, r_n \in R$ with

$$c = a_0 r_0 = \sum_{i=1}^{n} a_i r_i, \qquad \Rightarrow a_0 r_0 + \sum a_i(-r_i) = 0.$$

Thus from the property of a basis 2.3.5 (4):

$$r_0 = r_1 = \ldots = r_n = 0 \Rightarrow a_0 R \cap \sum_{\substack{a \in X \\ a \neq a_0}} aR = 0,$$

hence

$$F = \bigoplus_{a \in X} aR.$$

"(2)⇒(1)": Let φ: $R_R \cong A_i$ be the isomorphism which we are supposing to exist. We claim that $\{\varphi_i(1) | i \in I\}$ is a basis of F. From

$$A_i = \varphi_i(R) = \varphi_i(1 \cdot R) = \varphi_i(1)R$$

we have

$$F = \bigoplus_{i\in I} A_i = \bigoplus_{i\in I} \varphi_i(1)R,$$

thus $\{\varphi_i(1)|i \in I\}$ is a generating set of F. Let $I' \subset I$, I' be finite and

$$\sum_{i\in I'} \varphi_i(1)r_i = 0.$$

Then it follows from 2.4.2 for all $i \in I'$

$$\varphi_i(1)r_i = \varphi_i(r_i) = 0$$

and, since φ_i is an isomorphism, $r_i = 0$, thus $\{\varphi_i(1)|i \in I\}$ is in fact a basis of F. □

4.4.2 *Definition.* A module F, which satisfies the conditions of 4.4.1, is called a *free module.*

4.4.3 LEMMA. *Let I be a set. Then $R^{(I)}$ is a free R-module with a basis having the cardinality of I.*

Proof. We consider the family $(A_i|i \in I)$ with $A_i = R_R$ for all $i \in I$. Then it follows from 4.2.1 that

$$R^{(I)} = \coprod_{i\in I} A_i = \bigoplus_{i\in I} A_i' \quad \text{with} \quad R_R = A_i \overset{\varphi_i}{\cong} A_i'.$$

As shown previously it follows that $R^{(I)}$ is free and has $\{\varphi_i(1)|i \in I\}$ as a basis. □

It is pertinent to recall (see 4.2) that in the case $I = \{1, 2, \ldots, n\}$ we have

$$\varphi_i(1) = (0 \ldots 0\,1\,0 \ldots 0)$$

i.e. $\{\varphi_i(1)|i = 1, \ldots, n\}$ is then the "canonical basis of R^n".

4.4.4 COROLLARY. *Every module M_R is the epimorphic image of a free right R-module. If M_R is finitely generated, then M_R is an epimorphic image of a free right R-module with a finite basis.*

Proof. Let Y be a generating system of M. Then we consider the free module

$$R^{(Y)} = \bigoplus_{b\in Y} \varphi_b(1)R.$$

From

$$R^{(Y)} \ni \sum \varphi_b(1)r_b \mapsto \sum br_b \in M$$

and by virtue of the uniqueness of the representation by the basis in $R^{(Y)}$ an epimorphism is then defined. □

4.4.5 *Notation.* We have denoted the basis of $R^{(I)}$ by $\{\varphi_i(1) | i \in I\}$ but in the following we are not to be confined to this notation. Obviously it can also be denoted by any other set, which has the cardinality of I, e.g. by I itself.

$$M = \bigoplus_{i \in I} iR.$$

We point out another important property of free modules which later (in the case of projective modules) plays a fundamental role.

4.4.6 THEOREM. *If*

$$\varphi : A_R \to F_R$$

is an epimorphism and if F_R is free then φ splits. (*Definition see* 3.4.8).

Proof. Let Y be a basis of F_R and to every $b \in Y$ let $a_b \in A$ be chosen with $\varphi(a_b) = b$. Then the mapping

$$\varphi' : F \ni \sum br_b \mapsto \sum a_b r_b \in A$$

is an R-homomorphism (since Y is a basis).

Thus we have

$$\varphi\varphi'(\sum br_b) = \varphi(\sum a_b r_b) = \sum \varphi(a_b) r_b = \sum br_b,$$

thus $\varphi\varphi' = 1_F$ and consequently

$$A = \operatorname{Im}(\varphi') \oplus \operatorname{Ker}(\varphi). \qquad \square$$

4.5 FREE AND DIVISIBLE ABELIAN GROUPS

Every abelian group can be considered in a natural sense as a $\mathbb{Z}$-module, so that all module-theoretic concepts are applicable to abelian groups. Accordingly an abelian group is called free if it is free as a $\mathbb{Z}$-module, i.e. if it is a direct sum of copies of $\mathbb{Z}_{\mathbb{Z}}$.

If, in the following, the discourse concerns groups, then it is always to be additive abelian groups that are involved.

4.5.1 *Definition.* A group A is called *divisible* $:\Leftrightarrow$

$$\forall z \in \mathbb{Z}[z \neq 0 \Rightarrow Az = A].$$

4.5.2 LEMMA. *Every epimorphic image of a divisible group is divisible and consequently every factor group of a divisible group is divisible.*

Proof. Let A be divisible and let

$$\Phi: A \to B$$

be an epimorphism. Then we have for $0 \neq z \in \mathbb{Z}$,

$$Bz = \Phi(A)z = \Phi(Az) = \Phi(A) = B,$$

thus B is divisible. □

4.5.3 LEMMA. *The direct product and the direct sum of divisible groups are divisible.*

Proof. Let $(A_i | i \in I)$ be a family of divisible groups. Then we have for $0 \neq z \in \mathbb{Z}$

$$\left(\prod_{i\in I} A_i\right)z = \prod_{i\in I} (A_i z) = \prod_{i\in I} A_i,$$

$$\left(\bigoplus_{i\in I} A_i\right)z = \bigoplus_{i\in I} (A_i z) = \bigoplus_{i\in I} A_i,$$

as we immediately deduce from the definition of the direct product and the direct sum. □

Examples: $\mathbb{Q}$ and $\mathbb{Q}/\mathbb{Z}$ are divisible groups. $\mathbb{Z}$ is not divisible.

4.5.4 THEOREM. *Every abelian group can be mapped monomorphically into a divisible group.*

Proof. Let A be an abelian group. From 4.4.4 there is a free abelian group F and an epimorphism.

$$\Phi: F \to A.$$

If we put $\bar{x} := x + \mathrm{Ker}(\Phi)$ then

$$\Phi: F/\mathrm{Ker}(\Phi) \ni \bar{x} \mapsto \Phi(x) \in A$$

is an isomorphism (3.4.1). Let Y be a basis of $F = F_{\mathbb{Z}}$, then we consider

$$D = \mathbb{Q}^{(Y)} = \bigoplus_{b\in Y} b\mathbb{Q}.$$

Since $\mathbb{Q}_{\mathbb{Z}} \cong b\mathbb{Q}_{\mathbb{Z}}$, $b\mathbb{Q}_{\mathbb{Z}}$ is divisible and then from 4.5.3 so also is D. Since $F = \bigoplus b\mathbb{Z}$ F is a subgroup of D. Then $\mathrm{Ker}(\Phi)$ is also a subgroup of D and from 4.5.2 $\bar{D} := D/\mathrm{Ker}(\Phi)$ is also divisible.

Let now

$$\iota: F/\mathrm{Ker}(\Phi) \ni \bar{x} \mapsto \bar{x} \in \bar{D}$$

be the inclusion mapping, then $\iota\Phi'^{-1}$ is the desired monomorphism of A into the divisible group D. □

We now prove the theorem dual to 4.4.6 which shows that divisible groups are injective $\mathbb{Z}$-modules (definition in next chapter).

4.5.5 THEOREM. *If*

$$\varphi: D_{\mathbb{Z}} \to B_{\mathbb{Z}}$$

is a monomorphism and if $D_{\mathbb{Z}}$ is divisible, then φ splits (i.e. $\mathrm{Im}(\varphi)$ *is a direct summand in B).*

Proof. By 4.5.2 $\mathrm{Im}(\varphi)$ is divisible, so that without loss we can consider $D_{\mathbb{Z}}$ to be a submodule of $B_{\mathbb{Z}}$ and $\varphi = \iota$ to be the inclusion mapping. Let then

$$\Gamma := \{U \mid U \hookrightarrow B \wedge D \cap U = 0\}.$$

Since we have $U = 0 \in \Gamma$, $\Gamma \neq \varnothing$; since further the union of every totally ordered subset of Γ (under inclusion) is again an element of Γ, there is by reason of Zorn's Lemma a maximal element in Γ, which is again to be denoted by U. As a result we then have

$$D + U = D \oplus U \hookrightarrow B,$$

and it is to be shown that $B = D \oplus U$.

For an arbitrary $b \in B$ we consider the ideal $z_0\mathbb{Z}$ consisting of the $z \in \mathbb{Z}$ with $bz \in D + U$. Let $bz_0 = d + u$. Since D is divisible there is a d_0 with $d_0 z_0 = d \Rightarrow (b - d_0)z_0 = u$. Evidently $z_0\mathbb{Z}$ is then also the ideal of the $z \in \mathbb{Z}$ with $(b - d_0)z \in D + U$.
We claim that

$$D \cap (U + (b - d_0)\mathbb{Z}) = 0.$$

Assume

$$d_1 = u_1 + (b - d_0)z_1 \in D \cap (U + (b - d_0)\mathbb{Z}).$$

Then

$$(b - d_0)z_1 = d_1 - u_1 \in D + U$$

and so

$$z_1 = z_0 t,\ t \in \mathbb{Z}.$$

Then

$$(b - d_0)z_0 t = ut = d_1 - u_1.$$

Then

$$0 = d_1 - (u_1 + ut) \Rightarrow d_1 = 0.$$

From the maximality of U it follows that $(b - d_0)\mathbb{Z} \hookrightarrow U \Rightarrow b - d_0 \in U \Rightarrow b \in D + U$. Thus we have in fact, $B = D \oplus U$, which was to be shown. □

4.6 MONOID RINGS

As a further example of the application of free modules we introduce here the *monoid ring*. Let G be an arbitrary, multiplicatively written, monoid, i.e. G is a set with an operation $G \times G \ni (a, b) \to ab \in G$, which is associative and in which there exists a neutral element e. Further let R be an arbitrary ring.

In the sense of 4.4.5 let

$$GR := \bigoplus_{g \in G} gR,$$

where G is itself thus taken as a basis. We observe that then $g1 = g$ for $g \in G$, $1 \in R$ holds (g stands in the place of $\varphi_g(1)$ in the sense of 4.4.5).

By means of a definition of a multiplication GR is now to be made into a ring (with unit element).

4.6.1 *Definition*. Let T, T' be finite subsets of G. For

$$\sum_{g \in T} g r_g, \qquad \sum_{g' \in T'} g' r'_{g'} \in GR$$

let then

$$\left(\sum_{g \in T} g r_g\right)\left(\sum_{g' \in T'} g' r'_{g'}\right) := \sum_{\substack{g \in T \\ g' \in T'}} g g' r_g r'_{g'}.$$

This definition means: In GR for a product of elements from G the product is taken in G and for such a product of elements from R the product in R is taken; in other respects we calculate distributively and the elements from R and G are permutable.

Remark. On the right side the same monoid element in $\sum g g' r_g r'_g$ can occur many times in the form gg'; in general this is therefore not a representation by a basis. A representation by a basis ensues if by distributivity we collect together basic elements:

$$\sum_{\substack{g \in T \\ g' \in T'}} g g' r_g r'_{g'} = \sum_{b \in TT'} b s_b \quad \text{with} \quad s_b = \sum_{\substack{gg' = b \\ g \in T, g' \in T'}} r_g r'_{g'}.$$

For a finite monoid $G=\{g_1, \ldots, g_n\}$ the definition can be written in the following form:

$$\left(\sum_{i=1}^{n} g_i r_i\right)\left(\sum_{j=1}^{n} g_j r'_j\right) = \sum_{i,j=1}^{n} g_i g_j r_i r'_j = \sum_{k=1}^{n} g_k s_k \quad \text{with} \quad s_k = \sum_{\substack{g_i g_j = g_k \\ i,j=1,\ldots,n}} r_i r'_j.$$

It is easy to verify that, under the given definitions, GR is a ring. The associativity follows from that in G and in R and the distributivity follows from that in R. Let e be the identity (=neutral element) of G, then $e = e1$ is the unit element of GR, as follows immediately from 4.6.1.

4.6.2 *Definition.* GR is called the *monoid ring* of G with coefficients in R. If G is a group then GR is called the *group ring* of G with coefficients in R.

If we consider the subring eR of GR then the mapping

$$eR \ni er \mapsto r \in R$$

is a ring isomorphism, and eR is usually identified with R; for the 1-element of the group ring $e1$ (with $1 \in R$) we then write 1.

Finally we point out that by putting

$$r\sum g r_g := er \sum g r_g = \sum g r r_g, \qquad r \in R$$

GR becomes a left R-module, for which G is again a basis. Since GR is an eR-GR-bimodule, GR is also an R-GR-bimodule. If R is commutative GR is then an R-algebra (see 2.2.5).

Ring-theoretic, module-theoretic, group theoretic and—for deeper considerations—also arithmetic concepts are involved in the investigation of group rings. This many-sidedness makes this area particularly interesting and stimulating.

4.7 PUSHOUT AND PULLBACK

Let:

$$\alpha: A \to B, \qquad \varphi: A \to M$$

be two given homomorphisms with the same domain.

With respect to many problems, which we shall later encounter, the question arises as to whether we can incorporate these homomorphisms in a commutative square:

$$(*) \qquad \begin{array}{ccc} A & \xrightarrow{\alpha} & B \\ {\scriptstyle\varphi}\downarrow & & \downarrow{\scriptstyle\beta} \\ M & \xrightarrow{\psi} & N. \end{array}$$

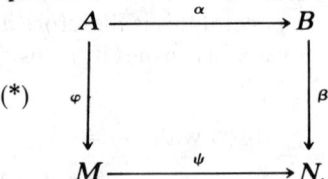

We wish to show that this is possible in a non-trivial way and indeed with a "universal" pair ψ, β i.e. with a pair such that over the pair every other commutative "completion of the square" of α, φ can be factorized.

Obviously the dual question also arises as to whether for a given ψ, β with the same codomain there exists a "universal" commutative "completion of the square". Here, too, the answer is positive.

In the first case the solution is called a pushout, in the second case a pullback.

4.7.1 *Definition.* Let the commutative diagram (*) be given.

(1) The pair (ψ, β) is called the *pushout* of the pair $(\varphi, \alpha) :\Leftrightarrow$ for every pair (ψ', β') with $\psi': M \to X$, $\beta': B \to X$ and $\psi'\varphi = \beta'\alpha$ there is precisely one $\sigma: N \to X$ with $\psi' = \sigma\psi$, $\beta' = \sigma\beta$.

(2) The pair (φ, α) is called the *pullback* of the pair $(\psi, \beta) :\Leftrightarrow$ for every pair (φ', α') with $\varphi': Y \to M$, $\alpha': Y \to B$ and $\psi\varphi' = \beta\alpha'$ there is precisely one $\tau: Y \to A$ with $\varphi' = \varphi\tau$, $\alpha' = \alpha\tau$.

We clarify the situation in the corresponding diagrams:

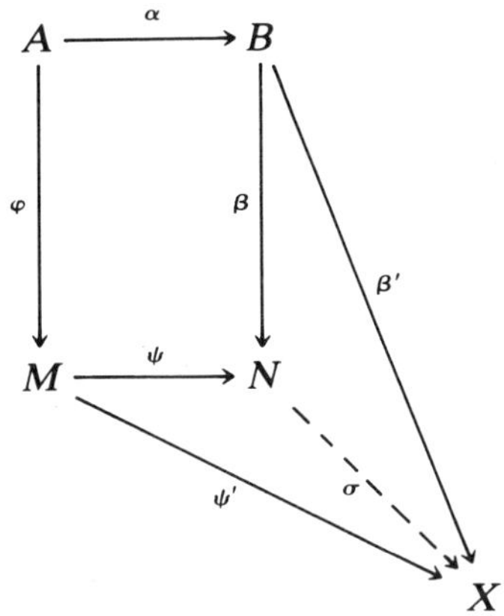

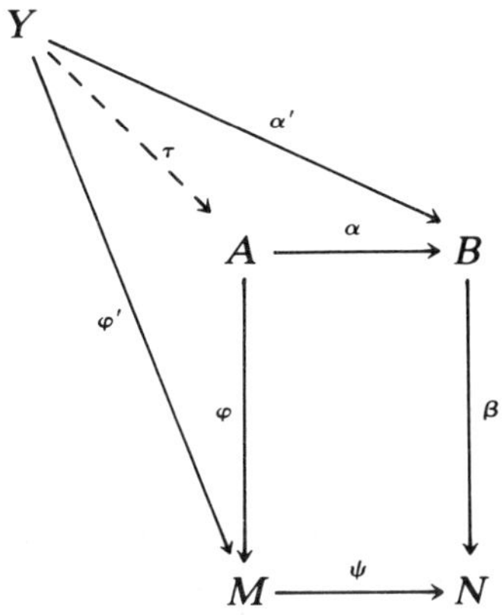

Before we prove the existence of pushouts and pullbacks, we establish their uniqueness.

4.7.2 *Remark. Pushouts and pullbacks are for given* (φ, α), (ψ, β) *resp., uniquely determined up to isomorphism.*

Proof. Let (ψ, β) and (ψ', β') be two pushouts for (φ, α). Then in addition to $\sigma: N \to X$ there also exists $\rho: X \to N$ with $\psi = \rho\psi'$, $\beta = \rho\beta'$. For $\rho\sigma: N \to N$ we have $\psi = \rho\psi' = \rho\sigma\psi$, $\beta = \rho\beta' = \rho\sigma\beta$. Accordingly from the prescribed uniqueness it follows that $\rho\sigma = 1_N$, correspondingly we obtain $\sigma\rho = 1_X$, thus

σ and ρ are isomorphisms inverse to one another. The statement for the pullback follows dually. □

In the following we denote elements from $M\oplus B$ by (b, m) and elements from $(M\oplus B)/U$ by $\overline{(m, b)}$.

4.7.3 THEOREM

(1) *Let the pair* (φ, α) *be given with*

$$\varphi: A \to M, \qquad \alpha: A \to B.$$

Let

$$N := (M\oplus B)/U \quad \textit{with} \quad U := \{(\varphi(a), -\alpha(a)) | a \in A\}$$

and let

$$\psi: M \ni m \mapsto \overline{(m, 0)} \in N, \qquad \beta: B \ni b \mapsto \overline{(0, b)} \in N,$$

then (ψ, β) *is a pushout of* (φ, α).

(2) *Let the pair* (ψ, β) *be given with*

$$\psi: M \to N, \qquad \beta: B \to N.$$

Let

$$A := \{(m, b) | m \in M \wedge b \in B \wedge \psi(m) = \beta(b)\}$$

and let

$$\varphi: A \ni (m, b) \mapsto m \in M, \qquad \alpha: A \ni (m, b) \mapsto b \in B,$$

then (φ, α) *is a pullback of* (ψ, β).

Proof. (1) First of all it is clear that U is a submodule, that N is a factor module of $M\oplus B$ and that ψ and β are homomorphisms with $\psi\varphi = \beta\alpha$. Let now ψ', β' be given as in 4.7.1. We define $\sigma: N \to X$ by $\sigma(\overline{(m, b)}) := \psi'(m) + \beta'(b)$. In order to prove that σ is a mapping it suffices to show that for $(m, b) \in U$ we have $\sigma(\overline{(m, b)}) = 0$:

$$\sigma(\overline{(\varphi(a), -\alpha(a))}) = \psi'\varphi(a) - \beta'\alpha(a) = 0$$

since $\psi'\varphi = \beta'\alpha$. It is again immediately clear that the mapping σ is a homomorphism and that $\sigma\psi = \psi'$, $\sigma\beta = \beta'$. It remains to show the uniqueness of σ. Suppose we also have $\psi' = \sigma_1\psi$, $\beta' = \sigma_1\beta$ for $\sigma_1: N \to X$. Then it follows that

$$(\sigma - \sigma_1)\psi = 0, \qquad (\sigma - \sigma_1)\beta = 0,$$

thus

$$0=(\sigma-\sigma_1)\psi(m)=(\sigma-\sigma_1)(\overline{(m,0)}),$$
$$0=(\sigma-\sigma_1)\beta(b)=(\sigma-\sigma_1)(\overline{(0,b)}).$$

Since $\{\overline{(m,0)},\overline{(0,b)}\,|\,m\in M\wedge b\in B\}$ is a generating set of N, for which $\sigma-\sigma_1$ is the zero mapping, it follows that $\sigma-\sigma_1=0$. This completes the proof for the pushout.

(2) The proof for the pullback proceeds dually. We merely put $\tau: Y\to A$:

$$\tau(y):=(\varphi'(y),\alpha'(y)),\qquad y\in Y$$

and establish the uniqueness of τ. Let $(\tau-\tau_1)(y)=(m,b)$, then it follows that

$$0=\varphi(\tau-\tau_1)(y)=\varphi(m,b)=m$$
$$0=\alpha(\tau-\tau_1)(y)=\alpha(m,b)=b,$$

thus $(\tau-\tau_1)(y)=0$, i.e. $\tau-\tau_1=0$. □

In the following we use the pushout and the pullback as they are already given explicitly in the theorem. The following theorem is of use for the definition of injective modules in the next chapter.

4.7.4 THEOREM. *Let* (ψ,β) *be a pushout of* (φ,α). *Then we have*:

(1) α *is* mono $\Rightarrow\psi$ *is* mono, α *is* epi $\Rightarrow\psi$ *is* epi;
φ *is* mono $\Rightarrow\beta$ *is* mono, φ *is* epi $\Rightarrow\beta$ *is* epi.

(2) *Let* α *be a monomorphism, then we have*: $\mathrm{Im}(\psi)$ *is a direct summand in* $N\Leftrightarrow$ *there exists a* $\kappa: B\to M$ *so that* $\varphi=\kappa\alpha$:

(4.7.5)

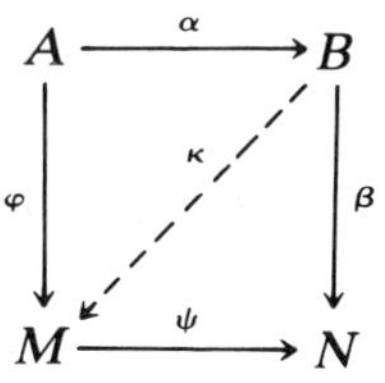

Proof. (1) Let α be a monomorphism and let $\psi(m)=\overline{(m,0)}=0$. Then there is $a\in A$ with $(m,0)=(\varphi(a),-\alpha(a))$. Hence $-\alpha(a)=0\Rightarrow a=0\Rightarrow m=\varphi(a)=0$.

Let α be an epimorphism and let $\overline{(m, b)} \in N$, then there is $a \in A$ with $b = -\alpha(a)$ and so

$$\psi(m - \varphi(a)) = \overline{(m - \varphi(a), 0)}$$
$$= \overline{(m - \varphi(a), 0)} + \overline{(\varphi(a), -\alpha(a))} = \overline{(m, b)},$$

thus ψ is also an epimorphism.

Correspondingly for the second line in (1).

(2) Let $\text{Im}(\psi)$ be a direct summand in N:

$$N = \text{Im}(\psi) \oplus N_0.$$

Since α is a monomorphism, from (1) ψ is also a monomorphism and consequently ψ induces an isomorphism $\psi_0: M \to \text{Im}(\psi)$. Let $\pi: N \to \text{Im}(\psi)$ be the projection arising from $N = \text{Im}\ \psi \oplus N_0$, then $\kappa := \psi_0^{-1} \pi \beta$ fulfils what we require:

$$\kappa\alpha(a) = \psi_0^{-1} \pi \beta \alpha(a) = \psi_0^{-1} \pi \psi \varphi(a) = \psi_0^{-1} \pi(\overline{(\varphi(a), 0)}) = \varphi(a).$$

Conversely let κ be given with $\varphi = \kappa\alpha$. Then we consider

$$\xi: N \ni \overline{(m, b)} \mapsto m + \kappa(b) \in M.$$

As $\xi(\overline{(\varphi(a), -\alpha(a))}) = \varphi(a) - \kappa\alpha(a) = 0$ this is a mapping and then also a homomorphism. As $\xi\psi(m) = \xi(\overline{(m, 0)}) = m$ we have $\xi\psi = 1_M$, thus we deduce as asserted: $N = \text{Im}(\psi) \oplus \text{Ker}(\xi)$. □

We come now to the dual theorem which leads on to the definition of projective modules.

4.7.6 THEOREM. *Let (φ, α) be pullback of (ψ, β). Then we have:*

(1)

$$\beta \textit{ is } \text{mono} \Rightarrow \varphi \textit{ is } \text{mono}, \ \beta \textit{ is } \text{epi} \Rightarrow \varphi \textit{ is } \text{epi},$$

$$\psi \textit{ is } \text{mono} \Rightarrow \alpha \textit{ is } \text{mono}, \ \psi \textit{ is } \text{epi} \Rightarrow \alpha \textit{ is } \text{epi}.$$

(2) *Let ψ be an epimorphism, then we have*: $\text{Ker}(\alpha)$ *is a direct summand in $A \Leftrightarrow$ there exists a $\kappa: M \to B$ so that $\beta = \psi\kappa$:*

(4.7.7)

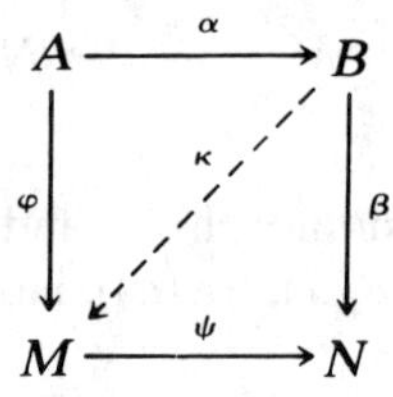

Proof. Since the proof proceeds dually to that of 4.7.4, we shall be brief.

(1) Let ψ be an epimorphism. Let $b \in B$, then there is an $m \in M$ with $\psi(m) = \beta(b)$. Thus it follows that $(m, b) \in A$ and $\alpha((m, b)) = b$, i.e. α is an epimorphism. Correspondingly in the other cases.

(2) Let $\mathrm{Ker}(\alpha)$ be a direct summand in A:

$$A = \mathrm{Ker}(\alpha) \oplus A_0.$$

Since, like ψ, α is also an epimorphism, $\alpha_0 := \alpha | A_0$ is an epimorphism. Let $\iota: A_0 \to A$ be thc inclusion, then $\kappa := \varphi\iota\alpha_0^{-1}$ fulfils what we require:

$$\psi\kappa(b) = \psi\varphi(\iota\alpha_0^{-1}(b)) = \beta\alpha\iota\alpha_0^{-1}(b) = \beta(b).$$

Conversely let κ with $\psi\kappa = \beta$ be given. Then we have for

$$\eta: B \ni b \mapsto (\kappa(b), b) \in A$$

$\alpha\eta = 1_B$ from which $A = \mathrm{Ker}(\alpha) \oplus \mathrm{Im}(\eta)$ follows. □

4.8 A CHARACTERIZATION OF GENERATORS AND COGENERATORS

In 3.3 we became acquainted with generators and cogenerators. For these a further characterization follows.

We preface these considerations by a lemma which is also itself of interest.

4.8.1 LEMMA. *Notations as in* 4.1

(a) *For every homomorphism* $\psi: \coprod_{i \in I} A_i \to M$ *we have*:

$$\mathrm{Im}(\psi) = \sum_{i \in I} \mathrm{Im}(\psi\eta_i).$$

(b) *For every homomorphism* $\psi: M \to \prod_{i \in I} A_i$ *we have*

$$\mathrm{Ker}(\psi) = \bigcap_{i \in I} \mathrm{Ker}(\pi_i\psi).$$

Proof. (a) As a consequence of the finiteness of the values of the elements of the coproduct every element from $\coprod_{i \in I} A_i$ may be written as a finite sum

$$\sum{}' \eta_i(a_i) \quad \text{with} \quad a_i \in A_i$$

Then it follows that

$$\psi(\Sigma' \eta_i(a_i)) = \Sigma' \psi\eta_i(a_i),$$

thus we have

$$\mathrm{Im}(\psi) \hookrightarrow \sum_{i\in I} \mathrm{Im}(\psi\eta_i).$$

Conversely if $m \in \sum_{i\in I} \mathrm{Im}(\psi\eta_i)$ then m may be represented as a finite sum

$$m = \Sigma' \psi\eta_i(a_i) = \psi(\Sigma' \eta_i(a_i)), \qquad a_i \in A_i.$$

From this it follows that $m \in \mathrm{Im}(\psi)$, thus we also have

$$\sum_{i\in I} \mathrm{Im}(\psi\eta_i) \hookrightarrow \mathrm{Im}(\psi).$$

(b) If $m \in \mathrm{Ker}(\psi)$, then it follows immediately that $m \in \mathrm{Ker}(\pi_i\psi)$ for every $i \in I$, thus we have

$$\mathrm{Ker}(\psi) \hookrightarrow \bigcap_{i\in I} \mathrm{Ker}(\pi_i\psi).$$

Conversely let $m \in \bigcap \mathrm{Ker}(\pi_i\psi)$. Then this implies that all components of $\psi(m)$ are equal to 0, thus we have $\psi(m) = 0$ and it follows that

$$\bigcap_{i\in I} \mathrm{Ker}(\pi_i\psi) \hookrightarrow \mathrm{Ker}(\psi).$$ □

We come now to the characterization of generators and cogenerators.

4.8.2 THEOREM

(a) *The following conditions are equivalent:*
   (1) *B_R is a generator.*
   (2) *Every direct sum of copies of B is a generator.*
   (3) *A direct sum of copies of B is a generator.*
   (4) *Every module M_R is an epimorphic image of a direct sum of copies of B.*

(b) *The following conditions are equivalent:*
   (1) *C_R is a cogenerator.*
   (2) *Every direct product of copies of C is a cogenerator.*
   (3) *A direct product of copies of C is a cogenerator.*
   (4) *Every module M_R can be mapped monomorphically into a direct product of copies of C.*

Proof

(a) (1)$\Leftrightarrow$(2), (1)$\Leftrightarrow$(3) follow from 3.3.2.

"(1)$\Rightarrow$(4)": Let

$$\coprod_{\varphi \in \mathrm{Hom}_R(B,M)} B_\varphi \quad \text{with} \quad B_\varphi = B \quad \text{for all} \quad \varphi \in \mathrm{Hom}_R(B, M),$$

then we consider the homomorphism

$$\psi: \coprod_{\varphi \in \mathrm{Hom}_R(B,M)} B_\varphi \to M,$$

which is defined by

$$\psi((b_\varphi)) := \sum_{\substack{b_\varphi \text{ component in } (b_\varphi) \\ b_\varphi \neq 0}} \varphi(b_\varphi).$$

Since in (b_φ) only finitely many $b_\varphi \neq 0$, the sum appearing on the right is meaningful. There then follows from 4.8.1

$$\mathrm{Im}(\psi) = \sum_{\varphi \in \mathrm{Hom}_R(B,M)} \mathrm{Im}(\varphi) = M,$$

thus ψ is an epimorphism.

"(4)$\Rightarrow$(1)": Conversely if there is an epimorphism

$$\psi: \coprod_{i \in I} B_i \to M, \quad \text{with} \quad B_i = B \quad \text{for all} \quad i \in I.$$

and if η_i is the ith inclusion of B in $\coprod B_i$, then we have $\psi\eta_i \in \mathrm{Hom}_R(B, M)$ as well as, from 4.8.1,

$$M = \mathrm{Im}(\psi) = \sum_{i \in I} \mathrm{Im}(\psi\eta_i) \hookrightarrow \sum_{\varphi \in \mathrm{Hom}_R(B,M)} \mathrm{Im}(\varphi) \hookrightarrow M,$$

thus

$$\sum_{\varphi \in \mathrm{Hom}_R(B,M)} \mathrm{Im}(\varphi) = M,$$

i.e. B is a generator.

(b) (1)$\Leftrightarrow$(2), (1)$\Leftrightarrow$(3) follow from 3.3.2.

"(1)$\Rightarrow$(4)": Let now

$$\prod_{\varphi \in \mathrm{Hom}_R(M,C)} C_\varphi \quad \text{with } C_\varphi = C \quad \text{for all} \quad \varphi \in \mathrm{Hom}_R(M, C),$$

then we consider the homomorphism

$$\psi: M \to \prod_{\varphi \in \mathrm{Hom}_R(M,C)} C_\varphi,$$

which is defined by

$$\psi(m) := (c_\varphi) \quad \text{with} \quad c_\varphi := \varphi(m) \quad \text{for all} \quad \varphi \in \mathrm{Hom}_R(M, C).$$

For $m \in \text{Ker}(\psi)$ it follows that

$$m \in \bigcap_{\varphi \in \text{Hom}_R(M,C)} \text{Ker}(\varphi) = 0,$$

thus ψ is also a monomorphism.

"(4)⇒(1)": Conversely if there is a monomorphism

$$\psi : M \to \prod_{i \in I} C_i, \quad \text{with} \quad C_i = C \quad \text{for all} \quad i \in I,$$

and if π_i is the ith projection, then we have $\pi_i\psi \in \text{Hom}_R(M, C)$ as well as, from 4.8.1,

$$\bigcap_{\varphi \in \text{Hom}_R(M,C)} \text{Ker}(\varphi) \hookrightarrow \bigcap_{i \in I} \text{Ker}(\pi_i\psi) = \text{Ker}(\psi) = 0,$$

thus

$$\bigcap_{\varphi \in \text{Hom}_R(M,C)} \text{Ker}(\varphi) = 0,$$

i.e. C is a cogenerator. □

EXERCISES

(1)

Show:

(a) For a homomorphism $\alpha : A \to B$ the following are equivalent:

(1) $\text{Ker}(\alpha)$ is a direct summand of A and $\text{Im}(\alpha)$ is a direct summand of B.

(2) There is a homomorphism $\beta : B \to A$ with $\alpha = \alpha\beta\alpha$.

(b) How is the equivalence simplified if α is a monomorphism or an epimorphism?

(2)

Give examples for a family of modules $(A_i \mid i \in I)$ and a module M with

$$\text{Hom}_R\left(\prod_{i \in I} A_i, M\right) \not\cong \prod_{i \in I} \text{Hom}_R(A_i, M)$$

$$\text{resp.} \quad \text{Hom}_R\left(M, \bigoplus_{i \in I} A_i\right) \not\cong \bigoplus_{i \in I} \text{Hom}_R(M, A_i)$$

($\not\cong$ means "not isomorphic as additive groups").

(3)

(a) Let $M_R \neq 0$ be a module with $M_R \cong M_R \oplus M_R$ and let $S := \text{End}(M_R)$. Show: For every $n \in \mathbb{N}$ a basis of S_S exists with n elements.

(b) Construct an example of a module $M \neq 0$ with $M \cong M \oplus M$ and for every $n \in \mathbb{N}$ a basis of S_S with n elements.

(4)

Show: A ring $R \neq 0$ is a skew field if and only if every right R-module is free.

(5)

Show: 4.4.6 holds also for direct summands of free modules.

(6)

(a) Show: If $\beta: B_R \to C_R$ is an epimorphism, if $\varphi: F_R \to C_R$ is a homomorphism and if F_R is a free module, then there is a $\varphi': F_R \to B_R$ with $\varphi = \beta\varphi'$.

(b) Show: (a) holds also if F_R is replaced by a direct summand of a free module.

(7)

(a) Show: If

$$\begin{array}{ccc} A & \xrightarrow{\alpha} & B \\ \downarrow{\scriptstyle\varphi} & & \downarrow{\scriptstyle\beta} \\ M & \xrightarrow{\psi} & N \end{array}$$

is a pushout of (φ, α) then there is an isomorphism $\eta: B/\mathrm{Im}(\alpha) \to N/\mathrm{Im}(\psi)$, so that

$$\begin{array}{ccc} B & \xrightarrow{\nu} & B/\mathrm{Im}(\alpha) \\ \downarrow{\scriptstyle\beta} & & \downarrow{\scriptstyle\eta} \\ N & \xrightarrow{\nu} & N/\mathrm{Im}(\psi) \end{array}$$

is commutative (ν is, as appropriate, the natural epimorphism).

(b) Formulate and prove the dual statements.

(8)

Let R be an integral domain with quotient field K. A module M is called *divisible* if, for every $0 \neq r \in R$, $Mr = M$ holds. Show:

(a) The class of divisible R-modules is closed on taking factor modules, products and coproducts.

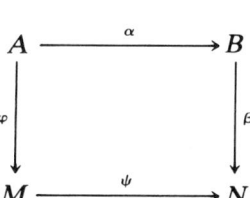

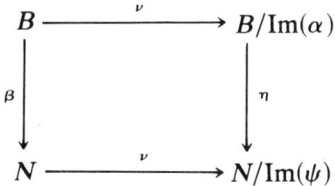

(b) The only divisible submodules of K_R are 0 and K.
(c) $R \neq K \Rightarrow$ every divisible, cyclic R-module is 0.

(9)

Let R be an integral domain. Show:

(a) M_R is then divisible if and only if to every cyclic ideal $A \hookrightarrow R$ and to every homomorphism $\varphi: A_R \to M_R$ there is a homomorphism $\varphi': R_R \to M_R$ with $\varphi'|A = \varphi$.

(b) If for fixed M_R and arbitrary N_R every homomorphism $\varphi: M_R \to N_R$ splits, then M_R is divisible.

(10)

(a) Let T be a divisible abelian group.

(1) Show: If finitely many, arbitrary elements are omitted from an arbitrary generating set of T over $\mathbb{Z}$, then the set of the remaining elements is still a generating set (see also 2.3.7).

(2) Show that T contains no maximal subgroup.

(b) Show: an abelian group which has no maximal subgroup is divisible.

(c) Give an example of a divisible abelian group, which contains a simple subgroup.

(11)

For an abelian group A the *torsion subgroup* $T(A)$ is defined by

$$T(A) := \{a \mid a \in A \wedge \exists z \in \mathbb{Z}[z \neq 0 \wedge az = 0]\}.$$

Show:

(a) A is divisible $\Rightarrow T(A)$ is a direct summand in A.

(b) A is divisible $\wedge\ T(A) = 0 \Rightarrow A$ is $\mathbb{Z}$-isomorphic to a direct sum of copies of $\mathbb{Q}_{\mathbb{Z}}$.

(Hint: A may be made into a $\mathbb{Q}$-vector space.)

(12)

Let p be a prime number and let

$$\mathbb{Q}_p := \left\{\frac{z}{p^n} \,\middle|\, z \in \mathbb{Z} \wedge n \in \mathbb{Z}\right\}.$$

Show:

$$\mathbb{Q}/\mathbb{Z} = \bigoplus_{\text{Prime } p} \mathbb{Q}_p/\mathbb{Z}.$$

(13)

Let G be a group, R a ring and GR the group ring of G with coefficients in R.

(a) Show: For $\sum g r_g$, $\sum g' r'_{g'} \in GR$ a new left GR-module structure is defined on GR by

$$(\sum g r_g) \circ (\sum g' r'_{g'}) := \sum g g' g^{-1} r_g r'_{g'}.$$

(b) Use (a) in order to give an example of a module which as well as being a left GR-module is also a right GR-module but which is not a GR-GR-bimodule.

(14)

A module M_R is called (von Neumann) *regular* if every cyclic submodule is a direct summand. Show:

(a) In a regular module every finitely generated submodule is a direct summand.

(b) R_R is regular$\Leftrightarrow R_R$ regular$\Leftrightarrow$to every $r \in R$ there is an $r' \in R$ with $r = rr'r$ (see also Chapter 2, Exercise 13).

(c) If R is regular then every free right R-module is regular.

(Hint: If $\{x_i | i \in I\}$ is a basis of F_R and if $x \in F_R$, consider the left ideal of R which is generated by the coefficients of x in the representation of the basis and which from (1) and (2) is of the form Re with $e^2 = e$; this is used to find a projection $F \to xR$.)

Chapter 5

Injective and Projective Modules

Injective and projective modules or, more generally, injective and projective objects in a category, play an important role in the later development of algebra. It is therefore advisable to become familiar with these concepts at the earliest opportunity in order to give due emphasis to the resulting point of view in further considerations. Here we shall present the general properties of injective and projective modules. We shall return many times to these concepts.

As a tool for the investigation of injective and projective modules we need large and small submodules as well as complements. These concepts are essentially needed also in other respects (as e.g. with respect to the radical and the socle) and are to be investigated here somewhat further than would be necessary for the considerations of this chapter.

5.1 BIG AND SMALL SUBMODULES

5.1.1 *Definition*

(a) A submodule A of a module M is called *small* (= *superfluous*) in M, notationally $A \overset{\circ}{\hookrightarrow} M$, respectively *large* (= *essential*) in M, notationally $A \overset{*}{\hookrightarrow} M :\Leftrightarrow$

$$\forall U \hookrightarrow M[A + U = M \Rightarrow U = M]$$

$$\text{resp. } \forall U \hookrightarrow M[A \cap U = 0 \Rightarrow U = 0].$$

(b) A right, left or two-sided ideal A of a ring R is called *small* resp. *large* in $R :\Leftrightarrow A$ is a small, resp. large, submodule of R_R, ${}_RR$ or ${}_RR_R$.

(c) A homomorphism $\alpha : A \to B$ is called *small*, resp. *large* $:\Leftrightarrow$

$$\operatorname{Ker}(\alpha) \overset{\circ}{\hookrightarrow} A \text{ resp. } \operatorname{Im}(\alpha) \overset{*}{\hookrightarrow} B.$$

Remark. We obtain immediately from the definition:

(1) $A \overset{\circ}{\hookrightarrow} M \Leftrightarrow \forall U \overset{\neq}{\hookrightarrow} M[A+U \overset{\neq}{\hookrightarrow} M]$.
(2) $A \overset{*}{\hookrightarrow} M \Leftrightarrow \forall U \hookrightarrow M, U \neq 0[A \cap U \neq 0]$.
(3) $M \neq 0 \wedge A \overset{\circ}{\hookrightarrow} M \Rightarrow A \neq M$.
(4) $M \neq 0 \wedge A \overset{*}{\hookrightarrow} M \Rightarrow A \neq 0$.

5.1.2 *Examples*

(1) For every module M we have: $0 \overset{\circ}{\hookrightarrow} M$, $M \overset{*}{\hookrightarrow} M$.

(2) A module is called *semisimple* if every submodule is a direct summand (see Chapter 8).

M is semisimple $\Rightarrow 0$ is the only small submodule of M and M is the only large submodule of M.

Proof. $A \hookrightarrow M \Rightarrow$ there exists $U \hookrightarrow M$ with $A \oplus U = M$. If $A \overset{\circ}{\hookrightarrow} M$ then $U = M$ and so $A = 0$. If $A \overset{*}{\hookrightarrow} M$ then $U = 0$ and so $A = M$. □

(3) Let R be a *local ring* (see Chapter 7), but not a skew field and let A be the two-sided ideal consisting of the non-invertible elements of R. Then $A \neq 0$ (since R is not a skew field) and A is the largest proper right, left or two-sided ideal of R (see 7.1.1). It follows therefore that A is small and (since $A \neq 0$) large in R_R, ${}_RR$ and ${}_RR_R$ resp.

Example of a local ring:

$$R := \mathbb{Z}/p^n\mathbb{Z}, \qquad A := p\mathbb{Z}/p^n\mathbb{Z}, \qquad p = \text{prime}.$$

(4) In a free $\mathbb{Z}$-module only the trivial submodule 0 is a small submodule.

Proof. Let

$$F = \bigoplus_{i \in I} x_i\mathbb{Z}$$

be a free $\mathbb{Z}$-module with basis $\{x_i \mid i \in I\}$, $A \hookrightarrow F$, $a \in A$ and let

$$a = x_{i_1}z_1 + \ldots + x_{i_m}z_m, \qquad z_i \in \mathbb{Z}$$

with $z_1 \neq 0$. Let $n \in \mathbb{Z}$ with GCD $(z_1, n) = 1$ and $n > 1$ (e.g. let n be a prime p not dividing z_1). Put

$$U = \bigoplus_{\substack{i \in I \\ i \neq i_1}} x_i\mathbb{Z} + x_{i_1}n\mathbb{Z},$$

then it follows that $a\mathbb{Z} + U = F$, hence certainly $A + U = F$ with $U \neq F$.

In particular the only small ideal of $\mathbb{Z}$ is 0. However every ideal $\neq 0$ is large in $\mathbb{Z}$, for if $a\mathbb{Z}$ and $b\mathbb{Z}$ are two ideals $\neq 0$ then we have $0 \neq ab \in a\mathbb{Z} \cap b\mathbb{Z}$.

(5) Every finitely generated submodule of $\mathbb{Q}_{\mathbb{Z}}$ is small in $\mathbb{Q}_{\mathbb{Z}}$. For the proof let $q_1, \ldots, q_n \in \mathbb{Q}$ and let $U \hookrightarrow \mathbb{Q}_{\mathbb{Z}}$ with

$$q_1\mathbb{Z} + \ldots + q_n\mathbb{Z} + U = \mathbb{Q},$$

then $\{q_1, \ldots, q_n\} \cup U$ is a generating set of $\mathbb{Q}$, consequently from 2.3.7 U is already a generating set of $\mathbb{Q}$, thus we have $U = \mathbb{Q}$.

We come now to simple deductions from the definition.

5.1.3 LEMMA

(a) $A \hookrightarrow B \hookrightarrow M \hookrightarrow N \wedge B \overset{\circ}{\hookrightarrow} M \Rightarrow A \overset{\circ}{\hookrightarrow} N$.

(b) $A_i \overset{\circ}{\hookrightarrow} M, i = 1, \ldots, n \Rightarrow \sum_{i=1}^{n} A_i \overset{\circ}{\hookrightarrow} M$.

(c) $A \overset{\circ}{\hookrightarrow} M \wedge \varphi \in \mathrm{Hom}_R(M, N) \Rightarrow \varphi(A) \overset{\circ}{\hookrightarrow} N$.

(d) *If $\alpha : A \to B$, $\beta : B \to C$ are small epimorphisms then $\beta\alpha : A \to C$ is also a small epimorphism.*

Proof. (a) Let $A + U = N$. Then $B + U = N$ and so $B + (U \cap M) = M$ (by the modular law). Hence $U \cap M = M$ (since $B \overset{\circ}{\hookrightarrow} M$) and so $M \hookrightarrow U$ and since by assumption $A \hookrightarrow M$ we deduce that $U = A + U = N$ which was to be shown.

(b) Proof by induction on n. For $n = 1$ the assertion holds by assumption. Let

$$A := A_1 + \ldots + A_{n-1} \overset{\circ}{\hookrightarrow} M,$$

and suppose we have for $U \hookrightarrow M$

$$A + A_n + U = M$$

$\Rightarrow A_n + U = M$, since $A \overset{\circ}{\hookrightarrow} M$ and so $U = M$, since $A_n \overset{\circ}{\hookrightarrow} M$.

(c) Let $\varphi(A) + U = N$ with $U \hookrightarrow N$. Then we have for arbitrary $m \in M$: $\varphi(m) = \varphi(a) + u$ with

$$a \in A, u \in U \Rightarrow \varphi(m - a) = u \Rightarrow m - a \in \varphi^{-1}(U) \Rightarrow m \in A + \varphi^{-1}(U)$$
$$\Rightarrow A + \varphi^{-1}(U) = M \Rightarrow M = \varphi^{-1}(U),$$

since $A \overset{\circ}{\hookrightarrow} M \Rightarrow \varphi(M) = \varphi\varphi^{-1}(U) = U \cap \mathrm{Im}(\varphi)$. Thus $\varphi(A) \hookrightarrow \varphi(M) \hookrightarrow U$ and hence $U = \varphi(A) + U = N$, which was to be shown.

(d) Let $\mathrm{Ker}(\beta\alpha) + U = A$ with $U \hookrightarrow A$, then, as $\mathrm{Ker}(\beta\alpha) = \alpha^{-1}(\mathrm{Ker}(\beta))$ it follows that

$$\alpha(\mathrm{Ker}(\beta\alpha)) + \alpha(U) = \mathrm{Ker}(\beta) + \alpha(U) = \alpha(A) = B.$$

Since by assumption we have $\mathrm{Ker}(\beta) \overset{\circ}{\hookrightarrow} B$, we obtain $\alpha(U) = B$ and consequently

$$\mathrm{Ker}(\alpha) + U = A.$$

As $\mathrm{Ker}(\alpha) \overset{\circ}{\hookrightarrow} A$ it follows therefore that $U = A$, which was to be shown. □

Before we direct our attention to dual properties with regard to large submodules, we have a further important statement for cyclic submodules which are not small.

5.1.4 LEMMA. *For $a \in M_R$ we have: aR is not small in $M \Leftrightarrow$ there is a maximal submodule $C \hookrightarrow M$ with $a \notin C$.*

Proof. "$\Leftarrow$": If C is a maximal submodule of M with $a \notin C$ then it follows that $aR + C = M$, thus aR is not small in M.

"$\Rightarrow$": Proof by the use of Zorn's Lemma. Let

$$\Gamma := \{B \mid B \underset{\neq}{\hookrightarrow} M \wedge aR + B = M\}.$$

Since aR is not small, there is a $B \in \Gamma$, i.e. $\Gamma \neq \varnothing$.

Let $\Lambda \neq \varnothing$ be a totally ordered (wrt $\hookrightarrow$) subset of Γ. Then

$$B_0 := \bigcup_{B \in \Lambda} B$$

is an upper bound of Λ. Suppose $a \in B_0$, then a must already be contained in a B; from which it would follow that $aR \hookrightarrow B$, thus

$$B = aR + B = M \quad ↯.$$

From $a \notin B_0$ it follows that $B_0 \underset{\neq}{\hookrightarrow} M$. Since $B \hookrightarrow B_0$ for $B \in \Lambda \Rightarrow$

$$aR + B_0 = M,$$

thus we have $B_0 \in \Gamma$, i.e. Λ has an upper bound in Γ. Zorn's Lemma implies then that Γ contains a maximal element C.

We claim that C is in fact a maximal submodule of M. Let $C \underset{\neq}{\hookrightarrow} U \hookrightarrow M$, then it follows that $U \notin \Gamma$, since C is maximal in Γ. From $M = aR + C \hookrightarrow aR + U \hookrightarrow M$ it follows that $aR + U = M$ and as $U \notin \Gamma$ we must have $U = M$ which was to be shown. □

We now direct our attention to large submodules. For these we have first of all the statements dual to 5.1.3.

5.1.5 LEMMA

(a) $A \hookrightarrow B \hookrightarrow M \hookrightarrow N \wedge A \overset{*}{\hookrightarrow} N \Rightarrow B \overset{*}{\hookrightarrow} M.$

(b) $A_i \overset{*}{\hookrightarrow} M, i = 1, \ldots, n \Rightarrow \bigcap_{i=1}^{n} A_i \overset{*}{\hookrightarrow} M.$

(c) $B \overset{*}{\hookrightarrow} N \wedge \varphi \in \mathrm{Hom}_R(M, N) \Rightarrow \varphi^{-1}(B) \overset{*}{\hookrightarrow} M.$

(d) *Let $\alpha : A \to B$, $\beta : B \to C$ be large monomorphisms. Then $\beta\alpha : A \to C$ is also a large monomorphism.*

Proof. (a) From $U \hookrightarrow M \wedge B \cap U = 0$ we obtain $A \cap U = 0$ and so $U = 0$, since $A \overset{*}{\hookrightarrow} N \wedge U \hookrightarrow M \hookrightarrow N$.

(b) Proof by induction on n. For $n = 1$ the assertion holds by assumption. Let now

$$A := \bigcap_{i=1}^{n-1} A_i \overset{*}{\hookrightarrow} M$$

and we have for $U \hookrightarrow M$: $A \cap A_n \cap U = 0 \Rightarrow A_n \cap U = 0$, since $A \overset{*}{\hookrightarrow} M$. Then $U = 0$ follows since $A_n \overset{*}{\hookrightarrow} M$.

(c) Let

$$U \hookrightarrow M \wedge \varphi^{-1}(B) \cap U = 0 \Rightarrow B \cap \varphi(U) = 0 \Rightarrow \varphi(U) = 0,$$

since $B \overset{*}{\hookrightarrow} N$. Then it follows that

$$U \hookrightarrow \mathrm{Ker}(\phi) = \varphi^{-1}(0) \hookrightarrow \varphi^{-1}(B) \Rightarrow U = \varphi^{-1}(B) \cap U = 0.$$

(d) Let $U \hookrightarrow C$ and $\mathrm{Im}(\beta\alpha) \cap U = 0$. Since β is a monomorphism it follows that

$$0 = \beta^{-1}(0) = \beta^{-1}(\mathrm{Im}(\beta\alpha)) \cap \beta^{-1}(U) = \mathrm{Im}(\alpha) \cap \beta^{-1}(U).$$

As $\mathrm{Im}(\alpha) \overset{*}{\hookrightarrow} B$ we deduce therefore that $\beta^{-1}(U) = 0$, thus this yields that $\mathrm{Im}(\beta) \cap U = 0$ and from $\mathrm{Im}(\beta) \overset{*}{\hookrightarrow} C$ it follows that $U = 0$, which was to be shown. □

The following criterion for large submodules is important for applications.

5.1.6 LEMMA. *Let $A \hookrightarrow M_R$, then we have*

$$A \overset{*}{\hookrightarrow} M_R \Leftrightarrow \forall m \in M,\ m \neq 0\ \exists r \in R[mr \neq 0 \wedge mr \in A].$$

Proof. "$\Rightarrow$": From $m \neq 0$ we have $mR \neq 0$ and so $A \cap mR \neq 0$, since $A \overset{*}{\hookrightarrow} M$ we have the assertion.

"$\Leftarrow$": Let $B \hookrightarrow M \wedge B \neq 0$. Then there is $m \in B$, $m \neq 0$. Let $mr \neq 0 \wedge mr \in A$, then $0 \neq mr \in A \cap B$ and so $A \overset{*}{\hookrightarrow} M$. □

5.1.7 COROLLARY. *Let $M = \sum_{i \in I} M_i$, $M_i \hookrightarrow M$, $A_i \overset{*}{\hookrightarrow} M_i$ for every $i \in I$ and let*

$$A := \sum_{i \in I} A_i = \bigoplus_{i \in I} A_i,$$

then it follows that

$$A \overset{*}{\rightarrow} M \quad \textit{and} \quad M = \bigoplus_{i \in I} M_i.$$

Proof. $A \overset{*}{\rightarrow} M$: Since every element from M lies in a sum of finitely many of the M_i, it is sufficient from 5.1.6 to establish the assertion for a finite index set I, say $I = \{1, \ldots, n\}$.

Proof by induction on n. The initial induction step $n = 1$ holds by assumption. Let the assertion be valid for $n-1$ summands, i.e., suppose

$$A_1 + \ldots + A_{n-1} \overset{*}{\rightarrow} M_1 + \ldots + M_{n-1}.$$

Let now

$$0 \neq m = m_1 + \ldots + m_{n-1} + m_n \quad \text{with} \quad m_i \in M_i.$$

If indeed $m_1 + \ldots + m_{n-1} = 0$ then $m = m_n \neq 0 \Rightarrow$ there is $r \in R$, $0 \neq mr = m_n r \in A_n$. Let therefore $m_1 + \ldots + m_{n-1} \neq 0$, then, by induction assumption, there is an $r \in R$ with

$$0 \neq (m_1 + \ldots + m_{n-1})r \in A_1 + \ldots + A_{n-1}.$$

If for this r we have further $m_n r = 0$ then we are finished. Thus let $m_n r \neq 0$. Then there is an $s \in R$ with $0 \neq m_n rs \in A_n$, and it follows that $mrs \in A_1 + \ldots + A_n$; from the directness of the sum of the A_i we have moreover $mrs \neq 0$. Therefore we have shown that $A \overset{*}{\rightarrow} M$.

$M = \bigoplus_{i \in I} M_i$: It is still also sufficient to assume that $I = \{1, \ldots, n\}$ and to suppose that

$$0 \neq m_n = m_1 + \ldots + m_{n-1} \in M_n \cap \sum_{i=1}^{n-1} M_i.$$

Then there exists an $r \in R$ with

$$0 \neq (m_1 + \ldots + m_{n-1})r \in \sum_{i=1}^{n-1} A_i,$$

thus

$$0 \neq m_n r = (m_1 + \ldots + m_{n-1})r \in M_n \cap \sum_{i=1}^{n-1} A_i.$$

Let then $s \in R$ with $0 \neq m_n rs \in A_n$, then it follows that

$$0 \neq m_n rs = (m_1 + \ldots + m_{n-1})rs \in A_n \cap \sum_{i=1}^{n-1} A_i$$

in contradiction to the assumption. □

5.1.8 COROLLARY. *Let* $M = \bigoplus_{i \in I} M_i$, $M_i \hookrightarrow M$, $A_i \overset{*}{\hookrightarrow} M_i$ *for every* $i \in I$. *Then we have*

$$A := \sum_{i \in I} A_i = \bigoplus_{i \in I} A_i \quad \textit{and} \quad A \overset{*}{\hookrightarrow} M.$$

Proof. From $M = \bigoplus M_i$ and $A_i \hookrightarrow M_i$ it follows that $A = \bigoplus A_i$. Then $A \overset{*}{\hookrightarrow} M$ follows from 5.1.7.

5.1.9 COROLLARY. *Let* $M = \bigoplus_{i \in I} M_i$ *and let* $B \hookrightarrow M$ *then the following conditions are equivalent*:

(1) $\forall i \in I[B \cap M_i \overset{*}{\hookrightarrow} M_i]$.

(2) $\bigoplus_{i \in I} (B \cap M_i) \overset{*}{\hookrightarrow} M$.

(3) $B \overset{*}{\hookrightarrow} M$.

Proof. "(1)⇒(2)": From 5.1.8.

"(2)⇒(3)": From $\bigoplus_{i \in I} (B \cap M_i) \overset{*}{\hookrightarrow} B$ and 5.1.5(a).

"(3)⇒(1)": Let $0 \neq m_i \in M_i$, then there is, from 5.1.6, an $r \in R$ with $0 \neq m_i r \in B$. But also $m_i r \in M_i$ and it follows that $0 \neq m_i r \in B \cap M_i$, thus (1) holds. □

5.2 COMPLEMENTS

We are here concerned with weakening the concept of the direct sum of two modules. A direct sum

$$A \oplus B = M$$

is, as we know, determined by the two conditions

$$A + B = M, \qquad A \cap B = 0,$$

which are weakened in the following way by the definition of complements.

5.2.1 *Definition*: Let $A \hookrightarrow M$.

(a) $A^{\cdot} \hookrightarrow M$ is called an *addition complement*, briefly *adco*, of A in M :⇔

(1) $A + A^{\cdot} = M$.

(2) $A^{\cdot}$ is minimal in $A + A^{\cdot} = M$ i.e.

$$\forall B \hookrightarrow M[(A + B = M \wedge B \hookrightarrow A^{\cdot}) \Rightarrow B = A^{\cdot}].$$

(b) $A' \hookrightarrow M$ is called an *intersection complement*, briefly *inco*, of A in M :⇔

(1) $A \cap A' = 0$.

(2) A' is maximal in $A \cap A' = 0$, i.e.

$$\forall C \hookrightarrow M[(A \cap C = 0 \wedge A' \hookrightarrow C) \Rightarrow A' = C].$$

First of all the preliminary remark has to be justified.

5.2.2 COROLLARY. *Let $A \hookrightarrow M$ and $B \hookrightarrow M$. Then we have: $A \oplus B = M \Leftrightarrow$ B is an adco and inco of A in M.*

Proof. "$\Leftarrow$": This follows directly from the definition.

"$\Rightarrow$": Let $A + C = M$ and $C \hookrightarrow B$. By the modular law it follows that $(A \cap B) + C = B$, and as $A \cap B = 0$ we deduce that $C = B$. Accordingly B is adco. Let now $A \cap C = 0$ and $B \hookrightarrow C \Rightarrow A \oplus C = M \Rightarrow B = C$ by the previous conclusion on interchanging the roles of B and C. Thus B is also an inco of A. □

The question now arises as to the uniqueness and existence of such complements. Already in the case $A \oplus B = M$ (with respect to fixed M and A) B is in general only uniquely determined up to isomorphism. With respect to complements even this is no longer the case (see the example in Exercise 6(d)); nevertheless a certain uniqueness statement does arise later.

Now we address ourselves to the question of the existence of complements. As $\mathbb{Z}_{\mathbb{Z}}$ shows, adcos need not exist: Let $n, m \in \mathbb{Z}$ with $(n, m) = 1$, then we have

$$n\mathbb{Z} + m\mathbb{Z} = \mathbb{Z}.$$

For $n \neq 0$, $n \neq \pm 1$ and $(n, q) = 1$, $q > 1$ yields $(n, qm) = 1$ as well as $qm\mathbb{Z} \hookrightarrow\neq m\mathbb{Z}$, thus there is no adco to $n\mathbb{Z}$.

On the other hand, examples of modules, possessing adcos, are easy to construct, such as artinian modules and semisimple modules. In contrast to adcos, incos always exist and these can moreover be chosen in a particular way.

5.2.3 LEMMA. *Let $A, B \hookrightarrow M$ with $A \cap B = 0$. Then there is an inco A' of A with $B \hookrightarrow A'$ and consequently an inco A'' of A' with $A \hookrightarrow A''$.*

Proof. By the use of Zorn's Lemma. Let

$$\Gamma = \{C \mid C \hookrightarrow M \wedge B \hookrightarrow C \wedge A \cap C = 0\},$$

then $\Gamma \neq \varnothing$ since $B \in \Gamma$. Since the union of every totally ordered subset in Γ lies evidently again in Γ, every totally ordered subset from Γ has an upper bound in Γ. By Zorn there is then a maximal element A' in Γ. With A' in the place of A and A in the place of B it follows that $A \hookrightarrow A''$. □

The fact, that there is indeed always an inco but not always an adco is of great significance for the entire theory of modules. For example, it follows from this that there is always an injective hull (definition later) but in general not always a projective cover. The reason for this stems from the fact that in the category of modules Zorn's lemma cannot be applied in the dual case.

Between the concepts small and adco, resp. large and inco, there exists an important connection which will now be explained.

5.2.4 LEMMA

(a) *Let* $M = A + B$, *then we have*:

$$B \text{ is adco of } A \text{ in } M \Leftrightarrow A \cap B \overset{\circ}{\hookrightarrow} B.$$

(b) *If* $A^{\cdot}$ *is* adco *of* A *in* M *and* $A^{\cdot\cdot}$ *is* adco *of* $A^{\cdot}$ *in* M *then* $A^{\cdot}$ *is also* adco *of* $A^{\cdot\cdot}$ *in* M.

(c) *If* $A^{\cdot}$ *is* adco *of* A *in* M *and* $A^{\cdot\cdot}$ *is* adco *of* $A^{\cdot}$ *in* M *with* $A^{\cdot\cdot} \hookrightarrow A^{\cdot}$ *then we have* $A/A^{\cdot\cdot} \overset{\circ}{\hookrightarrow} M/A^{\cdot\cdot}$.

Proof

(a) "$\Rightarrow$": Let $U \hookrightarrow B$ with $(A \cap B) + U = B$, then it follows that $M = A + B = A + (A \cap B) + U = A + U$. Since B is adco of A, it follows that $U = B$, hence we have $A \cap B \overset{\circ}{\hookrightarrow} B$.

(a) "$\Leftarrow$": Let $M = A + U$ with $U \hookrightarrow B$, then it follows that $B = (A \cap B) + U$. As $A \cap B \overset{\circ}{\hookrightarrow} B$ we deduce therefore that $B = U$, thus B is adco of A in M.

(b) By assumption we have $M = A^{\cdot\cdot} + A^{\cdot}$. Let $U \hookrightarrow A^{\cdot}$ with $M = A^{\cdot\cdot} + U$, then it follows that $A^{\cdot} = (A^{\cdot\cdot} \cap A^{\cdot}) + U$. As $M = A + A^{\cdot}$ we obtain $M = A + (A^{\cdot\cdot} \cap A^{\cdot}) + U$. From $A^{\cdot\cdot} \cap A^{\cdot} \overset{\circ}{\hookrightarrow} A^{\cdot\cdot}$ it follows that $A^{\cdot\cdot} \cap A^{\cdot} \overset{\circ}{\hookrightarrow} M$. We deduce that $M = A + (A^{\cdot\cdot} \cap A^{\cdot}) + U = A + U$. Since $A^{\cdot}$ is adco of A and $U \hookrightarrow A^{\cdot}$ it follows that $U = A^{\cdot}$. Thus $A^{\cdot}$ is adco of $A^{\cdot\cdot}$ in M.

(c) Let $(A/A^{\cdot\cdot}) + (U/A^{\cdot\cdot}) = M/A^{\cdot\cdot}$ with $A^{\cdot\cdot} \hookrightarrow U \hookrightarrow M$, then it follows that $A + U = M$. As $M = A^{\cdot\cdot} + A^{\cdot}$ and $A^{\cdot\cdot} \hookrightarrow U$ we have further that $U = A^{\cdot\cdot} + (A^{\cdot} \cap U)$. Hence we deduce that $M = A + U = A + A^{\cdot\cdot} + (A^{\cdot} \cap U) = A + (A^{\cdot} \cap U)$. Since $A^{\cdot}$ is adco of A, it follows that $A^{\cdot} \cap U = A^{\cdot}$, thus $A^{\cdot} \hookrightarrow U$ and we deduce that $M = A^{\cdot\cdot} + A^{\cdot} \hookrightarrow U \hookrightarrow M$, thus we have $U = M$ and it follows that $U/A^{\cdot\cdot} = M/A^{\cdot\cdot}$, which was to be shown. □

We come now to the dual statement.

5.2.5 LEMMA

(a) *Let* A *and* B *be submodules of* M *with* $0 = A \cap B$, *then we have*: B *is* inco *of* A *in* $M \Leftrightarrow (A + B)/B \overset{*}{\hookrightarrow} M/B$.

(b) *If A' is* inco *of A in M and A'' is* inco *of A' in M, then A' is also* inco *of A'' in M.*

(c) *If A' is* inco *of A in M and A'' is* inco *of A' in M with $A \hookrightarrow A''$ then we have $A \overset{*}{\hookrightarrow} A''$.*

Proof

(a) "$\Rightarrow$": Let $(A+B)/B \cap U/B = 0$ with $B \hookrightarrow U \hookrightarrow M$, then it follows that $(A+B) \cap U = B$. Hence we have $A \cap U \hookrightarrow B$, thus $A \cap U \hookrightarrow A \cap B$. Since B is inco of A, we have $A \cap B = 0$, thus also $A \cap U = 0$ and as $B \hookrightarrow U$ and B is inco of A it follows that $B = U$. Thus we have $U/B = B/B = 0$, which is to say $(A+B)/B$ is large in M/B.

(a) "$\Leftarrow$": Let now $A \cap U = 0$ with $B \hookrightarrow U \hookrightarrow M$. Let $x \in (A+B) \cap U$, then it follows that $x = a + b = u$ with $a \in A$, $b \in B$, $u \in U$, thus we have $a = u - b \in A \cap U = 0$ and consequently $a = 0$ and $x = b \in B$. Hence we have $(A+B) \cap U = B$ and it follows that $(A+B)/B \cap U/B = 0$. As $(A+B)/B \overset{*}{\hookrightarrow} M/B$ we must have $U/B = 0$, that is to say $B = U$ holds. Consequently B is inco of A in M.

(b) By assumption we have $A'' \cap A' = 0$. Let $A' \hookrightarrow U \hookrightarrow M$ with $A'' \cap U = 0$. From $(A'' + A')/A'' \overset{*}{\hookrightarrow} M/A''$ it follows that $A'' + A' \overset{*}{\hookrightarrow} M$ (by 5.1.5 (c)). Let $x \in (A'' + A') \cap (A \cap U)$, then it follows that $x = a'' + a' = a = u$ with $a'' \in A''$, $a' \in A'$, $a \in A$, $u \in U$. Hence we have $a'' = u - a' \in A'' \cap U = 0$, thus $a'' = 0$ and it now follows that $x = a' = a \in A' \cap A = 0$. Thus we have $(A'' + A') \cap (A \cap U) = 0$. As $A'' + A' \overset{*}{\hookrightarrow} M$ we must then have $A \cap U = 0$. Since A' is inco of A and $A' \hookrightarrow U$ was assumed, it follows that $A' = U$, which was to be shown.

(c) Let $U \hookrightarrow A''$ with $A \cap U = 0$. For $x \in A \cap (A' + U)$ we have $x = a = a' + u$ with $a \in A$, $a' \in A'$, $u \in U$. We deduce that $a - u = a' \in A'' \cap A' = 0$, thus $x = a = u \in A \cap U = 0$. Consequently we have $A \cap (A' + U) = 0$, thus $A' + U = A'$, thus $U \hookrightarrow A'$. By observing that $U \hookrightarrow A''$ it follows that $U \hookrightarrow A'' \cap A' = 0$ and so $U = 0$. We deduce that $A \overset{*}{\hookrightarrow} A$. □

5.3 DEFINITION OF INJECTIVE AND PROJECTIVE MODULES AND SIMPLE COROLLARIES

5.3.1 THEOREM

(a) *The following are equivalent for a module Q_R:*

(1) *Every monomorphism*

$$\xi : Q \to B$$

splits (i.e. $\operatorname{Im}(\xi)$ *is a direct summand in B).*

(2) *For every monomorphism* $\alpha: A \to B$ *and for every homomorphism* $\varphi: A \to Q$ *there is a homomorphism* $\kappa: B \to Q$ *with* $\varphi = \kappa\alpha$.
(3) *For every monomorphism* $\alpha: A \to B$

$$\mathrm{Hom}(\alpha, 1_Q): \mathrm{Hom}_R(B, Q) \to \mathrm{Hom}_R(A, Q)$$

is an epimorphism.

(b) *The following are equivalent for a module* P_R:
(1) *Every epimorphism*

$$\xi: B \to P$$

splits (i.e. $\mathrm{Ker}(\xi)$ *is a direct summand in* B).
(2) *For every epimorphism* $\beta: B \to C$ *and for every homomorphism* $\psi: P \to C$ *there is a homomorphism* $\lambda: P \to B$ *with* $\psi = \beta\lambda$.
(3) *For every epimorphism* $\beta: B \to C$

$$\mathrm{Hom}(1_P, \beta): \mathrm{Hom}_R(P, B) \to \mathrm{Hom}_R(P, C)$$

is an epimorphism.

DIAGRAM FOR (a), (2):

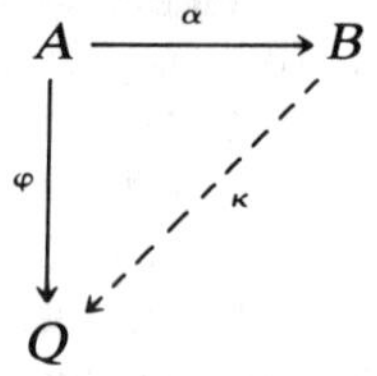

$\varphi = \kappa\alpha$ (i.e. the diagram is commutative).

DIAGRAM FOR (b), (2):

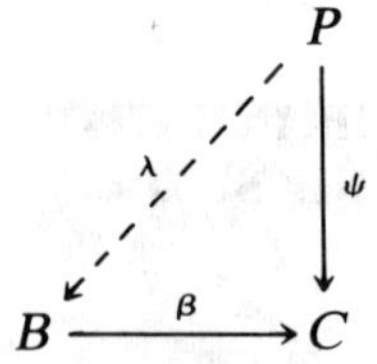

$\psi = \beta\lambda$ (i.e. the diagram is commutative).

Proof of 5.3.1. (a) "(1)⇒(2)": This follows from 4.7.4 since by assumption ψ splits in 4.7.4.

"(2)⇒(1)": By assumption there is a $\kappa : B \to Q$ so that the diagram

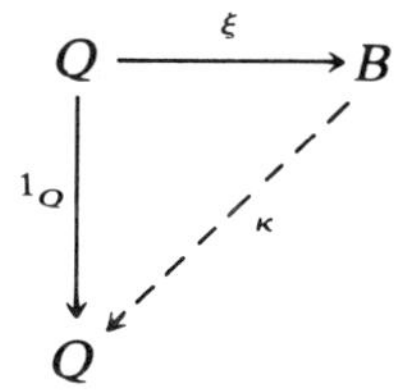

is commutative, i.e. we have $1_Q = \kappa\xi$, thus ξ splits by 3.4.11.

(a) "(2)⇔(3)": By considering the definition of Hom$(\alpha, 1_Q)$ (see 3.6) it is clear that (3) is an equivalent reformulation of (2).

(b) "(1)⇒(2)": This follows from 4.7.6, since, by assumption, α splits in 4.7.6.

(b) "(2)⇒(1)": By assumption there is a $\lambda : P \to B$ so that the diagram

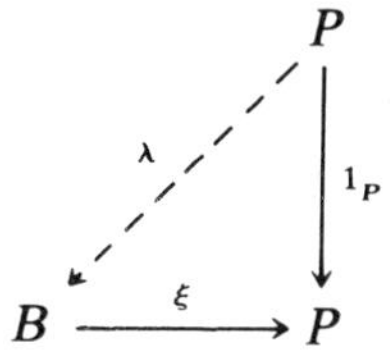

is commutative, i.e. we have $1_P = \xi\lambda$, thus ξ splits by 3.4.11.

(b) "(2)⇔(3)": Equivalent reformulation.

5.3.2 *Definition*

(a) A module Q_R, which satisfies the conditions of 5.3.1(a), is called an *injective R-module.*

(b) A module P_R, which satisfies the conditions of 5.3.1(b), is called a *projective R-module.*

This definition of an injective, resp. of a projective, module refers to the category of unitary right R-modules, since all monomorphisms $\alpha : A \to B$ resp. all epimorphisms $\beta : B \to C$ are allowed. Thus, appropriately, the issue concerns a categorical definition by means of universal mapping properties.

The question then arises whether we can also characterize injective and projective modules by means of "inner" properties. For projective modules this is—as we shall soon see—easily possible: An R-module is projective if and only if it is isomorphic to a direct summand of a free R-module. For injective modules there is in general no correspondingly simple characterization by inner properties. For $R = \mathbb{Z}$ we have however such a characterization: a $\mathbb{Z}$-module is injective if and only if it is divisible. The general case can be reduced to this one.

We come now first of all to some simple consequences of the definition.

5.3.3 COROLLARY

(a) *Q is injective* $\wedge\, Q \cong A \Rightarrow A$ *is injective.*

(b) *P is projective* $\wedge\, P \cong C \Rightarrow C$ *is projective.*

5.3.4 THEOREM

(a) *Let* $Q = \prod_{i\in I} Q_i$, *then we have*:

$$Q \text{ is injective} \Leftrightarrow \forall i \in I\ [Q_i \text{ is injective}].$$

(b) *Let* $P = \coprod_{i\in I} P_i$, *or* $P = \bigoplus_{i\in I} P_i$, *then we have*:

$$P \text{ is projective} \Leftrightarrow \forall i \in I\ [P_i \text{ is projective}].$$

Proof. Notation for injections and projections corresponding to the direct product and the direct sum as in Chapter 4.

(a) "$\Rightarrow$": Let Q be injective, and let $\alpha : A \to B$ be a monomorphism and let $\varphi: A \to Q_j$ for $j \in I$ be a homomorphism. For $\sigma\eta_j\varphi$ there then corresponds by assumption an $\omega : B \to Q$ with $\sigma\eta_j\varphi = \omega\alpha$:

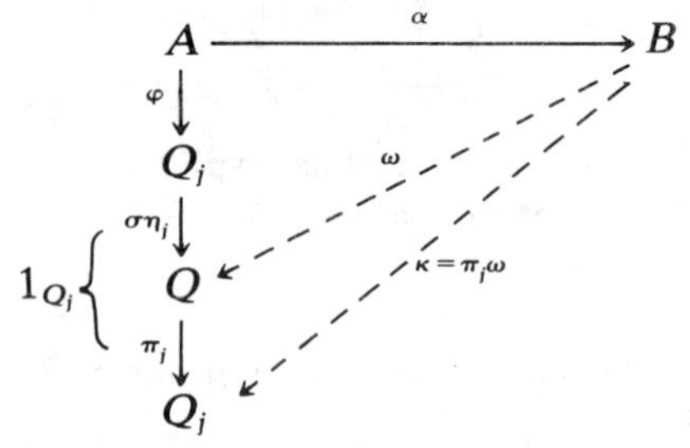

The desired homomorphism κ with $\varphi = \kappa\alpha$ is then $\kappa := \pi_j\omega$ since we have:

$$\varphi = 1_{Q_j}\varphi = (\pi_j\sigma\eta_j)\varphi = \pi_j(\sigma\eta_j\varphi) = \pi_j(\omega\alpha) = (\pi_j\omega)\alpha = \kappa\alpha.$$

(a) "$\Leftarrow$": Let now the monomorphism $\alpha : A \to B$ and the homomorphism $\varphi : A \to Q$ be given. To every $\pi_i\varphi$ there then exists by assumption a κ_i with $\pi_i\varphi = \kappa_i\alpha$. By 4.1.6 there is then a $\kappa : B \to Q$ with $\kappa_i = \pi_i\kappa$:

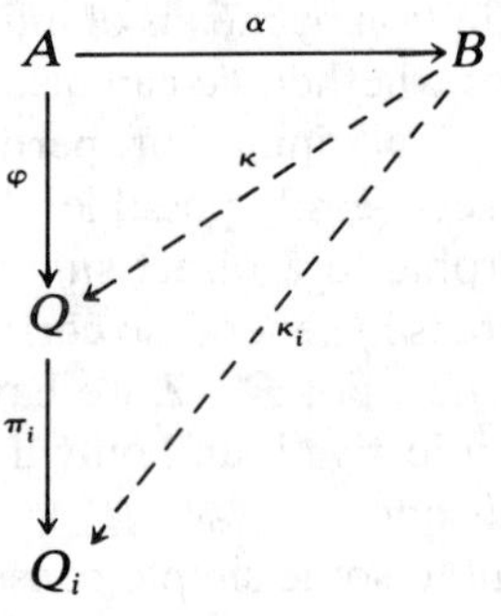

We claim that $\varphi = \kappa\alpha$. From $\pi_i\varphi = \kappa_i\alpha$ and $\kappa_i = \pi_i\kappa$ it follows that $\pi_i\varphi = \pi_i\kappa\alpha$, thus by 4.1.6 (uniqueness) $\varphi = \kappa\alpha$.

(b) By 5.3.3 it suffices to consider the case $P = \coprod_{i\in I} P_i$. The proof follows dually to (a); hence we can be brief.

(b) "⇒": The following situation is now given:

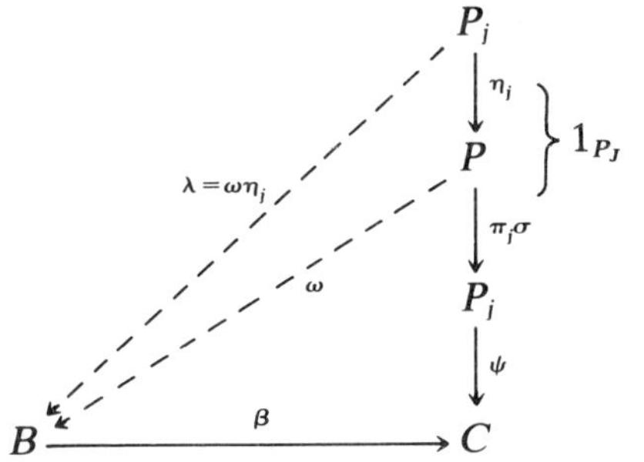

But ω exists with $\psi\pi_j\sigma = \beta\omega$, since P is projective, and $\lambda := \omega\eta_j$ yields the desired result since we have

$$\psi = \psi 1_{P_j} = \psi\pi_j\sigma\eta_j = (\psi\pi_j\sigma)\eta_j = (\beta\omega)\eta_j = \beta(\omega\eta_j) = \beta\lambda.$$

(b) "⇐": In the diagram

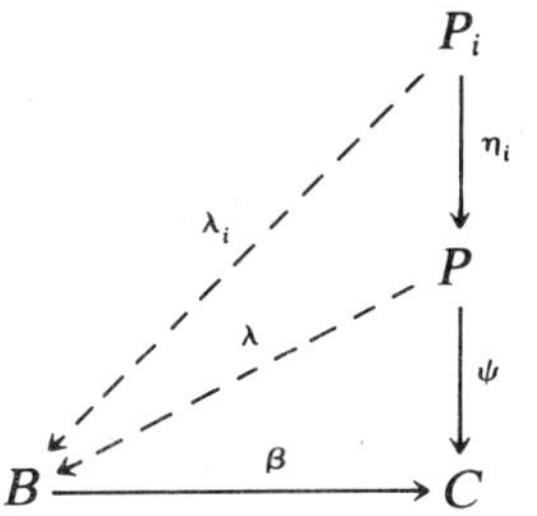

there exist λ_i with $\psi\eta_i = \beta\lambda_i$, by assumption, and λ with $\lambda_i = \lambda\eta_i$, since $P = \coprod P_i$. It remains to show that $\psi = \beta\lambda$. From $\psi\eta_i = \beta\lambda_i$ and $\lambda_i = \lambda\eta_i$ it follows that $\psi\eta_i = \beta\lambda\eta_i$, thus by 4.1.6 (uniqueness) $\psi = \beta\lambda$. □

In particular, according to this result, every direct summand of a projective module is again projective and—since for finite index sets direct sums are also direct products—every direct summand of an injective module is again injective.

5.4 PROJECTIVE MODULES

We are now in a position to produce the previously announced "inner" characterization of projective modules.

5.4.1 THEOREM. *A module is projective if and only if it is isomorphic to a direct summand of a free module.*

Proof. By 4.4.6 every free module is projective and by 5.3.4 and 5.3.3 so also is every module which is isomorphic to a direct summand of a free module. In order to show the converse, let P be a projective module and let

$$\xi : F \to P$$

be an epimorphism of a free module F onto P, existing by 4.4.4. Since P is projective, ξ splits:

$$F = \operatorname{Ker}(\xi) \oplus F_0$$

and F_0 is then isomorphic to P. □

By this theorem, to which there corresponds no dual theorem with respect to injective modules, the theory of projective modules is reduced to the question of the properties of free modules and of their direct summands.

Since, as is well known, every submodule of a free $\mathbb{Z}$-module is again free (see Exercise 10), we obtain the corollary.

COROLLARY. *Every projective $\mathbb{Z}$-module is free.*

An important lemma for the investigation of projective modules is the so-called *Dual-basis Lemma*, which serves in a certain manner with regard to arbitrary projective modules in the place of the basis property of free modules.

5.4.2 THEOREM (DUAL-BASIS LEMMA). *The following properties are equivalent:*

(1) *P_R is projective.*

(2) *To every family $(y_i \mid i \in I)$ of generators of P over R there exists a family $(\varphi_i \mid i \in I)$ of $\varphi_i \in P^* := \operatorname{Hom}_R(P, R)$ with*

(a) $\forall p \in P\ [\varphi_i(p) \neq 0$ *only for finitely many* $i \in I]$;

(b) $\forall p \in P\left[p = \sum_{\substack{i \in I \\ \varphi_i(p) \neq 0}} y_i \varphi_i(p)\right]$.

(3) *There exist families $(y_i \mid i \in I)$ with $y_i \in P$ and $(\varphi_i \mid i \in I)$ with $\varphi_i \in P^*$, so that* (a) *and* (b) *hold.*

Proof. "(1)⇒(2)": As established in 4.4 there is a free R-module F with a basis $\{x_i \mid i \in I\}$ and an epimorphism $\xi : F \to P$ with $\xi(x_i) = y_i$. Let

$$\pi_j : F \ni \sum x_i r_i \to r_j \in R, \qquad j \in I$$

(where we put $r_j = 0$ in the case that the index j does not appear in $\sum x_i r_i$), then we have for $a = \sum x_i r_i \in F$: $\pi_j(a) \neq 0$ for only finitely many $j \in I$ and $a = \sum x_i \pi_i(a)$.

Since P is projective, there exists $\lambda : P \to F$ so that $1_P = \xi\lambda$:

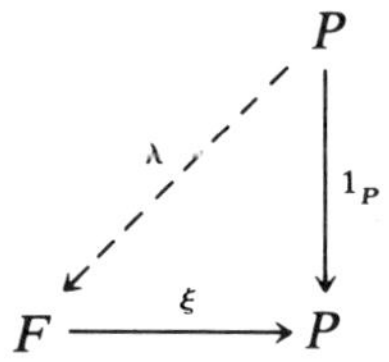

Define $\varphi_i := \pi_i\lambda$, $i \in I$, then we have $\varphi_i \in P^*$, and for $p \in P$ we have $\varphi_i(p) = \pi_i\lambda(p) \neq 0$ for only finitely many $i \in I$. Further we have for $p \in P$

$$p = \xi\lambda(p) = \xi\left(\sum x_i \pi_i(\lambda(p))\right) = \sum \xi(x_i)\pi_i\lambda(p) = \sum y_i \varphi_i(p),$$

thus (a) and (b) hold.

"(2)⇒(3)": Clear.

"(3)⇒(1)": From (b) $(y_i | i \in I)$ is a family of generators of P. Now let $\xi : F \to P$ be again an epimorphism as in the proof of (1)⇒(2). Further let $\tau : P \to F$ be defined by $\tau(p) := \sum x_i \varphi_i(p)$, then firstly τ is a mapping, since the $\varphi_i(p)$ are uniquely determined and by (a) almost all $\varphi_i(p)$ are equal to 0. Obviously τ is in fact an R-homomorphism. Then we have

$$\xi\tau(p) = \xi\left(\sum x_i \varphi_i(p)\right) = \sum y_i \varphi_i(p) = p,$$

thus $1_P = \xi\tau$, i.e. ξ splits and by 5.4.1 P is then projective.

5.5 INJECTIVE MODULES

In general a characterization of injective modules by "inner" properties is not possible in as simple a manner as in the case of projective modules. For $R = \mathbb{Z}$ there is nevertheless such a characterization and this has also considerable significance for the case of an arbitrary ring R, it is in fact used to show the existence of injective extensions.

5.5.1 THEOREM. *A $\mathbb{Z}$-module ($=$ abelian group) is injective if and only if it is divisible.*

Proof. Let $D_{\mathbb{Z}}$ be divisible, then 4.5.5 states that D is injective. Now let $Q_{\mathbb{Z}}$ be injective. Let $q_0 \in Q$, $0 \neq z_0 \in \mathbb{Z}$; if we consider the homomorphisms

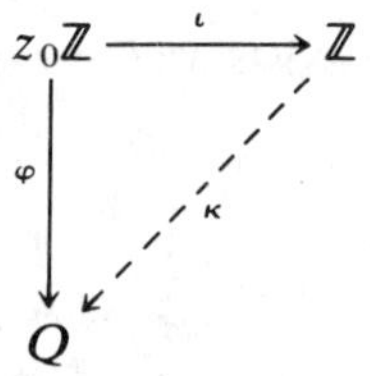

where ι is the inclusion mapping and φ is defined by $\varphi(z_0) := q_0$, then there is, since Q is injective, a κ with $\varphi = \kappa\iota$. Thus we have $\kappa(1)z_0 = \kappa(1z_0) = \kappa(z_0) = (\kappa\iota)(z_0) = \varphi(z_0) = q_0$. Since $q_0 \in Q$ was arbitrary, it follows therefore that $Qz_0 = Q$, i.e. Q is divisible. □

Let now R be again an arbitrary ring. Since every module is an epimorphic image of a free R-module and as free R-modules are projective, every module is an epimorphic image of a projective R-module. We address ourselves now to the dual question and wish to show that we can map every module monomorphically into an injective module.

5.5.2 LEMMA. *If D is a divisible ($=$ injective) $\mathbb{Z}$-module then* $\mathrm{Hom}_{\mathbb{Z}}(R, D)$ *is injective as a right R-module.*

Proof. Let $\alpha : A \to B$ be an R-monomorphism and let $\varphi : A \to \mathrm{Hom}_{\mathbb{Z}}(R, D)$ be an R-homomorphism. Let σ be the $\mathbb{Z}$-homomorphism defined by

$$\sigma : \mathrm{Hom}_{\mathbb{Z}}(R, D) \ni f \to f(1) \in D.$$

Then we consider the diagram

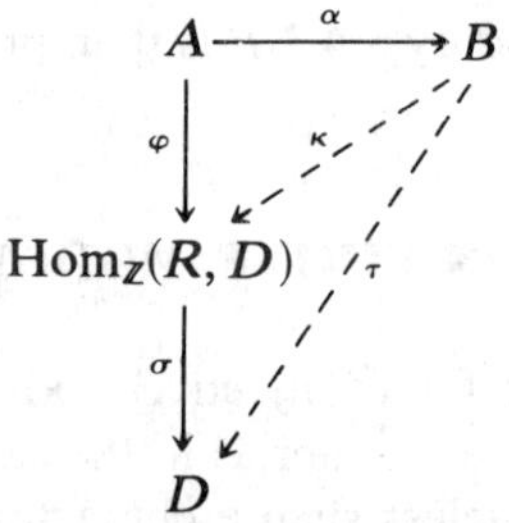

If we regard α and φ only as $\mathbb{Z}$-homomorphisms, then there is, since D is $\mathbb{Z}$-injective, a $\mathbb{Z}$-homomorphism $\tau : B \to D$ with $\sigma\varphi = \tau\alpha$. Now let $\kappa : B \to \mathrm{Hom}_{\mathbb{Z}}(R, D)$ be defined by

$$\kappa(b)(r) := \tau(br), \qquad b \in B, r \in R.$$

Then for fixed $b \in B$, obviously $\kappa(b) \in \mathrm{Hom}_{\mathbb{Z}}(R, D)$ and we have

$$\kappa(br_1)(r) = \tau(br_1r) = \kappa(b)(r_1r) = (\kappa(b)r_1)(r),$$

i.e. $\kappa(br_1) = \kappa(b)r_1$, thus κ is an R-homomorphism. Therefore we have

$$\begin{aligned}\kappa\alpha(a)(r) &= \tau(\alpha(a)r) = \tau(\alpha(ar)) = \tau\alpha(ar) = \sigma\varphi(ar)\\ &= \varphi(ar)(1) = (\varphi(a)r)(1) = \varphi(a)(r)\end{aligned}$$

and consequently $\kappa\alpha = \varphi$. □

5.5.3 THEOREM. *For every module there is a monomorphism into an injective module.*

Proof. Let M_R be given. By 4.5.4 there is a $\mathbb{Z}$-monomorphism

$$\mu : M \to D$$

into a divisible abelian group. By 5.5.2 $\mathrm{Hom}_{\mathbb{Z}}(R, D)_R$ is injective as an R-module. If we define

$$\rho : M \to \mathrm{Hom}_{\mathbb{Z}}(R, D)$$

by $\rho(m)(r) := \mu(mr)$, $m \in M$, $r \in R$, then ρ is evidently an R-homomorphism and, since μ is a monomorphism, even a monomorphism. □

5.5.4 COROLLARY. *Q_R is injective $\Leftrightarrow$ Q_R is isomorphic to a direct summand of a module of the form* $\mathrm{Hom}_{\mathbb{Z}}(R, D)_R$ *with D a divisible abelian group.*

Proof. "$\Rightarrow$": In proof of 5.5.3.
"$\Leftarrow$": By 5.5.2 and 5.3.4. □

Corollary 5.5.4 can be considered as an "inner" characterization of injective modules.

5.5.5 COROLLARY. *Every module is a submodule of an injective module.*

We formulate the proof as an independent lemma.

5.5.6 LEMMA. Let $\rho : M_R \to N_R$ be a monomorphism. Then there is a module N' with $M \hookrightarrow N'$ and an isomorphism $\tau : N' \to N$ so that $\rho = \tau\iota$, where ι is the inclusion mapping of M in N'.

Proof. Let D be a set of the same cardinality as the complement $N \backslash \rho(M)$ of $\rho(M)$ in N with $D \cap M = \varnothing$ and let $\beta : D \to N \backslash \rho(M)$ be an injective mapping. Then define a set $N' := M \cup D$ and let

$$\tau : N' \to N$$

be the bijective mapping defined by

$$\tau(m) := \rho(m), \qquad m \in M$$

$$\tau(d) := \beta(d), \qquad d \in D.$$

In order to make N' into an R-module containing M_R and to make τ into an R-module homomorphism, we put:

$$x + y := \tau^{-1}(\tau(x) + \tau(y)), \qquad x, y \in N'$$

$$xr := \tau^{-1}(\tau(x)r), \qquad r \in R.$$

As is immediately seen, all assertions are then satisfied. □

Then 5.5.5 follows from this lemma since $\mathrm{Hom}_{\mathbb{Z}}(R, D)_R$ and the isomorphic module N' are both injective.

5.6 INJECTIVE HULLS AND PROJECTIVE COVERS

Now that we have seen that every module can be mapped, on the one hand, monomorphically into an injective module and is, on the other hand, an epimorphic image of a projective module, we turn to the question whether in a certain sense there are "smallest" such modules.

5.6.1 *Definition.* Let M_R be given.

(a) A monomorphism $\eta : M \to Q$ is called an *injective hull* of $M :\Leftrightarrow Q$ is injective and η is a large monomorphism (see 5.1.1).

(b) An epimorphism $\xi : P \to M$ is called a *projective cover* of $M :\Leftrightarrow P$ is projective and ξ is a small epimorphism (s. 5.1.1).

If $\eta : M \to Q$ is an injective hull then, if no misunderstanding can arise, we designate Q simply as the injective hull of M without expressing the η. This holds correspondingly in the case of the projective cover.

With this interpretation we denote an injective hull of M also by $I(M)$ and a projective cover of M by $P(M)$. We note however that $I(M)$ and $P(M)$ are not thereby uniquely determined but only up to isomorphism (see 5.6.3).

Example. $\mathbb{Z}_{\mathbb{Z}} \xrightarrow{\iota} \mathbb{Q}_{\mathbb{Z}}$ is an injective hull of $\mathbb{Z}_{\mathbb{Z}}$, for ι is a monomorphism, $\mathbb{Q}_{\mathbb{Z}}$ is injective = (divisible) and by 5.1.6 $\mathbb{Z}_{\mathbb{Z}}$ is large in $\mathbb{Q}_{\mathbb{Z}}$.

5.6.2 Corollary

(a) *If* $\eta_i : M_i \to Q_i$ *for* $i = 1, 2, \ldots, n$ *is an injective hull of* M_i *then*

$$\bigoplus_{i=1}^{n} \eta_i : \bigoplus_{i=1}^{n} M_i \to \bigoplus_{i=1}^{n} Q_i$$

is an injective hull of $\bigoplus_{i=1}^{n} M_i$.

(b) *If* $\xi_i : P_i \to M_i$ *for* $i = 1, 2, \ldots, n$ *is a projective cover of* M_i, *then*

$$\bigoplus_{i=1}^{n} \xi_i : \bigoplus_{i=1}^{n} P_i \to \bigoplus_{i=1}^{n} M_i$$

is a projective cover of $\bigoplus_{i=1}^{n} M_i$.

Proof. (a) This follows from 5.1.7 and 5.3.4.

(b) This follows from 5.1.3 and 5.3.4. □

Two questions now arise immediately, namely the uniqueness and existence of hulls and covers. We begin with the question of the uniqueness and at once prove a somewhat more general result.

5.6.3 Theorem

(a) *Let* $\varphi : M_1 \to M_2$ *be an isomorphism, let* $\eta_1 : M_1 \to Q_1$ *be an injective hull and let* $\eta_2 : M_2 \to Q_2$ *be a monomorphism with* Q_2 *injective. Then there exists a split monomorphism*

$$\psi : Q_1 \to Q_2,$$

so that the diagram

$$\begin{array}{ccc} M_1 & \xrightarrow{\varphi} & M_2 \\ {\scriptstyle \eta_1}\downarrow & & \downarrow{\scriptstyle \eta_2} \\ Q_1 & \xrightarrow{\psi} & Q_2 \end{array}$$

is commutative and

$$\tilde{\eta}_2 : M_2 \ni m \to \eta_2(m) \in \operatorname{Im}(\psi)$$

is an injective hull of M_2. η_2 *is an injective hull of* M_2 *if and only if* ψ *is an isomorphism.*

(b) *Let $\varphi : M_1 \to M_2$ be an isomorphism, let $\xi_1 : P_1 \to M_1$ an epimorphism with P_1 projective and let $\xi_2 : P_2 \to M_2$ be a projective cover. Then there exists a split epimorphism*

$$\psi : P_1 \to P_2,$$

so that the diagram

$$\begin{array}{ccc} P_1 & \xrightarrow{\psi} & P_2 \\ {\scriptstyle \xi_1}\downarrow & & \downarrow{\scriptstyle \xi_2} \\ M_1 & \xrightarrow{\varphi} & M_2 \end{array}$$

is commutative and, if $P_1 = \mathrm{Ker}(\psi) \oplus P_0$ (note $P_0 \cong P_1/\mathrm{Ker}(\psi)$) $\hat{\xi}_1 := \xi_1 | P_0$ is a projective cover of M_1. ξ_1 is a projective cover of M_1 if and only if ψ is an isomorphism.

Proof. (a) In the commutative diagram

$$\begin{array}{ccc} M_1 & \xrightarrow{\eta_1} & Q_1 \\ {\scriptstyle \varphi}\downarrow & & \\ M_2 & \swarrow{\scriptstyle \psi} & \\ {\scriptstyle \eta_2}\downarrow & & \\ Q_2 & & \end{array}$$

ψ exists, since Q_2 is injective. Since $\eta_2 \varphi = \psi \eta_1$ is a monomorphism it follows that $\mathrm{Ker}(\psi) \cap \mathrm{Im}(\eta_1) = 0$. Since $\mathrm{Im}(\eta_1)$ is large in Q_1 it follows that $\mathrm{Ker}(\psi) = 0$, i.e. ψ is a monomorphism. Since Q_1 is injective, ψ splits and $\mathrm{Im}(\psi)$ is injective.

Since $\mathrm{Im}(\eta_2) \hookrightarrow \mathrm{Im}(\psi)$ the definition of $\tilde{\eta}_2$ is meaningful. $\tilde{\eta}_2$ is, along with η_2, a monomorphism and $\mathrm{Cod}(\tilde{\eta}_2) = \mathrm{Im}(\psi)$ is injective. It remains to be shown that $\mathrm{Im}(\tilde{\eta}_2) = \mathrm{Im}(\eta_2)$ is large in $\mathrm{Im}(\psi)$. Let

$$\tilde{\psi} : Q_1 \ni q \to \psi(q) \in \mathrm{Im}(\psi),$$

then $\tilde{\psi}$ is an isomorphism and we have

$$\tilde{\psi}\eta_1(M_1) = \tilde{\eta}_2 \varphi(M_1) = \tilde{\eta}_2(M_2).$$

Since $\eta_1(M_1)$ is large in Q_1, it follows therefore from 5.1.5(c) (with $\varphi = \tilde{\psi}^{-1}$), that $\tilde{\eta}_2(M_2)$ is also large in $\mathrm{Cod}(\tilde{\psi}) = \mathrm{Im}(\psi)$.

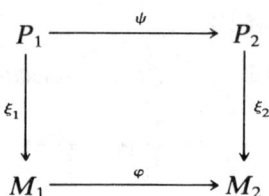

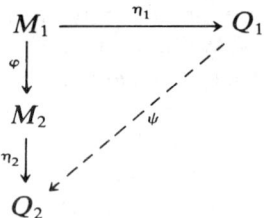

If η_2 is an injective hull of M_2 then $\mathrm{Im}(\eta_2) \overset{*}{\hookrightarrow} Q_2$ holds and as $\mathrm{Im}(\eta_2) \hookrightarrow \mathrm{Im}(\psi)$ it follows that $\mathrm{Im}(\psi) \overset{*}{\hookrightarrow} Q_2$. Since however $\mathrm{Im}(\psi)$ is a direct summand in Q_2, this is only possible with $\mathrm{Im}(\psi)=Q_2$, i.e. ψ is an isomorphism. Conversely if ψ is an isomorphism, then it follows that $\eta_2=\tilde{\eta}_2$, thus η_2 is now an injective hull of M_2.

(b) In the commutative diagram

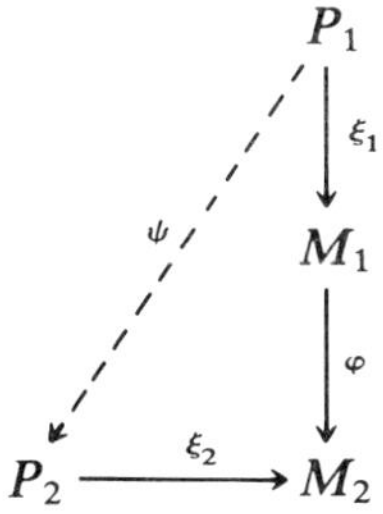

ψ exists, since P_1 is projective. Since $\varphi\xi_1=\xi_2\psi$ is an epimorphism it follows that $\mathrm{Im}(\psi)+\mathrm{Ker}(\xi_2)=P_2$. Since $\mathrm{Ker}(\xi_2)$ is small in P_2, it follows that $\mathrm{Im}(\psi)=P_2$, i.e. ψ is an epimorphism. Since P_2 is projective, ψ splits:

$$P_1=\mathrm{Ker}(\psi)\oplus P_0,$$

and $P_0\cong P_1/\mathrm{Ker}(\psi)$ is projective.

As $\mathrm{Ker}(\psi)\hookrightarrow \mathrm{Ker}(\xi_1)$, $\hat{\xi}_1 := \xi_1|P_0$ is, along with ξ_1, an epimorphism. Since P_0 is projective, it remains to be shown that $\mathrm{Ker}(\hat{\xi}_1)\overset{\circ}{\hookrightarrow} P_0$. Let

$$\hat{\psi}: P_0 \ni p \to \psi(p)\in P_2,$$

then $\hat{\psi}$ is an isomorphism and as $\varphi\hat{\xi}_1=\xi_2\hat{\psi}$ and since φ and $\hat{\psi}$ are isomorphisms

$$\mathrm{Ker}(\hat{\xi}_1)=\hat{\psi}^{-1}(\mathrm{Ker}(\xi_2)).$$

Since $\mathrm{Ker}(\xi_2)$ is small in P_2 it follows from 5.1.3(c), that $\mathrm{Ker}(\hat{\xi}_1)$ is small in $\hat{\psi}^{-1}(P_2)=P_0$.

If ξ_1 is a projective cover of M_1, then $\mathrm{Ker}(\xi_1)\overset{\circ}{\hookrightarrow} P_1$ holds and as $\mathrm{Ker}(\psi)\hookrightarrow\mathrm{Ker}(\xi_1)$ it follows that $\mathrm{Ker}\,(\psi)\overset{\circ}{\hookrightarrow} P_1$. Since however $\mathrm{Ker}(\psi)$ is a direct summand in P_1 this is only possible with $\mathrm{Ker}(\psi)=0$, i.e. ψ is an isomorphism. If conversely ψ is an isomorphism then it follows that $\xi_1=\hat{\xi}_1$, thus ξ_1 is now a projective cover of M_1. □

Once more we point out explicitly that from 5.6.3 the injective hull and the projective cover (if they exist) are uniquely determined up to isomorphism. For example if we put in the injective case $M=M_1=M_2$ and $\varphi=1_M$ then η_2 is an injective hull of M if and only if ψ is an isomorphism.

We come now to the question of the existence of projective covers and injective hulls. Whereas—as is shown subsequently—to every module there exists an injective hull, the dual statement does not hold. Thus there are modules which do not have projective covers. For example no $\mathbb{Z}$-module, which is not itself already projective (=free), has a projective cover, for as we have earlier shown in 5.1.2, the trivial submodule 0 is the only small submodule of a free $\mathbb{Z}$-module.

The interesting question then arises of characterizing the rings R for which every R-module has a projective cover. These are the perfect rings which are treated later.

5.6.4 THEOREM. *Every module has an injective hull. More precisely: If* $\mu: M \to Q$ *is a monomorphism into an injective module* Q *and if* $\mathrm{Im}(\mu)''$ *is an intersection complement of an intersection complement of* $\mathrm{Im}(\mu)$ *in* Q *with* $\mathrm{Im}(\mu) \hookrightarrow \mathrm{Im}(\mu)''$, *then*

$$\tilde{\mu}: M \to \mathrm{Im}(\mu)''$$

with $\tilde{\mu}(m) = \mu(m)$ *for all* $m \in M$ *(restriction of the codomain of* μ *to* $\mathrm{Im}(\mu)''$), *is an injective hull of* M.

Proof. Let $A := \mathrm{Im}(\mu)$. As shown in 5.2.5(c), A is large in A''. It remains to be shown that A'' is injective. To that end we prove that A'' is a direct summand of Q; since Q is injective, this follows then from 5.3.4 also for A''. We consider the diagram

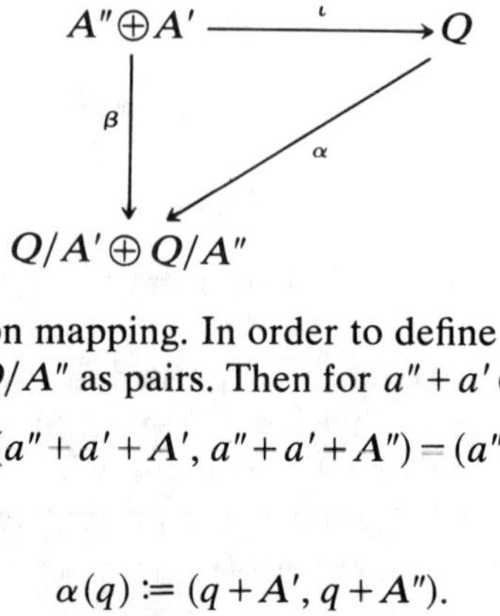

where ι is the inclusion mapping. In order to define α and β, we write the elements of $Q/A' \oplus Q/A''$ as pairs. Then for $a''+a' \in A'' \oplus A'$ let

$$\beta(a''+a') := (a''+a'+A',\, a''+a'+A'') = (a''+A',\, a'+A'')$$

and also

$$\alpha(q) := (q+A',\, q+A'').$$

Then the diagram is commutative, i.e. we have $\beta = \alpha\iota$ from which $\mathrm{Im}(\beta) \hookrightarrow \mathrm{Im}(\alpha)$ follows. As $A'' \cap A' = 0$ α and β are monomorphisms. Since Q is injective and α is a monomorphism, α splits.

We assert that $\mathrm{Im}(\beta)$ is large in $Q/A' \oplus Q/A''$. Since from 5.2.5 $A''+A/A' \overset{*}{\hookrightarrow} Q/A'$ and $A''+A'/A'' \overset{*}{\hookrightarrow} Q/A''$, the assertion follows by 5.1.7.

As $\mathrm{Im}(\beta) \hookrightarrow \mathrm{Im}(\alpha)$ then $\mathrm{Im}(\alpha)$ is also large in $Q/A' \oplus Q/A''$. Since α splits, it follows that $\mathrm{Im}(\alpha) = Q/A' \oplus Q/A''$, i.e. α is an isomorphism. To an arbitrary $q \in Q$ there is therefore a $q_1 \in Q$ with $(q+A', 0+A'') = (q_1+A', q_1+A'')$ from which firstly $q_1 \in A''$ and then $q \in A''+A'$ follow. Thus $A'' \oplus A' = Q$ holds. □

We now summarize, once again, how we have produced the injective hull for a module M_R:

(1) Embedding (monomorphism) of M as an abelian group in a divisible abelian group D.
(2) Embedding $\mu : M_R \to \mathrm{Hom}_Z(R, D)_R$, where the module $\mathrm{Hom}_Z(R, D)_R$ is injective.
(3) Let $\mathrm{Im}(\mu)''$ be an intersection complement of an intersection complement of $\mathrm{Im}(\mu)$ in $\mathrm{Hom}_Z(R, D)_R$ with $\mathrm{Im}(\mu) \hookrightarrow \mathrm{Im}(\mu)''$, then

$$\tilde{\mu} : M \ni m \to \mu(m) \in \mathrm{Im}(\mu)''$$

is an injective hull of M.

It is clear that, by this complicated construction, it is hardly to be expected in general that we can infer directly from the properties of M those of the injective hull of M. The question of which properties of M remain preserved or become lost on passing to the injective hull of M is in any event an interesting question which has been explored from different points of view and assumptions.

An injective hull, which is itself a "minimal injective extension" can also be characterized as a "maximal large (= essential) extension".

5.6.5 *Definition.* Let $\alpha : A \to B$ be a monomorphism.

(1) α is called a *large extension of* A $:\Leftrightarrow \mathrm{Im}(\alpha) \overset{*}{\hookrightarrow} B$.

(2) α is called a *maximal large extension* of A $:\Leftrightarrow \alpha$ is a large extension of A and every large extension of B is an isomorphism.

5.6.6 THEOREM. *Let $\gamma : M \to W$ be a monomorphism. Then we have: γ is a maximal large extension of M if and only if γ is an injective hull of M.*

Proof. Let $\eta : M \to Q$ be an injective hull of M, then we consider the commutative diagram

$$\begin{array}{ccc} M & \xrightarrow{\gamma} & W \\ \downarrow{\scriptstyle\eta} & \swarrow{\scriptstyle\varphi} & \\ Q & & \end{array}$$

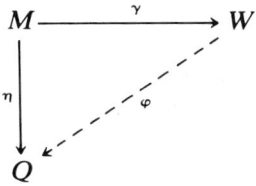

in which φ exists, since Q is injective. As $\mathrm{Im}(\gamma) \overset{*}{\hookrightarrow} W$ and $\mathrm{Im}(\gamma) \cap \mathrm{Ker}(\varphi) = 0$, we have $\mathrm{Ker}(\varphi) = 0$, i.e. φ is a monomorphism. As $\mathrm{Im}(\eta) \hookrightarrow \mathrm{Im}(\varphi)$ and $\mathrm{Im}(\eta) \overset{*}{\hookrightarrow} Q$ it follows that $\mathrm{Im}(\varphi) \overset{*}{\hookrightarrow} Q$. Let now γ be a maximal large extension, then it follows that φ is an isomorphism, thus γ is an injective hull.

The converse that every injective hull is a maximal large extension, follows from the fact, that every monomorphism

$$\alpha : Q \to B$$

with injective Q splits and proper direct summands are not large in a module containing them. □

We now direct our attention once again and briefly to the projective cover. As we know this need not exist. If we assume however that the corresponding addition complements exist then we can dualize Theorem 5.6.4. The intersection complements used in the proof of Theorem 5.6.4 exist by virtue of Zorn's lemma, whereas the dual addition complements exist only under appropriate assumptions. The exact formulation is not to be given here. Later, in the treatment of semiperfect and perfect modules, the question of the existence of projective covers will be thoroughly investigated.

5.7 BAER'S CRITERION

In order to establish whether a module Q is injective we have to test whether to every monomorphism $\alpha : A \to B$ and to every homomorphism $\varphi : A \to Q$ there is a homomorphism $\kappa : B \to Q$ with $\varphi = \kappa\alpha$. This prompts the question whether we can restrict the class of "test monomorphisms" $\alpha : A \to B$. This is in fact possible and indeed it suffices to consider all inclusions of right ideals $U \hookrightarrow R_R$.

5.7.1 THEOREM (BAER'S CRITERION). *A module Q_R is injective if and only if to every right ideal $U \hookrightarrow R_R$ and to every homomorphism $\rho : U \to Q$ there exists a homomorphism $\tau : R_R \to Q$ with $\rho = \tau\iota$, where ι is the inclusion mapping of U into R.*

Proof. That the condition is necessary for injectivity is clear. The converse proof, that it is sufficient, follows in two steps.

Step 1. Let $\alpha : A \to B$ be a monomorphism and let $\varphi \in \mathrm{Hom}_R(A, Q)$. Let $C \overset{\neq}{\hookrightarrow} B$ with $\mathrm{Im}(\alpha) \hookrightarrow C$ and let $\gamma : C \to Q$ with $\varphi(a) = \gamma\alpha(a)$ for all $a \in A$.

Assertion. There is a $C_1 \hookrightarrow B$ with $C \subsetneqq C_1$ and a $\gamma_1 : C_1 \to Q$ with $\gamma_1 | C = \gamma$ (hence also $\varphi(a) = \gamma_1\alpha(a)$).

To prove this assertion let $b \in B$, $b \notin C$; put $C_1 = C + bR$. If we had $C \cap bR = 0$ then immediately we could extend γ trivially to C_1. The difficulty stems from the fact that we can have $C \cap bR \neq 0$. Let

$$U = \{u | u \in R \wedge bu \in C\},$$

then U is obviously a right ideal in R and

$$\xi: U \ni u \to bu \in C$$

is an R-homomorphism. Let $\rho = \gamma\xi$, then we have $\rho : U \to Q$, and by assumption there is a $\tau : R \to Q$ with $\rho = \tau\iota$:

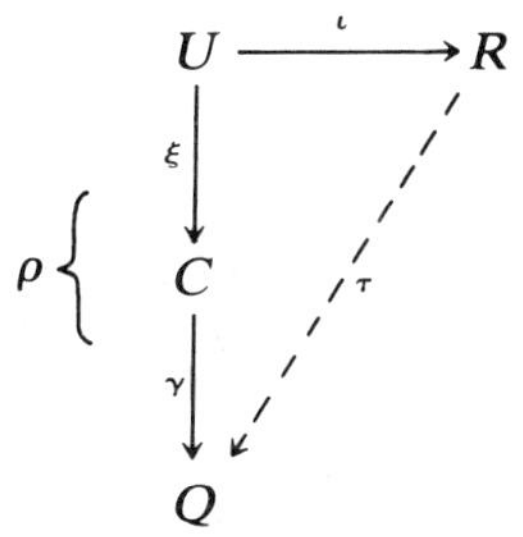

We now define $\gamma_1 : C_1 \to Q$ by

$$\gamma_1 : C + bR \ni c + br \to \gamma(c) + \tau(r) \in Q.$$

To establish that γ_1 is a mapping, let

$$c + br = c_1 + br_1, \qquad c, c_1 \in C, r, r_1 \in R.$$

Then

$$c - c_1 = b(r_1 - r) \in C \cap bR \Rightarrow r - r_1 \in U \Rightarrow \gamma\xi(r - r_1) = \tau(r - r_1)$$
$$\Rightarrow \gamma(c - c_1) = \gamma(b(r_1 - r)) = \gamma\xi(r_1 - r) = \tau(r_1 - r) \Rightarrow$$
$$\gamma(c) + \tau(r) = \gamma(c_1) + \tau(r_1).$$

Since γ and τ are R-homomorphisms, γ_1 is also an R-homomorphism and by the definition of γ_1 we have $\gamma_1 | C = \gamma$.

Step 2. Let $C_0 := \mathrm{Im}(\alpha)$ and let α_0 be the isomorphism of A onto C_0 induced by α. In addition let $\gamma_0 := \varphi\alpha_0^{-1}$, then we have $\varphi(a) = \gamma_0\alpha(a)$ for all $a \in A$. The homomorphism γ_0 is now extended to the whole of B with the help of Step 1 and Zorn's Lemma. For this let Γ be the set of all pairs (C, γ) with $\mathrm{Im}(\alpha) = C_0 \hookrightarrow C \hookrightarrow B$ and $\gamma : C \to Q$ with $\gamma | C_0 = \gamma_0$. As $(C_0, \gamma_0) \in \Gamma$ this set is not empty. An ordering is defined in Γ by

$$(C, \gamma) \leq (C_1, \gamma_1) :\Leftrightarrow \begin{cases} 1. & C \hookrightarrow C_1 \\ 2. & \gamma_1 | C = \gamma. \end{cases}$$

Now let Λ be a non-empty totally ordered subset of Γ and let

$$D := \bigcup_{(C,\gamma)\in\Lambda} C,$$

then we have $C_0 \hookrightarrow D \hookrightarrow B$. Further let

$$\delta : D \ni d \to \gamma(d) \in Q,$$

for $d \in C$ with $(c, \gamma) \in \Lambda$. Then by 2 this is a homomorphism with $\delta | C_0 = \gamma_0$. Consequently (D, δ) is an upper bound of Λ in Γ. Hence by Zorn's Lemma there exists a maximal element in Γ, which, from Step 1, must be equal to a (B, κ) with $\varphi = \kappa\alpha$. This completes the proof. □

Further, following this, we point out that this theorem remains valid if in it we replace R_R by an arbitrary generator (see exercise 21). The correspondingly dual assertion does not hold however.

An important application of Baer's criterion follows in the next chapter, where it is shown that a ring R is noetherian if and only if every direct sum of injective R-modules is again injective.

5.8 FURTHER CHARACTERIZATIONS AND PROPERTIES OF GENERATORS AND COGENERATORS

In 3.3 as well as in 4.8 we have introduced and characterized generators and cogenerators. These considerations are here to be carried forward. In particular a characterization of cogenerators is given which makes it possible for them to be constructed and so to demonstrate their existence. Moreover, in addition, a "minimal" cogenerator can be given.

5.8.1 THEOREM

(a) *The module B_R is a generator if and only if for every projective module P_R a direct sum of copies of B exists which contains a direct summand isomorphic to P.*

(b) *The module C_R is a cogenerator if and only if for every injective module Q_R a direct product of copies of C_R exists which contains a direct summand isomorphic to Q.*

Proof. (a) Let B_R be a generator. By 4.8.2(4) there exists an epimorphism of a direct sum of copies of B onto P. Since by 5.3.1 every epimorphism onto a projective module splits, the condition follows. The converse follows likewise from 4.8.2(4) if we observe that every direct summand of a module is an epimorphic image of the module and every module M, by 4.4.4, is an epimorphic image of a projective (indeed free) module.

(b) Dual to (a), in which 5.5.3 now appears in the place of 4.4.4. □

5.8.2 COROLLARY

(a) *Let P be a projective generator, then we have: The module B is then a generator if and only if there is a direct sum of copies of B which contains a direct summand isomorphic to P.*

(b) *Let Q be an injective cogenerator, then we have: The module C is then a cogenerator if and only if there is a direct product of copies of C which contains a direct summand isomorphic to Q.*

Proof. (a) If B is a generator, then the condition follows by 5.8.1. Conversely, if the condition is satisfied, i.e.

$$\coprod_{i\in I} B_i = P' \oplus L, \qquad B_i = B, \qquad P' \cong P,$$

then this module can evidently be mapped epimorphically onto P, thus is a cogenerator and by 4.8.2 this follows also for B itself.

(b) Dual to (a). □

A projective module P is defined by the fact that for every epimorphism $\beta: B \to C$ $\mathrm{Hom}(1_P, \beta)$ is also an epimorphism. A generator D can now conversely be characterized by the fact that for every epimorphism $\mathrm{Hom}(1_D, \beta)$, β is also an epimorphism. This holds correspondingly in the dual case.

5.8.3 THEOREM

(a) *The module D_R is a generator if and only if every homomorphism $\beta: B \to C$ for which $\mathrm{Hom}(1_D, \beta)$ is an epimorphism is itself an epimorphism.*

(b) *The module C_R is a cogenerator if and only if every homomorphism $\alpha: A \to B$ for which $\mathrm{Hom}(\alpha, 1_C)$ is an epimorphism is a monomorphism.*

Proof. (a) Let D be a generator, then we have

$$C = \sum_{\varphi \in \mathrm{Hom}_R(D,C)} \mathrm{Im}(\varphi).$$

Since $\mathrm{Hom}(1_D, \beta)$ is an epimorphism, there is to every $\varphi \in \mathrm{Hom}_R(D, C)$ a $\varphi' \in \mathrm{Hom}_R(D, B)$ with $\varphi = \beta\varphi'$. Then it follows that

$$C = \sum_{\varphi \in \mathrm{Hom}_R(D,C)} \mathrm{Im}(\varphi) = \sum_{\varphi' \in \mathrm{Hom}_R(D,B)} \mathrm{Im}(\beta\varphi') \hookrightarrow \mathrm{Im}(\beta) \hookrightarrow C,$$

thus $\mathrm{Im}(\beta) = C$, i.e. β is an epimorphism.

To prove the converse let M_R be arbitrary. We define

$$B := \coprod_{\varphi \in \mathrm{Hom}_R(D,M)} D_\varphi$$

with $D_\varphi = D$ for every $\varphi \in \mathrm{Hom}_R(D, M)$ and also $\beta : B \to M$ by

$$\beta((d_\varphi)) = \sum_{d_\varphi \neq 0} \varphi(d_\varphi).$$

Then for $\varphi_0 \in \mathrm{Hom}_R(D, M)$, $\varphi_0 = \beta\eta_{\varphi_0}$ obviously holds, where $\eta_{\varphi_0} : D \to \coprod D_\varphi$ is the canonical monomorphism. Hence it follows that $\mathrm{Hom}_R(1_D, \beta)$ is an epimorphism. By assumption β is then an epimorphism. Consequently we have

$$M = \mathrm{Im}(\beta) = \sum_{\varphi \in \mathrm{Hom}_R(D,M)} \mathrm{Im}(\varphi),$$

thus D is a generator.

(b) Since the proof runs dually, we can be brief. If C is a cogenerator, then the assertion follows from the relations

$$0 = \bigcup_{\varphi \in \mathrm{Hom}_R(A,C)} \mathrm{Ker}(\varphi) = \bigcap_{\varphi' \in \mathrm{Hom}_R(B,C)} \mathrm{Ker}(\varphi'\alpha)$$

$$= \bigcap_{\varphi'} \overset{-1}{\alpha}(\mathrm{Ker}(\varphi')) \hookleftarrow \overset{-1}{\alpha}(0) = \mathrm{Ker}(\alpha) \hookleftarrow 0,$$

thus $\mathrm{Ker}(\alpha) = 0$, i.e. α is a monomorphism.

To show the converse, let M_R be arbitrary. We define

$$B := \prod_{\varphi \in \mathrm{Hom}_R(M,C)} C_\varphi \qquad \text{with } C_\varphi = C \text{ for every } \varphi \in \mathrm{Hom}_R(M, C),$$

as well as $\alpha : M \to B$ by

$$\alpha(m) := (\varphi(m)), \qquad m \in M.$$

Then, for $\varphi_0 \in \mathrm{Hom}_R(M, C)$,

$$\varphi_0 = \pi_{\varphi_0}\alpha$$

where $\pi_{\varphi_0} : \prod C_\varphi \to C_{\varphi_0} = C$ is the canonical epimorphism. Hence $\mathrm{Hom}(\alpha, 1_C)$ is an epimorphism and by assumption α is then a monomorphism. Consequently we have

$$0 = \mathrm{Ker}(\alpha) = \bigcap_{\varphi \in \mathrm{Hom}_R(M,C)} \mathrm{Ker}(\varphi),$$

thus C is a cogenerator. □

5.8.4 COROLLARY

(a) *Let P_R be a projective generator and let $\beta : B_R \to C_R$ be a homomorphism. Then*:

$$\beta \text{ is an epimorphism} \Leftrightarrow \mathrm{Hom}(1_P, \beta) \text{ is an epimorphism.}$$

(b) *Let Q_R be an injective cogenerator and let $\alpha : A_R \to B_R$ be a homomorphism. Then*:

$$\alpha \text{ is a monomorphism} \Leftrightarrow \mathrm{Hom}(\alpha, 1_Q) \text{ is an epimorphism.}$$

Proof. "$\Rightarrow$" This holds since P is projective, resp. Q is injective.
"$\Leftarrow$" By 5.8.3. □

We now direct our attention in particular to cogenerators. Let E_R be a simple module and $I(E_R)$ an injective hull of E, for which we assume $E \hookrightarrow I(R)$. Let C_R be a cogenerator, then as

$$\bigcap_{\varphi \in \mathrm{Hom}_R(I(E), C)} \mathrm{Ker}(\varphi) = 0$$

there must be given a homomorphism $\varphi \in \mathrm{Hom}_R(I(E), C)$ with $E \not\hookrightarrow \mathrm{Ker}(\varphi)$. Since E is simple it follows that $E \cap \mathrm{Ker}(\varphi) = 0$. Since E is large in $I(E)$ it follows therefore that $\mathrm{Ker}(\varphi) = 0$, i.e. φ is a monomorphism. Obviously (by 5.6.3)

$$\varphi' : E \ni x \to \varphi(x) \in \mathrm{Im}(\varphi)$$

is also an injective hull of E where the module $\mathrm{Im}(\varphi)$, isomorphic to $I(E)$, is an injective submodule of C. As an injective submodule it is in fact a direct summand of C. We have therefore established that the cogenerator C to every simple module E contains an injective hull. Henceforth it is crucial that this property is characteristic for cogenerators.

5.8.5 THEOREM

(a) *The module C_R is a cogenerator if and only if for every simple module it contains an injective hull.*

(b) *Let $\{E_j \mid j \in J\}$ be a system of representatives for the classes of isomorphic simple R-modules, and let $I(E_j)$ be an injective hull of E_j. Then*

$$C_0 := \coprod_{j \in J} I(E_j)$$

is a cogenerator.

(c) *A module C_R is a cogenerator if and only if it possesses a submodule isomorphic to C_0.*

Proof. (a) We have previously determined that a cogenerator for every simple module E contains an injective hull. Conversely let this now be satisfied for C. Let $0 \neq m \in M$, then mR is finitely generated and has therefore by 2.3.12 a maximal submodule A. Then $E := mR/A$ is simple. Let

$$\gamma : E \to I(E)$$

be an injective hull of E with $I(E) \hookrightarrow C$, which exists by assumption. Since γ is a monomorphism, $\gamma(m+A) \neq 0$. Let $\nu : mR \to mR/A$ be the natural epimorphism then $\gamma\nu(m) = \gamma(m+A) \neq 0$. Since $I(E)$ is injective, there is a $\gamma' : M \to I(E)$ so that the diagram

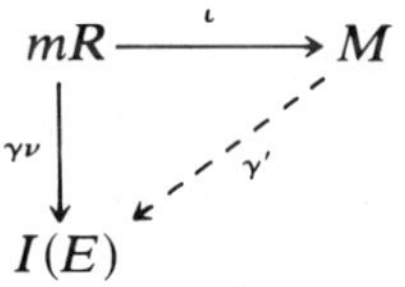

is commutative. Consequently we have $\gamma'(m) \neq 0$. Then let φ be defined by

$$\varphi : M \ni x \to \gamma'(x) \in C,$$

thus it follows that $\varphi(m) \neq 0$, i.e. $m \notin \operatorname{Ker}(\varphi)$. Thus altogether we have

$$\bigcap_{\varphi \in \operatorname{Hom}_R(M,C)} \operatorname{Ker}(\varphi) = 0,$$

i.e., C is in fact a cogenerator.

(b) From consideration of 5.6.3 it follows by (a) that $C_0 = \coprod_{j \in J} I(E_j)$ is a cogenerator.

(c) If we have $C_1 \hookrightarrow C$ with $C_1 \cong C_0$ then it follows by (a) that C is a cogenerator. Now conversely let C be a cogenerator, then there is by (a) to every E_j an injective hull $Q_j \hookrightarrow C$, which by 5.6.3 is isomorphic to $I(E_j)$; let $\gamma_j : I(E_j) \cong Q_j$. We assert that

$$\sum_{j \in J} Q_j = \bigoplus_{j \in J} Q_j,$$

where the sum is to be taken in C. Let $E_j' := \gamma_j(E_j)$, then E_j' is isomorphic to E_j and $E_j' \overset{*}{\hookrightarrow} Q_j$. To prove the assertion it suffices by 5.1.7 to show that

$$\sum_{j \in J} E_j' = \bigoplus_{j \in J} E_j'.$$

If we suppose that this sum is not direct then there is a finite sub-sum $\neq 0$ which is not direct. Of all finite sub-sums $\neq 0$, which are not direct, let (with respect to new indices) $E_1' + \ldots + E_n'$ be one with the smallest n.

Then we have

$$E_2' + \ldots + E_n' = E_2' \oplus \ldots \oplus E_n',$$

and also $E_1' \cap (E_2' \oplus \ldots \oplus E_n') \neq 0$. Since E_1' is simple it follows that

$$E_1' \hookrightarrow E_2' \oplus \ldots \oplus E_n'.$$

Let π_i be the projection of $E_2' \oplus \ldots \oplus E_n'$ onto E_i', $i = 2, \ldots, n$, then there is an $i_0 \in \{2, \ldots, n\}$ with $\pi_{i_0}(E_1') \neq 0$. Since E_1' and E_{i_0}' are simple, it follows that $E_1' \cong \pi_{i_0}(E_1') = E_{i_0}'$, thus also $E_1 \cong E_1' \cong E_{i_0}' \cong E_{i_0}$, in contradiction to the assumption concerning $\{E_j \mid j \in J\}$. Thus we have in fact

$$\sum_{j \in J} Q_j = \bigoplus_{j \in J} Q_j,$$

and hence the isomorphisms $\gamma_j: I(E_j) \cong Q_j$ can be assembled to give an isomorphism

$$\gamma: \coprod_{j \in J} I(E_j) \to \bigoplus_{j \in J} Q_j$$

(see 4.3.1) for which we have $\gamma((a_j)) = \sum' \gamma_j(a_j)$, where $(a_j) \in \coprod I(E_j)$, $\sum' \gamma_j(a_j) \in \bigoplus Q_j$. □

Condition (b) of this theorem provides us with a "minimal" cogenerator C_0 which is injective for finite J (for this see also Exercise (28)). For an arbitrary J, C_0 is injective in the case that R_R is noetherian (see 6.5.1).

In the general case to obtain an injective cogenerator, we take an arbitrary injective module containing $\coprod_{j \in J} E_j$ as for example $I\left(\coprod_{j \in J} E_j\right)$.

Later we shall be closely concerned with injective cogenerators, where the preceding theorem is essentially used.

5.8.6 *Example.* For $R = \mathbb{Z}$, $\mathbb{Q}/\mathbb{Z}$ is an injective cogenerator. Since $\mathbb{Q}/\mathbb{Z}$ is divisible $\mathbb{Q}/\mathbb{Z}$ is firstly injective. For an arbitrary ring R every cyclic R-module $M = mR$ is isomorphic to a module R/A with $A_R \hookrightarrow R_R$ ($A = \mathrm{Ker}(\alpha)$, if $\alpha: R \ni r \mapsto mr \in M$). Consequently every cyclic and thereby in particular every simple $\mathbb{Z}$-module is isomorphic to a module of the form $\mathbb{Z}/n\mathbb{Z}$ (with $n \in \mathbb{Z}$). Since for $n \neq 0$

$$\mathbb{Z}/n\mathbb{Z} \ni \bar{z} \mapsto \frac{z}{n} + \mathbb{Z} \in \mathbb{Q}/\mathbb{Z}$$

is obviously a monomorphism, $\mathbb{Q}/\mathbb{Z}$ contains for every simple $\mathbb{Z}$-module an isomorphic copy. Consequently $\mathbb{Q}/\mathbb{Z}$ is also a cogenerator.

EXERCISES

(1)

Let $A \hookrightarrow B \hookrightarrow M$. Show:

(a) $B \overset{\circ}{\hookrightarrow} M \Leftrightarrow B/A \overset{\circ}{\hookrightarrow} M/A \wedge A \overset{\circ}{\hookrightarrow} M$.

(b) $A \overset{*}{\hookrightarrow} M \Leftrightarrow A \overset{*}{\hookrightarrow} B \wedge B \overset{*}{\hookrightarrow} M$.

(c) Let $A^{\cdot}$ be adco of A in M. Show for $T \hookrightarrow M$: $T \overset{\circ}{\hookrightarrow} M \Rightarrow T \cap A^{\cdot} \overset{\circ}{\hookrightarrow} A$.

(d) Let A' be inco of A in M. Show for $T \hookrightarrow M$: $T \overset{*}{\hookrightarrow} M \Rightarrow (T+A')/A' \overset{*}{\hookrightarrow} M/A'$.

(2)

Let A and B be submodules of M.

(a) Show: $A+B=M \wedge A \cap B \overset{\circ}{\hookrightarrow} B \Rightarrow B$ is adco of A in M.

(b) Show: $A \cap B = 0 \wedge (A+B)/B \overset{*}{\hookrightarrow} M/B \Rightarrow B$ is inco of A in M.

(3)

Show: For $A \hookrightarrow M$ the following properties are equivalent:

(a) $A \overset{\circ}{\hookrightarrow} M$.

(b) For every generating set $(x_i | i \in I)$ of M and every family $(a_i | i \in I)$ with $a_i \in A$, $(x_i - a_i | i \in I)$ is also a generating set of M.

(c) There is a generating set $(x_i | i \in I)$ of M so that for every family $(a_i | i \in I)$ with $a_i \in A$, $(x_i - a_i | i \in I)$ is also a generating set of M. (Note the special case $M_R = R_R$ with 1 as generating family.)

(d) If from a generating set of M all elements of A are omitted then a generating set of M is again obtained.

(4)

For $m \in M_R$ let

$$r_R(m) = \{r | r \in R \wedge mr = 0\}.$$

Show

$$\mathrm{Si}(M) := \{m | m \in M \wedge r_R(m) \overset{*}{\hookrightarrow} R_R\}$$

is a submodule of M_R ($\mathrm{Si}(M)$ is called the *singular* submodule of M).

(5)

Show:

(a) Let R be a commutative ring, $A \hookrightarrow R_a$ and $A^{\cdot}$ an adco of A in R. Then there is a $B \hookrightarrow A$ with $A^{\cdot} \oplus B = R$.

(b) If R is an integral domain and $A \overset{\neq}{\hookrightarrow} R_R$ has an adco in R then A is small in R.

(6)

A submodule $X \hookrightarrow M_R$ is called *closed* in M if from $X \overset{*}{\hookrightarrow} U \hookrightarrow M$ it always follows that $X = U$. Show:

(a) For X the following are equivalent:

(i) X is closed in M.

(ii) X is an intersection complement of a submodule of M, i.e. there is an $A \hookrightarrow M$ with $X = A'$.

(iii) From $X \hookrightarrow V \overset{*}{\hookrightarrow} M$ it always follows that $V/X \overset{*}{\hookrightarrow} M/X$, i.e. the natural mapping $\nu : M \to M/X$ "contains" large submodules.

(b) Additionally let $M \overset{*}{\hookrightarrow} Q_R$ with Q_R injective (thus injective hull of M). Then we have: X is closed in M if and only if there is a direct summand $Y \hookrightarrow Q$ with $Y \cap M = X$.

(c) Every submodule in M_R is a direct summand ($= \dot{M_R}$ semisimple) if and only if every submodule is closed.

(d) Construct an example in which a closed submodule is not a direct summand (say in $M_{\mathbb{Z}} := (\mathbb{Z}/8\mathbb{Z}) \oplus (\mathbb{Z}/2\mathbb{Z})$).

(e) If R is an integral domain then in every R-module M the torsion submodule

$$T(M) := \{m \mid m \in M \wedge mr = 0 \quad \text{for an } r \in R,\ r \neq 0\}$$

is closed.

(7)

For $R = \mathbb{Z}$, i.e. in the category of abelian groups, the closed submodules are to be characterized. Show for $X \hookrightarrow M_{\mathbb{Z}}$:

(a) If X is an intersection complement of A in M and if we have $m \in M$, $m \notin X$, $mp \in X$ for a prime number p, then there is an $x \in X$ with $mp = xp$. (Show that $(m\mathbb{Z} + X)/X$ is simple and $(X + A)/X$ is large in M/X so that it follows that $m \in X + A$.)

(b) If $X \overset{*}{\underset{\neq}{\hookrightarrow}} U \hookrightarrow M$, then there is a $u \in U$ and a prime number p with $u \notin X$, $up \in X$. (Show that U/X has a simple subgroup.)

(c) A subgroup X is closed in M if and only if for every prime p we have: $Xp = X \cap Mp$.

(d) A subgroup X is closed in M if and only if:

$$\operatorname{Soc}(M/X) = (\operatorname{Soc}(M) + X)/X.$$

Here $\operatorname{Soc}(M)$ is the sum of all simple subgroups of M.

(8)

Let $0 \neq e \neq 1$ be an idempotent ($e = e^2$) from the centre of R. Show: The right R-module eR is projective but not free.

(9)

Let $\beta_1 : P_1 \to M$, $\beta_2 : P_2 \to M$ be epimorphisms and P_1, P_2 projective. Show:

$$P_1 \oplus \mathrm{Ker}(\beta_2) \cong P_2 \oplus \mathrm{Ker}(\beta_1).$$

(10)

Let $U \hookrightarrow F$ where F is a free right R-module with a basis $(e_i \mid i \in I)$. Let the set I be well-ordered (by $\leq$) and to every $j \in I$ let there be defined:

$$F_j = \bigoplus_{i<j} e_i R, \qquad \bar{F}_j = \bigoplus_{i \leq j} e_i R, \qquad U_j = U \cap F_j, \qquad \bar{U}_j = U \cap \bar{F}_j,$$

and also $A_j = \pi_j(\bar{U}_j)$ where π_j is the jth projection of F onto R. Show:

(a) If the right ideal A_j is projective, then there is a $V_j \cong A_j$ with $\bar{U}_j = U_j \oplus V_j$.

(b) If there is for every $i \in I$ a V_i with $\bar{U}_i = U_i \oplus V_i$ then it follows that

$$U = \bigoplus_{i \in I} V_i.$$

(To show that $X = \sum V_i$ coincides with U one shows that the set $\{i \mid i \in I \wedge U_i \not\hookrightarrow X\}$ is empty.)

(c) If in R every right ideal is projective then every submodule of a free right R-module is a direct sum of right ideals (up to isomorphism).

(d) Over a principal ideal domain every submodule of a free module is again free.

(11)

Let $(T_i \mid i \in I)$ be a family of rings and let

$$R := \prod_{i \in I} T_i.$$

R itself becomes a ring, if addition and multiplication are defined componentwise, R is then called the *ring product* of the family $(T_i \mid i \in I)$. Let

$$A := \coprod_{i \in I} T_i,$$

then obviously we have $A \subset R$. For $k \in I$ denote by e_k that element from A whose kth component is 1 and whose remaining components are 0. Show:

(a) A is a two-sided ideal in R with $A_R = \bigoplus_{i \in I} e_i R$ and $A_R \not\stackrel{*}{\hookrightarrow} R_R$.

(b) A_R is projective.

(c) $\mathrm{Hom}_R((R/A)_R, R_R) = 0$ and $(R/A)_R$ is not projective for infinite I.

(d) For $j \in I$ we have: $(e_j R)_R$ is injective $\Leftrightarrow (T_j)_{T_j}$ is injective.

(e) If all $(T_j)_{T_j}$ are injective and I is infinite then A_R is a direct sum of injective modules but A_R is not itself injective.

(f) R_R is injective $\Leftrightarrow \forall_j \in I\ [(T_j)_{T_j}$ is injective].

(12)

Let R be an integral domain with quotient field $K \neq R$. Show:

(a) $\mathrm{Hom}_R(K, R) = 0$.

(b) K_R is not projective.

(c) If a projective module P_R has a finitely generated large submodule, then P is itself finitely generated.

(d) Every (as R-module) projective ideal is finitely generated.

(Hint: For (c) use the Dual Basis Lemma.)

(13)

Show in the category of abelian groups (i.e. $R = \mathbb{Z}$):

(a) If P is projective ($=$free) and if A, B are two direct summands in P then $A \cap B$ is also a direct summand in P.

(b) If Q is injective ($=$divisible) and if A, B are two direct summands in Q then $A + B$ is also a direct summand in Q.

(14)

Show: A module P is projective if and only if to every epimorphism $\beta: Q \to C$ with injective Q and to every homomorphism $\varphi: P \to C$ there is a homomorphism $\varphi': P \to Q$ with $\varphi = \beta\varphi'$.

(15)

In the following let $I(M)$ always be an injective hull of M with $M \hookrightarrow I(M)$.

(a) Show: Every endomorphism φ of $I(M)$ with $\varphi(m) = m$ for all $m \in M$ is an automorphism.

(b) Show that the following conditions are equivalent:

(1) Every endomorphism φ of $I(M)$ with $\varphi(m) = m$ for all $m \in M$ is the identity of $I(M)$;

(2) $\mathrm{Hom}_R(I(M)/M, I(M)) = 0$.

(16)

Let always $M \hookrightarrow I(M)$, and $X \hookrightarrow I(X)$ resp. Let M be called *X-determined* if $\mathrm{Hom}_R(I(M)/M, I(X)) = 0$. Show:

(a) M is X-determined $\Leftrightarrow$ to every homomorphism $\varphi: M \to X$ there is only one homomorphism $\varphi': I(M) \to I(X)$ with $\varphi(m) = \varphi'(m)$ for all $m \in M$.

(b) M is injective $\Leftrightarrow \forall X \in M_R$ [M is X-determined].

(c) $\forall x \in X[\{r \mid r \in R \wedge xr = 0\} \overset{*}{\hookrightarrow} R_R \Rightarrow x = 0] \Leftrightarrow \forall M \in M_R$ [M is X-determined].

(d) M is $\prod_{i \in I} X_i$-determined $\Leftrightarrow \forall i \in I$ [M is X_i-determined].

(e) $M_1 \oplus M_2$ is X-determined $\Leftrightarrow M_1$ and M_2 are X-determined.

(f) $M_1 \oplus M_2$ is $M_1 \oplus M_2$–determined$\Leftrightarrow M_i$ is M_j-determined for $i, j = 1, 2$.

(g) Let C be a cogenerator and X an arbitrary module. Precisely the injective modules are $C \oplus X$-determined. In particular we have: $C \oplus X$ is $C \oplus X$-determined$\Leftrightarrow C$ and X are injective.

(17)

Show: If Q_1, Q_2 are injective and $\mu_1 : Q_1 \to Q_2$, $\mu_2 : Q_2 \to Q_1$ are monomorphisms, then we have: $Q_1 \cong Q_2$. (Hint: Without loss we can assume that $Q_2 \hookrightarrow Q_1$, $\mu_1 : Q_1 \to Q_1$ and μ_2 is an inclusion mapping. Let $Q_1 = Q_2 \oplus A$, then let

$$B := A + \mu_1(A) + \mu_1^2(A) + \mu_1^3(A) + \dots$$

and let C be an injective hull of $B \cap Q_2 = \mu_1(B)$ in Q_2. By using the homomorphism $B \ni b \mapsto \mu_1(b) \in C$ it may be shown that $A \oplus C \cong C$.)

(18)

Let $S := \operatorname{End}(M_R)$ where M is considered as an S-R-bimodule ${}_SM_R$. Show:

(a) Let $x \in M$, let xR be simple and let xR be contained in an injective submodule of M_R. Then Sx is a simple left S-module.

(b) Let $x, y \in M$, $xR \cong yR$ and let xR be contained in an injective submodule of M_R. Then Sx is isomorphic to a submodule of Sy.

(c) Let $x, y \in M$, $xR \cong yR$ and as well let xR and yR be contained in injective submodules of M. Then it follows that $I(Sx) \cong I(Sy)$.

(19)

For an integral domain R show:

Every divisible torsion-free R-module is injective.

(M_R is *divisible* $:\Leftrightarrow \forall r \in R, r \neq 0\ [Mr = M]$;

M_R is *torsion-free* $:\Leftrightarrow \forall m \in M, m \neq 0 \forall r \in R, r \neq 0\ [mr \neq 0]$.)

(20)

Let R be an integral domain with quotient field K. In the lattice $\operatorname{Lat}(K_R)$ of R-submodules of K_R a multiplication is defined:

$$U \cdot V := \left\{ \sum_{i=1}^{n} u_i v_i \,\middle|\, u_i \in U \wedge v_i \in V \wedge n \in \mathbb{N} \right\}.$$

This multiplication is commutative and associative and has R as unit element. Show:

(a) For $0 \neq U \hookrightarrow K_R$ the following are equivalent:

(1) There is a $V \hookrightarrow K_R$ with $U \cdot V = R$.

(2) U_R is projective and finitely generated.
(3) U_R is projective.
(Hint: Use the Dual Basis Lemma.)

(b) If $0 \neq U_R \hookrightarrow R_R$ holds then the three conditions are further equivalent to

(4) For all divisible M_R the mapping

$$\mathrm{Hom}(\iota, 1_M) : \mathrm{Hom}_R(R, M) \to \mathrm{Hom}_R(U, M)$$

is surjective.

(c) The following are equivalent for R:
(1) Every ideal is projective.
(2) Every divisible R-module is injective.

An integral domain with property (c) (1) is called a *Dedekind ring*. In particular every principal ideal domain is a Dedekind ring.

(21)

Let M_R be called X_R-*injective*$:\Leftrightarrow$for every monomorphism $\alpha : A \to X$

$$\mathrm{Hom}(\alpha, 1_M) : \mathrm{Hom}_R(X, M) \to \mathrm{Hom}_R(A, M)$$

is surjective. Show:

(a) Let $\xi_1 : X_1 \to X$ be a monomorphism and let $\xi_2 : X \to X_2$ be an epimorphism with $\mathrm{Im}(\xi_1) = \mathrm{Ker}(\xi_2)$. If M is X-injective then M is also X_1- and X_2-injective.

(b) Let M be X-injective and let M_1 be large in M. Then we have: M_1 is X-injective$\Leftrightarrow$for every $\varphi \in \mathrm{Hom}_R(X, M)$ we have $\mathrm{Im}(\varphi) \hookrightarrow M_1$.

(c) If M is X_i-injective for every X_i of the family $(X_i \mid i \in I)$ then M is also $\left(\coprod_{i \in I} X_i\right)$-injective. (Hint: Use (b) with an injective hull of M.)

(d) Let M be X-injective and let X be a generator, then M is injective. (Generalization of Theorem 5.7.1.)

(22)

Let M_R be called Y_R-*projective* $:\Leftrightarrow$ for every epimorphism $\beta : Y \to B$

$$\mathrm{Hom}(1_M, \beta) : \mathrm{Hom}_R(M, Y) \to \mathrm{Hom}_R(M, B)$$

is surjective. Show:

(a) Let $R = \mathbb{Z}$. $\mathbb{Q}_{\mathbb{Z}}$ is $\mathbb{Z}_{\mathbb{Z}}$-projective, but not $\mathbb{Z}^{\mathbb{N}}$-projective $\left(\mathbb{Z}^{\mathbb{N}} = \prod_{n \in \mathbb{N}} \mathbb{Z}_n\right.$ with $\mathbb{Z}_n = \mathbb{Z}$ for all $\left. n \in \mathbb{N}\right)$.

(b) If every simple right R-module is X-projective, then X is semisimple ($=$ sum of its simple submodules).

(23)

Show:

(a) If R is a Dedekind ring (see Example 20) with quotient field $K \neq R$ then K/R is an injective cogenerator of M_R.

(b) If R is a principal ideal domain and with exactly one maximal ideal $pR \neq 0$, then the K/R-projective modules (see Example 22) are precisely the torsion-free R-modules.

(Hint: Use the following two facts concerning R:

(1) If an R-module M is not torsion-free, it has a direct summand which is isomorphic to K/R or $R/(p^n)$ for an $n \geq 1$.

(2) The R-modules $A_n := R/(p^n)$ have the following property: $A_n \hookrightarrow B \wedge B/A_n$ torsion-free $\Rightarrow A_n$ is a direct summand in B.)

(24)

Let $S := \mathrm{End}(C_R)$ and consider C as an S-R-bimodule ${}_SC_R$. For $U \subset C$ let

$$l_S(U) := \{s \mid s \in S \wedge a(U) = 0\}$$

and also for $T \subset S$

$$r_C(T) := \{c \mid c \in C \wedge t(c) = 0 \quad \text{for all} \quad t \in T\},$$

then $l_S(U)$ is a left ideal of S and $r_C(T)$ is a submodule of C_R. Let the other annihilators be analogously formed.

Show for a cogenerator C_R:

(a) $B \hookrightarrow C_R \Rightarrow r_C l_S(B) = B$.

(b) $A \hookrightarrow R_R \Rightarrow r_R l_C(A) = A$.

(c) ${}_SS$ is a cogenerator $\Rightarrow C_R$ is injective.

(Hint: Let $\eta : C \to I(C)$ be an injective hull; by using the left ideal $L \hookrightarrow {}_SS$,

$$L := \{\lambda\eta \mid \lambda \in \mathrm{Hom}_R(I(C), C)\}$$

we show that η splits.)

(d) If R is a cogenerator on both sides then R is injective on both sides.

(25)

Let Q_R be injective, $S := \mathrm{End}(Q_R)$ and let $U \hookrightarrow Q$, $V \hookrightarrow Q$. Show:

(a) $l_S(U \cap V) = l_S(U) + l_S(V)$.

(b) $l_S r_Q(I) = I$ for all finitely generated left ideals $I \hookrightarrow {}_SS$.

(26)

Let $S = \mathrm{End}(M_R)$. For $U \hookrightarrow M$ we define in S the right ideal

$$\lambda_S(U) := \{s \mid s \in S \wedge \mathrm{Im}(s) \hookrightarrow U\},$$

for $T \subset S$ we define in M the R-submodule

$$\rho_M(T) := \sum_{s \in T} \operatorname{Im}(s).$$

Show:

(a) M_R is a generator $\Rightarrow \rho_M \lambda_S(U) = U$ for all $U \hookrightarrow M_R$.

(b) M_R is projective $\Rightarrow \lambda_S(U+V) = \lambda_S(U) + \lambda_S(V)$ for all $U, V \hookrightarrow M_R$.

(c) M_R is projective $\Rightarrow \lambda_S \rho_M(I) = I$ for all finitely generated $I \hookrightarrow S_S$.

(27)

Let $R = K[x, y]$ be the polynomial ring in the indeterminates x and y over a field K. For fixed $n \in \mathbb{N}$ let A denote the ideal generated by the elements

$$\{x^i y^{n+1-i} \mid 0 \leq i \leq n+1\}$$

and let $S := R/A$. Show:

(a) S_S is not injective.

(b) The R-module

$$M := \left(\sum_{i=0}^{n} x^i y^{n-i} R \right) \Big/ (x^{n+1}R + y^{n+1}R)$$

is also an S-module (i.e. $MA = 0$) and possesses exactly one simple submodule E_S.

(c) The inclusion $E_S \hookrightarrow M_S$ is a maximally large extension.

(d) M_S is an injective cogenerator.

(28)

Let the cogenerator C_0, introduced in 5.8.6, be injective and let D be also a cogenerator of M_R to which in every cogenerator of M_R there is an isomorphic submodule.

Show: $C_0 \cong D$.

Chapter 6

Artinian and Noetherian Modules

One of the starting points in the historical development of "non-commutative" rings and of modules over such rings was the theory of algebras over a field K. The algebras themselves, their ideals as well as modules over such algebras are also K-vector spaces. Consequently it was possible to draw upon the theory of vector spaces for much of what was done in the initial stage of the development. If a finiteness assumption is needed then it is clear that finite dimension is required of the underlying K-vector spaces.

The further development aimed, as far as possible, at removing the assumption of an algebra. If we only have a ring (which is not an algebra), then certainly in such a case we do not have the linear theory available and in particular the question arises as to a substitute for the finiteness condition of an algebra which is now no longer applicable.

Here, above all, Emmy Noether provided the appropriate notions and interpretations and thereby sowed the seeds for the further development. As finiteness assumptions she introduced maximal and minimal condition which can also be formulated as chain conditions. In other parts of algebra these have turned out to be just as significant and natural. These conditions are now about to be provided so that in the following considerations we can always refer back to them. From the considerations of this chapter the investigation of artinian and noetherian modules is not in any way concluded but, as further concepts and lemmas present themselves, we shall return to the theme many times.

In order to avoid misunderstanding it is to be emphasized that in the following it is a question of finite or countable chains of submodules with inclusion as order relation.

6.1. DEFINITIONS AND CHARACTERIZATIONS

6.1.1 *Definitions*

(1) A module $M = M_R$ is called *noetherian* resp. *artinian* :$\Leftrightarrow$ every non-empty set of submodules possesses (with respect to inclusion as ordering) a maximal resp. minimal element.

(2) A ring R is called *right noetherian* resp. *artinian* :$\Leftrightarrow$ R_R is noetherian resp. artinian.

(3) A chain of submodules of M

$$\ldots \hookrightarrow A_{i-1} \hookrightarrow A_i \hookrightarrow A_{i+1} \hookrightarrow \ldots$$

(finite or infinite) is called *stationary* :$\Leftrightarrow$ the chain contains only finitely many different A_i.

Remarks. (a) These properties are obviously preserved under isomorphism.

(b) A noetherian resp. artinian module is also called a module with *maximal* resp. *minimal condition.*

6.1.2 THEOREM. *Let $M = M_R$ and let $A \hookrightarrow M$.*

(I) *The following properties are equivalent:*

(1) *M is artinian.*

(2) *A and M/A are artinian.*

(3) *Every descending chain $A_1 \hookleftarrow A_2 \hookleftarrow A_3 \hookleftarrow \ldots$ of submodules of M is stationary.*

(4) *Every factor module of M is finitely cogenerated.*

(5) *In every set $\{A_i \mid i \in I\} \neq \emptyset$ of submodules $A_i \hookrightarrow M$ there is a finite subset $\{A_i \mid i \in I_0\}$ (i.e. finite $I_0 \subset I$) with*

$$\bigcap_{i \in I} A_i = \bigcap_{i \in I_0} A_i.$$

(II) *The following properties are equivalent:*

(1) *M is noetherian.*

(2) *A and M/A are noetherian.*

(3) *Every ascending chain $A_1 \hookrightarrow A_2 \hookrightarrow A_3 \hookrightarrow \ldots$ of submodules of M is stationary.*

(4) *Every submodule of M is finitely generated.*

(5) *In every set $\{A_i \mid i \in I\} \neq \emptyset$ of submodules $A_i \hookrightarrow M$ there is a finite subset $\{A_i \mid i \in I_0\}$ (i.e. finite $I_0 \subset I$) with*

$$\sum_{i \in I} A_i = \sum_{i \in I_0} A_i.$$

(III) *The following properties are equivalent*:

(1) *M is artinian and noetherian.*

(2) *M is of finite length* (*see* 3.5.1).

Proof. (I) "(1)⇒(2)": Since every non-empty set of submodules of A is also such a set of M there is therefore a minimal element, thus A is artinian. Let $\nu: M \to M/A$ and let $\{\Omega_i \mid i \in I\}$ be a non-empty set of submodules of M/A.

We claim that if $\nu^{-1}(\Omega_{i_0})$ is minimal in $\{\nu^{-1}(\Omega_i) \mid i \in I\}$ then Ω_{i_0} is minimal in $\{\Omega_i \mid i \in I\}$. Suppose $\Omega_i \hookrightarrow \Omega_{i_0}$. Then $\nu^{-1}(\Omega_i) \hookrightarrow \nu^{-1}(\Omega_{i_0})$ and so, from the minimality of Ω_{i_0}:

$$\nu^{-1}(\Omega_i) = \nu^{-1}(\Omega_{i_0}),$$

and we have $\Omega_i = \nu\nu^{-1}(\Omega_i) = \nu\nu^{-1}(\Omega_{i_0}) = \Omega_{i_0}$. This follows also directly from 3.1.13.

(I) "(2)⇒(3)": Let $A_1 \hookleftarrow A_2 \hookleftarrow A_3 \hookleftarrow \ldots$ be a descending chain of $A_i \hookrightarrow M$ and let again $\nu: M \to M/A$. Let

$$\Gamma := \{A_i \mid i = 1, 2, 3, \ldots\}, \qquad \nu(\Gamma) := \{\nu(A_i) \mid i = 1, 2, 3, \ldots\},$$

$$\Gamma_A := \{A_i \cap A \mid i = 1, 2, 3, \ldots\}.$$

Since Γ is not empty, $\nu(\Gamma)$ and Γ_A are not empty. By assumption there is therefore a minimal element in $\nu(\Gamma)$, say $\nu(A_l)$ and a minimal element in Γ_A, say $A_m \cap A$. Let $n := \text{Max}(l, m)$, then we have

$$\nu(A_n) = \nu(A_{n+i}), \qquad A_n \cap A = A_{n+i} \cap A, \quad i = 0, 1, 2, \ldots$$

We claim that $A_n = A_{n+1}$, $i = 0, 1, 2, \ldots$ so that the given chain is in fact stationary.

From $\nu(A_n) = \nu(A_{n+i})$ we have

$$A_n + A = \nu^{-1}\nu(A_n) = \nu^{-1}\nu(A_{n+i}) = A_{n+i} + A,$$

i.e. $A_n + A = A_{n+i} + A$. Further we have $A_n \cap A = A_{n+i} \cap A$ as by assumption $A_n \hookleftarrow A_{n+i}$. By the modular law it now follows that

$$A_n = (A_n + A) \cap A_n = (A_{n+i} + A) \cap A_n = A_{n+i} + (A \cap A_n)$$
$$= A_{n+i} + (A \cap A_{n+i}) = A_{n+i}.$$

(I) "(3)⇒(1)": Indirect proof. Suppose the non-empty set Λ of submodules contains no minimal element. For every $U \in \Lambda$ there is then a $U' \in \Lambda$ with $U' \subsetneq U$. For every U let such a fixed U' be chosen (Axiom of Choice). For arbitrary $U_0 \in \Lambda$

$$U_0 \supsetneq U_0' \supsetneq U_0'' \supsetneq \ldots$$

is then an infinite, properly descending chain in contradiction to (3).

(I) "(4)⇔(5)": This follows immediately from 3.1.11 (if in (5) we write $U := \bigcap_{i\in I} A_i$).

(I) "(1)⇒(5)": By (1) in the set of all intersections of any finitely many of the A_i, $i\in I$ there is a minimal element; let this be $D := \bigcap_{i\in I_0} A_i$.

By the minimality of D it follows for every $j\in I$: $D\cap A_j = D$ and hence $D \hookrightarrow \bigcap_{j\in I} A_j$, thus $D = \bigcap_{j\in I} A_j$.

(I) "(5)⇒(3)": Let $A_1 \hookleftarrow A_2 \hookleftarrow A_3 \hookleftarrow \dots$ be given, then from (5) there is an n with

$$\bigcap_{i=1,2,3,\dots} A_i = \bigcap_{i=1,\dots,n} A_i$$

and consequently we have $A_n = A_i$ for $i \geqslant n$.

(II) The proof follows dually to the artinian case up to (4)⇔(5): this equivalence was shown in 2.3.13 (with M in the place of $\sum_{i\in I} A_i$ in (5)).

(III) "(1)⇒(2)": Since M is noetherian by (II) every submodule is noetherian. Consequently there is in every submodule $A \hookrightarrow M$ (including M itself), $A \neq 0$, a maximal submodule A'. To every such A let a fixed submodule A' be chosen. Then consider the chain

$$M \hookleftarrow\!\!\!\!\!\neq\; M' \hookleftarrow\!\!\!\!\!\neq\; M'' \hookleftarrow\!\!\!\!\!\neq\; M''' \hookleftarrow\!\!\!\!\!\neq\; \dots$$

Since M is artinian this must break off and then it represents a composition series, i.e. M is of finite length.

(III) "(2)⇒(1)": Let $\mathsf{A} := A_1 \hookrightarrow A_2 \hookrightarrow A_3 \hookrightarrow \dots$ be an ascending chain of submodules of M.

Let l be the length of M (= the length of a composition series of M).

We claim that in A at most $l+1$ different A_i occur. Suppose there were more than $l+1$, then a subchain of A of the form

$$A_{i_1} \hookrightarrow\!\!\!\!\!\neq\; A_{i_2} \hookrightarrow\!\!\!\!\!\neq\; \dots \hookrightarrow\!\!\!\!\!\neq\; A_{i_{l+2}}$$

would exist. This could be refined to a composition series of M (see 3.5.3) and consequently M would have to have a length $\geqslant l+1$. But if A has only finitely many different A_i then A is stationary, thus M is noetherian. Analogously it follows that M is artinian. □

The condition (I), (3) resp. (II), (3) in 6.1.2 is called the *descending* resp. *ascending chain condition.* Thus by 6.1.2 we conclude that a module satisfies the minimal resp. the maximal condition if and only if it satisfies the descending resp. ascending chain condition. This statement remains valid if we consider not all submodules but only the finitely generated submodules resp. the cyclic submodules resp. the direct summands of a module. Thus by way of example we have: a module then satisfies the minimal conditions for finitely generated submodules (that is to say, in every non-empty set of finitely generated submodules there is a minimal one), if and only if the descending chain condition for finitely generated submodules is satisfied (that is to say, every descending chain of finitely generated submodules is stationary). The easy proof of this and of the other corresponding statements is left to the reader as an exercise. The reader will also notice that in place of the set of all finitely generated submodules resp. all cyclic submodules resp. all direct summands an arbitrary set of submodules can appear. Of course only the three given cases are of interest here.

6.1.3 COROLLARY

(1) *If M is a finite sum of noetherian* resp. *artinian submodules then M is noetherian*, resp. *artinian.*

(2) *If R is a right noetherian* resp. *artinian ring and $M = M_R$ is finitely generated then M is noetherian*, resp. *artinian.*

(3) *Every factor ring of a right noetherian* resp. *artinian ring is again right noetherian* resp. *artinian.*

Proof. (1) let $M = \sum_{i=1}^{n} A_i$ with $A_i \hookrightarrow M$. We obtain the proof by means of induction on the number n of the summands. For $n = 1$ the assertion coincides with the assumption. Let the assumption be valid for $n-1$ and let

$$M = \sum_{i=1}^{n} A_i \qquad \text{with } A_i \text{ noetherian resp. artinian for all } i.$$

Then

$$L := \sum_{i=1}^{n-1} A_i \qquad \text{is noetherian resp. artinian.}$$

By the First Isomorphism Theorem (3.4.3) we have

$$M/A_n = (L + A_n)/A_n \cong L/L \cap A_n.$$

From 6.1.2 whenever L is noetherian resp. artinian so also is $L/L \cap A_n$ and hence also M/A_n. Since A_n is also noetherian resp. artinian the assertion now follows from 6.1.2.

(2) For $x \in M$ consider the mapping

$$\varphi_x : R \ni r \mapsto xr \in M.$$

This is immediately a homomorphism of R_R into M_R. From the Homomorphism Theorem we deduce that

$$R/\mathrm{Ker}(\varphi_x) \cong \mathrm{Im}(\varphi_x) = xR$$

as right R-modules. If R_R is artinian resp. noetherian then it follows from this by 6.1.2 that this also holds for xR. If $x_1, \ldots, x_n$ is a generating set of M_R then the assertion follows from Corollary (1) as

$$M = \sum_{i=1}^{n} x_i R.$$

(3) Let $A \hookrightarrow {}_R R_R$, then whenever R_R is noetherian resp. artinian so also is $(R/A)_R$. Since $(R/A)A = 0$ the submodules of $(R/A)_R$ coincide with the right ideals of R/A from which the assertion follows. □

6.2 EXAMPLES

(1) Every finite-dimensional vector space is of finite length. In order to see this let V_K be a vector space over the skew field K and let $\{x_1, \ldots, x_n\}$ be a basis of V_K. Then

$$0 \subsetneq x_1 K \subsetneq x_1 K + x_2 K \subsetneq \ldots \subsetneq x_1 K + \ldots + x_n K = V$$

is a composition series of V since from

$$(x_1 K + \ldots + x_{i+1} K)/(x_1 K + \ldots + x_i K) \cong x_{i+1} K \cong K$$

every factor is simple.

(2) Every finite-dimensional algebra R over a field K is on both sides of finite length since every right or left ideal is a subspace of R considered as a K-vector space.

(3) A vector space V_K of infinite dimension is neither artinian nor noetherian. Let $\{x_i \mid i \in \mathbb{N}\}$ be a set of linearly independent elements, then we may consider the chains

$$\sum_{i=1}^{\infty} x_i K \supsetneq \sum_{i=2}^{\infty} x_i K \supsetneq \sum_{i=3}^{\infty} x_i K \supsetneq \ldots$$

and

$$x_1 K \subsetneq x_1 K + x_2 K \subsetneq x_1 K + x_2 K + x_3 K \subsetneq \ldots$$

neither of which is stationary.

(4) $\mathbb{Z}_{\mathbb{Z}}$ is noetherian but not artinian. Since every ideal is a principal ideal and so finitely generated $\mathbb{Z}_{\mathbb{Z}}$ is noetherian.

Since

$$\mathbb{Z} \supsetneq 2\mathbb{Z} \supsetneq 2^2\mathbb{Z} \supsetneq \ldots$$

is not stationary, $\mathbb{Z}_{\mathbb{Z}}$ is not artinian.

Remark. In $\mathbb{Z}$ we have a ring which (on each side) is in fact noetherian but not artinian. The converse situation is with respect to a ring (with a unit element!) not possible! Actually we shall show later that every artinian ring is also noetherian. In order to obtain an example of an artinian module, which is not noetherian (see next example), we can in consequence not refer to a ring.

(5) Let p be a prime number and let

$$\mathbb{Q}_p := \left\{ \frac{a}{p^i} \middle| a \in \mathbb{Z} \wedge i \in \mathbb{N} \right\}$$

i.e. the set of the rational numbers, whose denominator is a power of p (including $p^0 = 1$). Then $\mathbb{Q}_p$ is a subgroup of $\mathbb{Q}$ (as additive group) and $\mathbb{Z} \hookrightarrow \mathbb{Q}_p$.

ASSERTION. *$\mathbb{Q}_p/\mathbb{Z}$ is artinian but not noetherian as a $\mathbb{Z}$-module.*

Proof. Let $\left|\frac{1}{p^i} + \mathbb{Z}\right)$ be the $\mathbb{Z}$-submodule of $\mathbb{Q}_p/\mathbb{Z}$ generated by $\frac{1}{p^i} + \mathbb{Z} \in \mathbb{Q}_p/\mathbb{Z}$. Then

$$0 \hookrightarrow \left|\frac{1}{p} + \mathbb{Z}\right) \hookrightarrow \left|\frac{1}{p^2} + \mathbb{Z}\right) \hookrightarrow \left|\frac{1}{p^3} + \mathbb{Z}\right) \hookrightarrow \dots$$

is a properly ascending chain for

$$\frac{1}{p^{i+1}} \notin \left|\frac{1}{p^i} + \mathbb{Z}\right),$$

thus $\mathbb{Q}_p/\mathbb{Z}$ is not noetherian. In order to show that $\mathbb{Q}_p/\mathbb{Z}$ is artinian we show that in the chain given above all proper submodules of $\mathbb{Q}_p/\mathbb{Z}$ occur. In every non-empty set of submodules there is evidently then a smallest submodule (not only a minimal one!).

We consider firstly:

$$(*) \qquad (a, p) = 1 \Rightarrow \left|\frac{a}{p^i} + \mathbb{Z}\right) = \left|\frac{1}{p^i} + \mathbb{Z}\right).$$

As $(a, p) = 1$ (coprime) there are $b, c \in \mathbb{Z}$ with

$$ab + p^i c = 1 \Rightarrow \frac{ab}{p^i} - \frac{1}{p^i} = -c \in \mathbb{Z}$$

thus

$$\frac{ab}{p^i} + \mathbb{Z} = \frac{1}{p^i} + \mathbb{Z} \Rightarrow \left|\frac{1}{p^i} + \mathbb{Z}\right) \hookrightarrow \left|\frac{a}{p^i} + \mathbb{Z}\right).$$

Since on the other hand $|a/p^i + \mathbb{Z}) \hookrightarrow |1/p^i + \mathbb{Z})$, the assertion follows.

Let now $B \hookrightarrow \mathbb{Q}_p/\mathbb{Z}$, then there are two distinct cases.

Case 1. For every $n \in \mathbb{N}$ there is an $i \in \mathbb{N}$ with $i \geqslant n$ and an $a/p^i + \mathbb{Z} \in B$ with $(a, p) = 1$ (i.e. there are elements of arbitrarily high order in B).

From $(*)$ it then obviously follows that $B = \mathbb{Q}_p/\mathbb{Z}$ for every $z/p^n + \mathbb{Z} \in B$.

Case 2. There is a maximal $i \in \mathbb{N}$ for which there exists an $a/p^i + \mathbb{Z} \in B$ with $(a, p) = 1$ (i.e. there is no element of arbitrarily high order in B). From $(*)$ it then follows that

$$\left|\frac{a}{p^i} + \mathbb{Z}\right) = \left|\frac{1}{p^i} + \mathbb{Z}\right) = B. \qquad \square$$

(6) Example of a ring which is artinian and noetherian on one side, thus of finite length, but which is neither artinian nor noetherian on the other side. Let R and K be fields and let R be an infinite dimensional extension field of K. Example: $\mathbb{R}$ and $\mathbb{Q}$.

Let S be the ring of all matrices of the form

$$\begin{pmatrix} k & r_1 \\ 0 & r_2 \end{pmatrix} \qquad \text{with } k \in K, r_1, r_2 \in R.$$

As we see immediately, S is a ring with unit element

$$\begin{pmatrix} 1 & 0 \\ 0 & 1 \end{pmatrix}.$$

S is neither left artinian nor left noetherian. Let $\{x_i \mid i \in \mathbb{N}\}$ be a set of elements from R which are linearly independent over K. Let

$$s_i := \begin{pmatrix} 0 & x_i \\ 0 & 0 \end{pmatrix}, \qquad i \in \mathbb{N},$$

then we have

$$\begin{pmatrix} k & r_1 \\ 0 & r_2 \end{pmatrix} s_i = \begin{pmatrix} 0 & kx_i \\ 0 & 0 \end{pmatrix},$$

thus it follows that the left ideal generated by s_i is

$$Ss_i = \begin{pmatrix} 0 & Kx_i \\ 0 & 0 \end{pmatrix}.$$

Then

$$Ss_i \subsetneqq Ss_1 + Ss_2 \subsetneqq Ss_1 + Ss_2 + Ss_3 \subsetneqq \ldots$$

is a properly ascending chain of left ideals and

$$\sum_{i=1}^{\infty} Ss_i \supsetneqq \sum_{i=2}^{\infty} Ss_i \supsetneqq \sum_{i=3}^{\infty} Ss_i \supsetneqq \dots$$

is a properly descending chain of left ideals.

To see that S_S is of finite length a composition series of S_S will be explicitly given. For this it is useful to have the product of two elements of S in view:

$$\begin{pmatrix} h & a_1 \\ 0 & a_2 \end{pmatrix}\begin{pmatrix} k & r_1 \\ 0 & r_2 \end{pmatrix} = \begin{pmatrix} hk & hr_1 + a_1 r_2 \\ 0 & a_2 r_2 \end{pmatrix}$$

Let now

$$A_1 := \begin{pmatrix} 0 & 1 \\ 0 & 0 \end{pmatrix} S = \begin{pmatrix} 0 & R \\ 0 & 0 \end{pmatrix}, \qquad A_2 := \begin{pmatrix} 0 & 0 \\ 0 & 1 \end{pmatrix} S = \begin{pmatrix} 0 & 0 \\ 0 & R \end{pmatrix},$$

then these right ideals are obviously simple (since R is a field) and we have $A_1 \cap A_2 = 0$.

Then it follows that $A_1 + A_2 / A_1 \cong A_2$ is also simple.

We claim that $0 \hookrightarrow A_1 \hookrightarrow A_1 + A_2 \hookrightarrow S$ is a composition series of S_S. It remains only to be shown that $A_1 + A_2$ is maximal in S. Let

$$\begin{pmatrix} h & a_1 \\ 0 & a_2 \end{pmatrix} \notin A_1 + A_2,$$

then it follows that $h \neq 0$. For

$$B := A_1 + A_2 + \begin{pmatrix} h & a_1 \\ 0 & a_2 \end{pmatrix} S$$

we then have

$$\left(\begin{pmatrix} h & a_1 \\ 0 & a_2 \end{pmatrix} + \begin{pmatrix} 0 & 0 \\ 0 & 1 - a_2 \end{pmatrix}\right)\begin{pmatrix} h^{-1} & -h^{-1} a_1 \\ 0 & 1 \end{pmatrix} = \begin{pmatrix} 1 & 0 \\ 0 & 1 \end{pmatrix} \in B,$$

thus $B = S$.

6.3 THE HILBERT BASIS THEOREM

The Hilbert Basis Theorem can be considered as the principle of construction for certain noetherian rings. It has important applications in algebraic geometry.

6.3.1 THEOREM. *Let R be a right noetherian ring. Then the polynomial ring $R[x]$ (in which x commutes with the elements from R) is right noetherian.*

COROLLARY. *$R[x_1, \ldots, x_n]$ is right noetherian.*

Proof of the theorem. We show that every right ideal A from $R[x]$ is finitely generated. For the proof we assume $A \neq 0$. We obtain the proof in three steps.

Step 1. Let $P(x) = x^n r_n + x^{n-1} r_{n-1} + \ldots + r_0 \in R[x]$ with $r_n \neq 0$, then r_n is called the highest coefficient of $P(x)$; the highest coefficient of the zero polynomial from $R[x]$ is put equal to 0. Let $A_0 :=$ the set of the highest coefficients of polynomials in A.

ASSERTION. $A_0 \hookrightarrow R_R$.

Proof. Let $a, b \in A_0$, $a \neq 0$, $b \neq 0$ then there are

$$P_1(x) = x^m a + x^{m-1} a_{m-1} + \ldots \in A,$$

$$P_2(x) = x^n b + x^{n-1} b_{n-1} + \ldots \in A.$$

Let further $r_1, r_2 \in R$ with $ar_1 + br_2 \neq 0$, it follows that $P_1(x) x^n r_1 + P_2(x) x^m r_2 \in A$, thus $ar_1 + br_2 \in A_0$ and consequently $A_0 \hookrightarrow R_R$.

Since R_R is noetherian, A_0 is finitely generated. Let $a_1, \ldots, a_k$ be a generating set of A_0, where all $a_i \neq 0$, then there are $P_1(x), \ldots, P_k(x) \in A$ with $a_1, \ldots, a_k$ as the highest coefficients (in the given sequence). By multiplication by powers of x it can be arranged that all $P_i(x)$ have the same degree, say n; which we now assume. Let now

$$B := \sum_{i=1}^{k} P_i(x) R[x],$$

then B is finitely generated and we have $B \hookrightarrow A$.

Step 2. Now let $F(x) \in A$.

ASSERTION. *$F(x)$ can be written in the form*

$$F(x) = G(x) + H(x)$$

where $G(x) \in B$ and $H(x) = 0$ or the degree of $H(x)$ is $\leqslant n$.

Proof. If $F(x) = 0$ or if the degree of $F(x)$ is $\leqslant n$ the assertion holds with $F(x) = H(x)$. Thus let the degree of $F(x)$ be $t > n$. If b is the highest coefficient of $F(x)$ then b can be represented in the form

$$b = a_1 r_1 + \ldots + a_k r_k, \qquad r_i \in R.$$

The polynomial

$$F_1(x) := F(x) - (\sum P_i(x) r_i) x^{t-n}$$

then has a degree $\leqslant t-1$ or $F_1(x) = 0$. Thus putting

$$G_1(x) := (\sum P_i(x) r_i) x^{t-n}$$

we then have

$$F(x) = G_1(x) + F_1(x),$$

where $G_1(x) \in B$. In the case that the degree of $F_1(x) \geqslant n$ we may decompose $F_1(x)$ correspondingly:

$$F_1(x) = G_2(x) + F_2(x)$$

with $G_2(x) \in B$ and $F_2(x) = 0$ or the degree of $F_2(x) \leqslant t-2$. From this it follows that

$$F(x) = G_1(x) + G_2(x) + F_2(x)$$

with $G_1(x) + G_2(x) \in B$ and $F_2(x) = 0$ or the degree of $F_2(x) \leqslant t-2$.

After at most $t-n$ steps (i.e. using induction) the desired decomposition is obtained

$$(*) \qquad F(x) = G(x) + H(x).$$

Since $F(x) \in A$ and $G(x) \in B \hookrightarrow A$ it follows that

$$H(x) = F(x) - G(x) \in A \cap (R + xR + \ldots + x^n R).$$

Step 3. Now consider the right R-module

$$A \cap (R + xR + \ldots + x^n R).$$

This is an R-submodule of the finitely generated right R-module $R + xR + \ldots + x^n R$, over the right noetherian ring R. By 6.1.3 and 6.1.2 this is then also finitely generated. Let, say,

$$A \cap (R + xR + \ldots + x^n R) = \sum_{j=1}^{l} Q_j(x) R.$$

ASSERTION

$$A = \sum_{i=1}^{k} P_i(x) R[x] + \sum_{j=1}^{l} Q_j(x) R[x].$$

Since $P_i(x), Q_j(x) \in A$ the right side is contained in A and from $(*)$ A is also contained in the right side. This completes the proof. □

6.4 ENDOMORPHISMS OF ARTINIAN AND NOETHERIAN MODULES

First let $M = M_R$ be an arbitrary module and let φ be an endomorphism of M, i.e. a homomorphism of M into itself. Then φ^n, $n \in \mathbb{N}$, is also an endomorphism of M and we have

$$\mathrm{Im}(\varphi) \hookleftarrow \mathrm{Im}(\varphi^2) \hookleftarrow \mathrm{Im}(\varphi^3) \hookleftarrow ..$$

$$\mathrm{Ker}(\varphi) \hookrightarrow \mathrm{Ker}(\varphi^2) \hookrightarrow \mathrm{Ker}(\varphi^3) \hookrightarrow \dots$$

In case M is artinian resp. noetherian, the first, resp. second of these chains must be stationary. This yields interesting corollaries.

6.4.1 THEOREM. *Let φ be an endomorphism of M.*

(1) *M is artinian* $\Rightarrow \exists n_0 \in \mathbb{N}\ \forall n \geq n_0\ [M = \mathrm{Im}(\varphi^n) + \mathrm{Ker}(\varphi^n)]$.

(2) *M is artinian $\wedge$ φ is a monomorphism $\Rightarrow \varphi$ is an automorphism.*

(3) *M is noetherian* $\Rightarrow \exists n_0 \in \mathbb{N}\ \forall n \geq n_0\ [0 = \mathrm{Im}(\varphi^n) \cap \mathrm{Ker}(\varphi^n)]$.

(4) *M is noetherian $\wedge$ φ is an epimorphism $\Rightarrow \varphi$ is an automorphism.*

Proof. (1) By the preceding remark there is an $n_0 \in \mathbb{N}$ with $\mathrm{Im}(\varphi^{n_0}) = \mathrm{Im}(\varphi^n)$ for $n \geq n_0$. For $n \geq n_0$ it then follows that $\mathrm{Im}(\varphi^n) = \mathrm{Im}(\varphi^{2n})$. Let $x \in M$, then $\varphi^n(x) \in \mathrm{Im}(\varphi^n) = \mathrm{Im}(\varphi^{2n}) \Rightarrow$ there exists $y \in M$ with $\varphi^n(x) = \varphi^{2n}(y) \Rightarrow \varphi^n(x - \varphi^n(y)) = 0 \Rightarrow k := x - \varphi^n(y) \in \mathrm{Ker}(\varphi^n) \Rightarrow x = \varphi^n(y) + k \in \mathrm{Im}(\varphi^n) + \mathrm{Ker}(\varphi^n)$, which was to be shown.

(2) If φ is a monomorphism then obviously so also is φ^n for every $n \in \mathbb{N}$, i.e. $\mathrm{Ker}(\varphi^n) = 0$. Then it follows from (1) that $M = \mathrm{Im}(\varphi^{n_0})$, thus also $M = \mathrm{Im}(\varphi)$, for $\mathrm{Im}(\varphi^{n_0}) \hookrightarrow \mathrm{Im}(\varphi)$. Consequently φ is an epimorphism, thus an automorphism.

(3) Here there is an $n_0 \in \mathbb{N}$ with $\mathrm{Ker}(\varphi^{n_0}) = \mathrm{Ker}(\varphi^n)$ for $n \geq n_0$. For $n \geq n_0$ it then follows that $\mathrm{Ker}(\varphi^n) = \mathrm{Ker}(\varphi^{2n})$. Let $x \in \mathrm{Im}(\varphi^n) \cap \mathrm{Ker}(\varphi^n)$, then there is a $y \in M$ with $x = \varphi^n(y)$, and we have

$$0 = \varphi^n(x) = \varphi^{2n}(y).$$

Consequently we have $y \in \mathrm{Ker}(\varphi^{2n}) = \mathrm{Ker}(\varphi^n)$, from which we have $x = \varphi^n(y) = 0$, thus we obtain $0 = \mathrm{Im}(\varphi^n) \cap \mathrm{Ker}(\varphi^n)$.

(4) If φ is an epimorphism then so also is φ^n for every $n \in \mathbb{N}$, i.e. $\mathrm{Im}(\varphi^n) = M$. From (3) it then follows that $0 = \mathrm{Ker}(\varphi^{n_0})$, thus since $\mathrm{Ker}(\varphi) \hookrightarrow \mathrm{Ker}(\varphi^{n_0})$ we have also $\mathrm{Ker}(\varphi) = 0$. Consequently φ is a monomorphism, thus also an automorphism. □

6.4.2 COROLLARY. *Let M be a module of finite length and let φ be an endomorphism of M. Then we have*

(5) $$\exists n_0 \in \mathbb{N}\ \forall n \geq n_0\ [M = \mathrm{Im}(\varphi^n) \oplus \mathrm{Ker}(\varphi^n)].$$

(6) *φ is an automorphism $\Leftrightarrow \varphi$ is an epimorphism $\Leftrightarrow \varphi$ is a monomorphism.*

Proof. (5) For n_0 we now take the maximum of the numbers n_0 in (1) and (3). (6) follows from (2) and (4). □

By means of 6.4.2 well known properties of finite-dimensional vector spaces are generalized.

6.5 A CHARACTERIZATION OF NOETHERIAN RINGS

We give here a characterization of noetherian rings which is of fundamental significance for a comprehensive theory of modules over noetherian rings. The proof is based essentially on Baer's Criterion.

6.5.1 THEOREM. *The following conditions are equivalent for a ring R:*

(1) *R_R is noetherian.*
(2) *Every direct sum of injective right R-modules is injective.*
(3) *Every countable direct sum of injective hulls of simple right R-modules is injective.*

Proof. "(1)$\Rightarrow$(2)": Let $Q := \bigoplus_{i \in I} Q_i$ be an internal or external direct sum of injective right R-modules Q_i. By Baer's Criterion 5.7.1 it suffices for the proof of injectivity to show that for every right ideal $U \hookrightarrow R_R$ and every homomorphism $\rho : U \to Q$ there exists a homomorphism $\tau : R \to Q$ with $\rho = \tau\iota$, where $\iota : U \to R$ is the inclusion mapping. Since R_R is noetherian, U is finitely generated:

$$U = \sum_{i=1}^{n} u_i R.$$

The images $\rho(u_i)$, $i = 1, \ldots, n$, of the u_i under ρ have components different from zero for only finitely many of the Q_i, say for the Q_i with $i \in I_0$, where I_0 is a finite subset of I.

Let

$$\iota_0 : \bigoplus_{i \in I_0} Q_i \to \bigoplus_{i \in I} Q_i$$

be the inclusion mapping and let ρ_0 be the homomorphism induced by the restriction of the domain of ρ to $\bigoplus_{i \in I_0} Q_i$. Then we have $\rho = \iota_0 \rho_0$.

Since I_0 is finite, $\bigoplus_{i \in I_0} Q_i$ is injective and there exists a homomorphism τ_0 so that the following diagram is commutative:

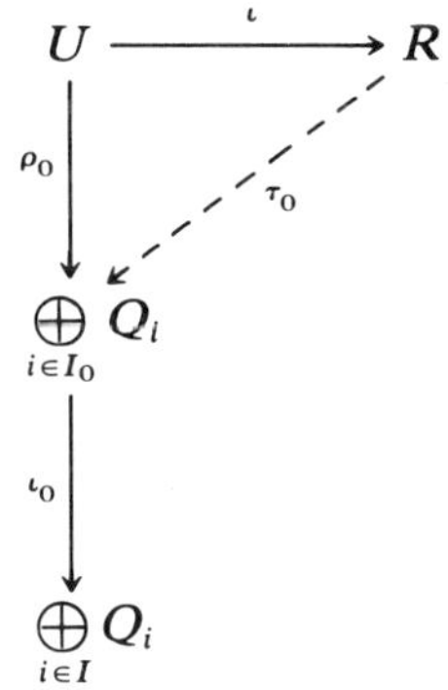

Consequently we have $\rho = \iota_0 \rho_0 = \iota_0 \tau_0 \iota = \tau \iota$ if we put $\tau := \iota_0 \tau_0$.

"(2)$\Rightarrow$(3)": (3) is a special case of (2).

"(3)$\Rightarrow$(1)": The proof is obtained indirectly.

Let R_R be non-noetherian, then there is a properly ascending chain of right ideals of R:

$$A := A_1 \subsetneqq A_2 \subsetneqq A_3 \subsetneqq \dots$$

Then

$$A := \bigcup_{i=1}^{\infty} A_i$$

is also a right ideal of R and to every $a \in A$ there is an $n_a \in \mathbb{N}$ so that $a \in A_i$ for all $i \geqslant n_a$. For every $i = 1, 2, 3, \dots$ let $c_i \in A$, $c_i \notin A_i$. In the cyclic module $(c_i R + A_i)/A_i$ by 2.3.12 there exists a maximal submodule N_i/A_i; then

$$E_i := ((c_i R + A_i)/A_i)/(N_i/A_i)$$

is a simple right R-module. Let ν_i: $(c_i R + A_i)/A_i \to E_i$ denote the natural epimorphism. Let $I(E_i)$ be the injective hull of E_i with $E_i \hookrightarrow I(E_i)$ and let $\iota_i : E_i \to I(E_i)$ be the inclusion mapping. Then there exists a commutative diagram

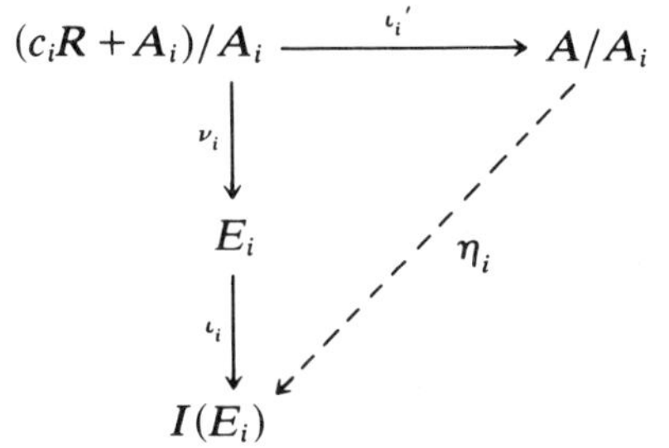

where ι_i' is the corresponding inclusion mapping and we have $\eta_i(\bar{c}_i)=\iota_i\nu_i(\bar{c}_i)\neq 0$ for $i=1,2,3,\ldots$.

We now define

$$\alpha: A\ni a\mapsto \sum_{i=1}^{n_a}\eta_i(a+A_i)\in\bigoplus_{i=1}^{\infty}I(E_i),$$

in which $\eta_i(a+A_i)$ is thus the ith component of $\alpha(a)$. Since $a\in A_i$ for $i\geqslant n_a$, $\alpha(a)$ lies in fact in the direct sum. (If we consider $\bigoplus I(E_i)$ as an external direct sum then we put $\alpha(a)=(\eta_i(a+A_i))$. Since by assumption $\bigoplus_{i=1}^{\infty}I(E_i)$ is injective, there is a β so that the diagram

$$\begin{array}{ccc} A & \xrightarrow{\iota} & R \\ {\scriptstyle\alpha}\downarrow & \swarrow{\scriptstyle\beta} & \\ \bigoplus_{i=1}^{\infty} I(E_i) & & \end{array}$$

is commutative. Let b_i be the ith component of $\beta(1)$ in $\bigoplus I(E_i)$, then there is an $n\in\mathbb{N}$ with $b_i=0$ for $i\geqslant n$. Since $\alpha(a)=\beta(a)=\beta(1)a$, $a\in A$ it follows that $\eta_i(a+A_i)=b_ia$, thus $\eta_i(a+A_i)=0$ for $i\geqslant n$ and all $a\in A$. But $\eta_n(c_n+A_n)\neq 0$ by the definition of η_i, contradiction! Hence 6.5.1 is completely proved. □

Remark. If we are only interested in 6.5.1 (1)⇔(2), then the proof can be simplified. We need (3)⇒(1) for a later theorem. The simplification in the proof of (2)⇒(1) as opposed to that of (3)⇒(1) will be indicated briefly. The proof now follows directly by first starting from an arbitrary chain of right ideals

$$A_1\hookrightarrow A_2\hookrightarrow A_3\hookrightarrow\ldots$$

Again let

$$A:=\bigcup_{i=1}^{\infty}A_i.$$

Now let η_i be the inclusion mappings

$$\eta_i: A/A_i\ni a+A_i\mapsto a+A_i\in I(A/A_i)$$

and let

$$\alpha: A\to\bigoplus_{i=1}^{\infty}I(A/A_i)$$

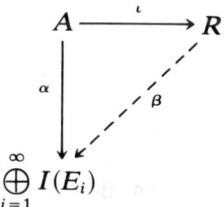

be defined by

$$\alpha(a) := \sum_{i=1}^{n_a} (a + A_i), \qquad a \in A.$$

Then it follows that $\eta_i = 0$ for $i \geqslant n$ and consequently $A = A_i$ for $i \geqslant n$.

If R is an arbitrary ring and $\eta_i : M_i \to I(M_i)$, $i = 1, \ldots, n$, are finitely many injective hulls of R-modules, then

$$\bigoplus_{i=1}^{n} \eta_i : \bigoplus_{i=1}^{n} M_i \to \bigoplus_{i=1}^{n} I(M_i)$$

is also an injective hull. If R_R is now noetherian then it follows from 6.5.1 and 5.1.7 that the corresponding result also holds for an arbitrary index set.

6.5.2. COROLLARY. *Let R_R be noetherian and let $(M_i \mid i \in I)$ be a family of right R-modules. If*

$$\eta_i : M_i \to I(M_i)$$

is any injective hull of M_i then

$$\bigoplus_{i \in I} \eta_i : \bigoplus_{i \in I} M_i \to \bigoplus_{i \in I} I(M_i)$$

is an injective hull of $\bigoplus_{i \in I} M_i$.

6.6 DECOMPOSITION OF INJECTIVE MODULES OVER NOETHERIAN AND ARTINIAN RINGS

In order to explain the issues to follow we need some definitions.

6.6.1 *Definitions*

(a) M_R is called *directly decomposable* resp. *directly indecomposable*: $\Leftrightarrow M_R = 0$ or there is a direct summand of M different from 0 and M resp. $M_R \neq 0$ and there is no direct summand of M different from 0 and M. (See 2.4.3.)

(b) Let $U \subsetneq M_R$. M is called *irreducible* (meet-irreducible) over $U :\Leftrightarrow$ for arbitrary submodules $A, B \hookrightarrow M$ with $U \subsetneq A$, $U \subsetneq B$ we have $U \neq A \cap B$.

(c) M is called *irreducible* (meet-irreducible) $:\Leftrightarrow M$ is irreducible over 0.

One of the fundamental questions of the theory of modules concerns the decomposition of a module into a direct sum of submodules. The utmost possible such decomposition is then obviously achieved if all submodules of the decomposition are themselves indecomposable. In this connection there arise three questions:

(1) Under what assumptions does a module admit a decomposition into a direct sum of directly indecomposable submodules?
(2) Is such a decomposition (if it exists) uniquely determined?
(3) What properties do directly indecomposable modules have?

Questions (1) and (3) are answered here for injective modules over noetherian and artinian rings. An answer to question (2) is given in the next chapter by the Krull–Remak–Schmidt Theorem.

We begin by investigating directly indecomposable, injective modules for which, first of all, the ring R is arbitrary.

6.6.2 THEOREM. *Let Q_R be injective, $Q_R \neq 0$. Then the following conditions are equivalent*:
(1) *Q is directly indecomposable.*
(2) *Q is the injective hull of every submodule $\neq 0$.*
(3) *Every submodule $\neq 0$ of Q is irreducible.*
(4) *Q is the injective hull of an irreducible submodule.*

Proof. "(1)⇒(2)": Let $U \hookrightarrow Q$, $U \neq 0$ and let $I(U) \hookrightarrow Q$ be the injective hull of U. Since $U \neq 0$ we also have $I(U) \neq 0$. Since $I(U)$ as an injective module is a direct summand of Q it follows that $I(U) = Q$.

"(2)⇒(3)": Let $M \hookrightarrow Q$ and let $A, B \hookrightarrow M$, $A \neq 0$, $B \neq 0$. Since Q is an injective hull of A, A is large in Q and it follows that $A \cap B \neq 0$.

"(3)⇒(4)": As an irreducible submodule we may take Q itself.

"(4)⇒(1)": Let Q be an injective hull of the irreducible submodule $M \neq 0$ of Q. Suppose $Q = A \oplus B$, $A \neq 0$, $B \neq 0$. Since M is large in Q it follows that $M \cap A \neq 0$, $M \cap B \neq 0$. Since M is irreducible, it follows that $(M \cap A) \cap (M \cap B) \neq 0$ in contradiction to $A \cap B = 0$. Thus Q is directly indecomposable. □

6.6.3 COROLLARIES
(a) *The injective hull of a simple R-module is directly indecomposable.*
(b) *A directly indecomposable, injective module Q contains at most one simple submodule.*
(c) *If R_R is artinian then every directly indecomposable, injective module $Q_R \neq 0$ is the injective hull of a simple R-module.*

Proof. (a) Every simple module is irreducible.

(b) Let E, E_1 be simple submodules of Q. From $E \overset{*}{\hookrightarrow} Q$ it follows that $E \cap E_1 \neq 0$, thus $E = E \cap E_1 = E_1$.

(c) Let $0 \neq q \in Q$, then by 6.1.3 qR is artinian. Thus a simple submodule E exists in $qR \hookrightarrow Q$. By the theorem Q is an injective hull of E. □

We come now to the following interesting theorem which yields a new characterization of noetherian resp. artinian rings.

6.6.4 THEOREM

(a) *The following conditions are equivalent*:

(1) R_R *is noetherian.*

(2) *Every injective module* Q_R *is a direct sum of directly indecomposable submodules.*

(b) *The following conditions are equivalent*:

(1) R_R *is artinian.*

(2) *Every injective module* Q_R *is a direct sum of injective hulls of simple* R*-modules.*

By 6.6.3(a) the injective hulls of simple R-modules appearing in the characterization of artinian rings are likewise directly indecomposable. From the theorem we have in particular: If R_R is noetherian but not artinian then there is a directly indecomposable injective R-module which contains no simple submodule.

The proof of the theorem is now only indicated for noetherian rings in the direction (1)⇒(2). In order to obtain (1)⇒(2) for artinian rings, we need the fact that every right artinian ring is also right noetherian, which will be proved in Chapter 9. For the proof of (2)⇒(1), further lemmas are required, and in particular the fact of the uniqueness (up to isomorphism) of the decomposition of a semisimple module into a direct sum of simple modules. As soon as the necessary lemmas become available we shall obtain the complete proof (in 9.5). Thus now we prove only

6.6.5 PROPOSITION. *If* R_R *is noetherian then every injective module* Q_R *is a direct sum of directly indecomposable submodules. If, moreover,* R_R *is artinian (it is shown later: artinian* $R_R \Rightarrow$ *noetherian* R_R*) then every one of the directly indecomposable summands is an injective hull of a simple* R*-module.*

For the proof of 6.6.5 we need two lemmas which are also of interest.

6.6.6 LEMMA. *Let* Γ *be a set of submodules of a module* M_R. *Then among all subsets* Λ *of* Γ *with*

$$(*) \qquad \sum_{U \in \Lambda} U = \bigoplus_{U \in \Lambda} U$$

there is a maximal set Λ_0.

Proof. By the help of Zorn's Lemma. Let

$$G := \{\Lambda \mid \Lambda \subset \Gamma \wedge (*) \text{ is satisfied}\},$$

then G is ordered by inclusion and $G \neq \varnothing$ for $\varnothing \in G$ $\left(\text{since } 0 = \sum_{U \in \varnothing} U = \bigoplus_{U \in \varnothing} U\right)$. Let H be a totally ordered subset from G and let

$$\Omega := \bigcup_{\Lambda \in H} \Lambda,$$

then $\Omega \subset \Gamma$. Assertion: $\Omega \in G$, i.e., $(*)$ is satisfied for Ω. Suppose that were not the case, then the sum of the submodules from Ω would thus not be direct. Consequently there must be in fact a finite subsum of the sum which is not direct. But finitely many submodules from Ω lie already in a $\Lambda \in H$ (since H is a totally ordered subset) so that their sum is direct. Consequently we have in fact $\Omega \in G$ and so Ω is an upper bound of H in G. Consequently by Zorn there exists a maximal element Λ_0 in G. □

6.6.7 COROLLARY

(a) *For every module M_R there is a maximal set of directly indecomposable, injective submodules whose sum is direct.*

(b) *For every module M_R there is a maximal set of simple submodules whose sum is direct.*

Proof. This follows from 6.6.6 if $\Gamma =$ set of directly indecomposable, injective submodules in case (a) and if $\Gamma =$ set of simple submodules in case (b). □

6.6.8 LEMMA. *If R_R is noetherian then every module $M_R \neq 0$ contains an irreducible submodule $\neq 0$.*

Proof. We show that every finitely generated submodule $B \hookrightarrow M$, $B \neq 0$, which is noetherian by 6.1.3, contains an irreducible submodule $\neq 0$. Let $\{X \mid X \underset{\neq}{\hookrightarrow} B \wedge X \text{ is inco in } B\}$ be the set of proper submodules of B which are intersection complements of a submodule of B in B. This set is not empty since 0 is an inco of B. Since B is noetherian there is a maximal element X_0 in this set. Let X_0 be an inco of $U_0 \hookrightarrow B$. Clearly then $U_0 \neq 0$.

We claim that every submodule $0 \neq C \hookrightarrow U_0$ is large in U_0 and consequently U_0 is irreducible. Suppose, for $L \hookrightarrow U_0$ we have $C \cap L = 0$, then it follows that $C \cap (X_0 + L) = 0$. From the maximality of X_0 and as $C \neq 0$ (thus $C' \neq B$) it follows that $X_0 + L = X_0$, thus $L \hookrightarrow X_0$ and consequently $L \hookrightarrow U_0 \cap X_0 = 0$. From $C \cap L = 0$ it follows therefore that $L = 0$, i.e. $C \overset{*}{\hookrightarrow} U_0$. □

Proof of 6.6.5. Consider a maximal set of directly indecomposable, injective submodules of Q, whose sum is direct (6.6.7). Let this direct sum be $Q_0 := \bigoplus_{i \in I} Q_i$. Since all the Q_i are injective by 6.5.1 Q_0 is injective. Consequently Q_0 is a direct summand of Q:

$$Q = Q_0 \oplus Q_1.$$

Suppose $Q_1 \neq 0$, then Q_1 contains an irreducible submodule $M \neq 0$ (6.6.8). Let $I(M)$ be an injective hull of M in Q_1, then $I(M)$ is a direct summand in Q_1, $Q_1 = I(M) \oplus Q_2$, and by 6.6.2 $I(M)$ is directly indecomposable. But then $Q_0 = \bigoplus_{i \in I} Q_i$ would not have been maximal, since $Q_0 \oplus I(M)$ is also a direct sum of directly indecomposable, injective submodules of Q. This contradiction means that already $Q = Q_0 = \bigoplus_{i \in I} Q_i$ holds.

If R_R is not only noetherian but also artinian then by 6.6.3 all $Q_i \neq 0$ are injective hulls of simple submodules. □

EXERCISES

(1)

Let R_n be the ring of all $n \times n$ square matrices with coefficients from R.

Show: R_n is right artinian resp. noetherian $\Leftrightarrow R$ is right artinian resp. noetherian.

(2)

Show: Every right artinian ring without zero divisors is a skew field.

(3)

Let $L := k(t_1, t_2, t_3, \ldots)$ be the field of rational functions in the indeterminates $t_1, t_2, t_3, \ldots$ with coefficients in the field k. The elements of L are then quotients of polynomials $\dfrac{P_1(t_i)}{P_2(t_i)}$ (with $P_2(t_i) \neq 0$).

Let $K := k(t_1^2, t_2^2, t_3^2, \ldots)$, then K is a subfield of L.

(a) Show
$$\tau : L \ni \frac{P_1(t_i)}{P_2(t_i)} \mapsto \frac{P_1(t_i^2)}{P_2(t_i^2)} \in K$$
is a ring isomorphism.

(b) Show: The product set $R = L \times L$ becomes by the definitions

$$(l_1, l_2) + (m_1, m_2) := (l_1 + m_1, l_2 + m_2),$$

$$(l_1, l_2)(m_1, m_2) := (l_1 m_1, l_1 m_2 + l_2 \tau(m_1)),$$

a ring with a unit element.

(c) Show: ${}_R R$ has length 2 (i.e. it has a composition series of the form $0 \hookrightarrow A \hookrightarrow R$).

(d) Show: R_R is neither artinian nor noetherian.

(4)

A ring is called a principal right ideal ring :$\Leftrightarrow$ every right ideal is principal (= cyclic). Let R be a principal right and left ideal ring without zero divisors and let $A \hookrightarrow R_R$, $A \neq 0$. Show: $(R/A)_R$ is artinian.

(5)

(a) If a module M_R satisfies the maximal conditions for finitely generated submodules then it is already noetherian.

(b) Give an example of a module M_R which satisfies the maximal condition for cyclic submodules but which is not noetherian.

(c) Show that for an abelian group $M = M_{\mathbb{Z}}$ the following are equivalent:

(1) M satisfies the minimal condition for cyclic subgroups.

(2) $T(M) = M$, i.e. $\forall m \in M\ \exists z \in \mathbb{Z}, z \neq 0\ [mz = 0]$.

(3) M satisfies the minimal condition for finitely generated subgroups.

(6)

Let A, B be rings and ${}_A M_B$ an A-B-bimodule. Then define

$$R := \left\{ \begin{pmatrix} a & m \\ 0 & b \end{pmatrix} \middle| \ a \in A, m \in M, b \in B \right\} \quad \text{with componentwise addition}$$

and

$$\begin{pmatrix} a_1 & m_1 \\ 0 & b_1 \end{pmatrix} \begin{pmatrix} a_2 & m_2 \\ 0 & b_2 \end{pmatrix} := \begin{pmatrix} a_1 a_2 & a_1 m_2 + m_1 b_2 \\ 0 & b_1 b_2 \end{pmatrix}.$$

The unit element of this ring is then $\begin{pmatrix} 1 & 0 \\ 0 & 1 \end{pmatrix}$.

Show:

(a) R_R is noetherian (resp. artinian) $R_R \Leftrightarrow A_A$, B_B, M_B are noetherian (resp. artinian).

(b) ${}_R R$ is noetherian (resp. artinian)$\Leftrightarrow {}_A A$, ${}_B B$, ${}_A M$ are noetherian (resp. artinian). (Hint: Consider the ring homomorphism $\rho : R \to A \times B$ with

$\rho\begin{pmatrix} a & m \\ 0 & b \end{pmatrix} = (a, b)$ and show for the kernel $K := \text{Ker}(\rho)$ that K_R and M_B (resp. ${}_RK$ and ${}_AM$) have isomorphic submodule lattices).

(7)

Show:

(a) Let $M = U \oplus U_1 = V \oplus V_1$ with $U \hookrightarrow V$. Then U has a direct complement in M, which contains V_1 (i.e. $M = U \oplus W$ with $V_1 \hookrightarrow W$), and V has a direct complement in M, which is contained in U_1.

(b) M_R satisfies the maximal condition for direct summands if and only if it satisfies the minimal condition for direct summands.

(c) Let M_R satisfy the maximal condition for direct summands. Show that for $\varphi \in \text{End}_R(M)$ the following are equivalent:

(1) φ is left invertible (i.e. split monomorphism).
(2) φ is right invertible (i.e. split epimorphism).
(3) φ is invertible (i.e. isomorphism).

(8)

Give an example of a ring R and a module M_R which does not have finite length and with the property that for every $\varphi \in \text{End}(M_R)$ there holds:

(a) $\exists n_0 \in \mathbb{N}\, \forall n \geqslant n_0\, [M = \text{Im}(\varphi^n) \oplus \text{Ker}(\varphi^n)]$; and

(b) φ is an automorphism $\Leftrightarrow \varphi$ is an epimorphism $\Leftrightarrow \varphi$ is a monomorphism.

(Hint: For M_R use a direct sum of infinitely many non-isomorphic simple R-modules).

(9)

Show: If B_R is artinian and $B_R \neq 0$ then there is an indecomposable factor module $\neq 0$ of B.

(M_R is called indecomposable if $M_R \neq 0$ and the sum of any two proper submodules is again a proper submodule of M_R.)

(10)

Show that for a commutative ring R the following statements are equivalent:

(1) For every $x \in R$ the series $xR \hookleftarrow x^2R \hookleftarrow x^3R \hookleftarrow \ldots$ is stationary.

(2) For every cyclic module M_R the injective endomorphisms are already automorphisms.

(3) Every prime ideal in R is already a maximal ideal.

(Hint: For (3)$\Rightarrow$(1) consider the multiplicative subset

$$S_x := \{x^n(1 - xr) \mid n \geqslant 0, r \in R\}.)$$

(11)

For a module M_R show the following are equivalent:

(1) Every set of submodules, whose sum is direct, is finite.

(2) Every submodule satisfies the maximal condition for direct summands.

(3) Every sequence $U_1 \hookrightarrow U_2 \hookrightarrow U_3 \hookrightarrow \ldots$ with $U_i \hookrightarrow M$ and U_i a direct summand in U_{i+1} is stationary.

(4) Every sequence $M \hookleftarrow U_1 \hookleftarrow U_2 \hookleftarrow U_3 \hookleftarrow \ldots$ with U_{i+1} a direct summand in U_i is stationary.

(5) Every submodule has a finitely generated large submodule.

(6) M satisfies the maximal condition for incos (= intersection complements).

(7) M satisfies the minimal condition for incos.

(8) The injective hull of M satisfies the maximal condition for direct summands.

(12)

As in Chapter 5, Exercise 4 let the *singular submodule* of a module M_R be defined by

$$\mathrm{Si}(M) := \{m \in M \mid \boldsymbol{r}_R(m) \overset{*}{\hookrightarrow} R_R\}.$$

Show that for a ring R with $\mathrm{Si}(R_R) = 0$ the following are equivalent:

(1) $I(R_R)$ satisfies the maximal condition for direct summands.

(2) For every family $(Q_i \mid i \in I)$ with Q_i injective and $\mathrm{Si}(Q_i) = 0$ $\coprod_{i \in I} Q_i$ is injective.

(Hint: Use the equivalent statements in Exercise 11 and show firstly with respect to (2)$\Rightarrow$(1) that in an ascending sequence

$$A_1 \hookrightarrow A_2 \hookrightarrow \ldots \hookrightarrow R_R$$

of intersection complements from $\mathrm{Si}(R_R) = 0$ it follows that $\mathrm{Si}(R/A_i) = 0$).

(13)

(a) Show that the following are equivalent for a module M_R:

(1) $M^{(I)}$ is injective for every index set I.

(2) $M^{(\mathbb{N})}$ is injective.

(3) M is injective and R satisfies the maximal condition for right ideals which are annihilators of subsets of M.

(b) Show that the following are equivalent for a ring R:

(1) R_R noetherian.

(2) For every injective module Q_R, $Q^{(\mathbb{N})}$ is also injective.

Chapter 7

Local Rings: Krull–Remak–Schmidt Theorem

In Chapter 6 it was shown that every injective module over a noetherian ring is a direct sum of directly indecomposable submodules. The question arises as to whether and in what sense such a decomposition is uniquely determined. This question is answered by the Krull–Remak–Schmidt Theorem. The proof of the Krull–Remak–Schmidt Theorem assumes that the endomorphism rings of the direct summands are local rings. Hence we have, first of all, to introduce local rings and then to state sufficient conditions in order that the endomorphism ring of a directly indecomposable module is local.

7.1 LOCAL RINGS

An element r of a ring R is called *right* resp. *left invertible*, if there is an $r' \in R$ with $rr' = 1$ resp. $r'r = 1$, and r' is then called the *right inverse* resp. *left inverse* of R. If we have $rr' = r'r = 1$ then r is said to be *invertible* and r' is said to be the *inverse* of r. If there are a right and a left inverse of r then these are equal and consequently there is then an inverse of r (see 2.5.4). As examples show there are right resp. left invertible elements which are not invertible.

We now have to consider rings in which the set of all non-invertible elements have a particular structure. For convenience we assume always that $R \neq 0$.

7.1.1 THEOREM. *Let A be the set of all non-invertible elements of R, then the following properties are equivalent*:

(1) *A is additively closed* ($\forall a_1, a_2 \in A[a_1 + a_2 \in A]$)

(2) *A is a two-sided ideal.*
(3r) *A is the largest proper right ideal.*
(3l) *A is the largest proper left ideal.*
(4r) *In R there exists a largest proper right ideal.*
(4l) *In R there exists a largest proper left ideal.*
(5r) *For every $r \in R$ either r or $1-r$ is right invertible.*
(5l) *For every $r \in R$ either r or $1-r$ is left invertible.*
(6) *For every $r \in R$ either r or $1-r$ is invertible.*

Proof. "(1)⇒(2)": We show first that every right resp. left invertible element is invertible. Let $bb' = 1$.

Case 1. $b'b \notin A$. Then there is $s \in R$ with $1 = sb'b$. Hence

$$b' = sb'bb' = sb'$$

and so

$$1 = b'b,$$

which was to be shown.

Case 2. $b'b \in A$. Then $1 - b'b \notin A$ must hold, since otherwise

$$1 - b'b + b'b = 1 \in A \quad ↯.$$

Let now

$$1 = s(1 - b'b).$$

Then

$$b' = s(1-b'b)b' = s(b' - b'bb') = s(b' - b') = 0$$

in contradiction to $bb' = 1$.

Since A, by assumption, is additively closed, we require only to show:

$$\forall a \in A \forall r \in R[ar \in A \wedge ra \in A].$$

Suppose $ar \notin A$, then there is $s \in R$ with $ars = 1$. By the preliminary remark (with $a = b$ and $rs = b'$) it follows that $rsa = 1$ in contradiction to $a \in A$. Analogously for ra.

"(2)⇒(3r)": Since $A \hookrightarrow {}_RR_R$ we have $A \hookrightarrow R_R$. Since $1 \notin A$, $A \neq R$. Let

$$B \underset{\neq}{\hookrightarrow} R_R \wedge b \in B.$$

Then

$$bR \hookrightarrow B \underset{\neq}{\hookrightarrow} R_R$$

so b has no right inverse. Therefore b has no inverse; hence $b \in A$ and so $b \hookrightarrow A$.

"$(3r) \Rightarrow (4r)$": Clear.

"$(4r) \Rightarrow (5r)$": Let C be a largest proper right ideal (which is then uniquely determined). Let $r \in R$; suppose r and $1-r$ are not right invertible. Then

$$rR \underset{\neq}{\hookrightarrow} R_R \wedge (1-r)R \underset{\neq}{\hookrightarrow} R_R,$$

hence

$$rR \hookrightarrow C \wedge (1-r)R \hookrightarrow C$$

and so

$$1 \in rR + (1-r)R \hookrightarrow C \Rightarrow C = R \quad ↯.$$

"$(5r) \Rightarrow (6)$": It suffices to show that every right invertible element is invertible. Let $bb' = 1$.

Case 1. $b'b$ right invertible, hence there is $s \in R$ with $1 = b'bs$ so $b = bb'bs = bs$ therefore $1 = b'b$.

Case 2. $1 - b'b$ right invertible, hence there is $s \in R$ with $1 = (1-b'b)s$ so

$$b = b(1-b'b)s = bs - bb'bs = 0$$

in contradiction to $bb' = 1$.

"$(6) \Rightarrow (1)$": Suppose, for $a_1, a_2 \in A$ that $a_1 + a_2$ is invertible, then there is $s \in R$ with $(a_1 + a_2)s = 1$; hence $a_1 s = 1 - a_2 s$. Since $(6) \Rightarrow (5r)$ holds we can (as shown in the proof $(5r) \Rightarrow (6)$) use the fact that every right invertible element is invertible. Hence it follows from $a \in A \wedge r \in R$ that $ar \in A$ (for if $ar \notin A$ then ar right invertible and so a right invertible, i.e. $a \notin A$ ↯). Then it follows that $a_1 s \in A \wedge a_2 s \in A$; in contradiction to which we obtain from $a_2 s \in A$ by (6)

$$a_1 s = 1 - a_2 s \notin A \quad ↯.$$

Analogously we obtain the left-sided assertions. □

7.1.2 *Definition.* A ring, which satisfies the equivalent properties of 7.1.1, is called a *local ring.*

7.1.3 COROLLARY. *Let R be a local ring and A the ideal of the non-invertible elements of R. Then we have*

(1) *R/A is a skew field.*

(2) *Every left resp. right invertible element is invertible.*

(3) *Every non-zero ring, which is the image of a local ring under a surjective ring homomorphism, is itself local.*

In particular: every isomorphic image of a local ring is local.

Proof. (1) Every element not contained in A has an inverse.

(2) This is contained in the proof of 7.1.1.

(3) Let $\sigma: R \to S$ be a surjective ring homomorphism. We show that 7.1.1 (6) is satisfied for S. Let $s \in S$, then there is $r \in R$ with $\sigma(r) = s$ and consequently $\sigma(1-r) = \sigma(1) - \sigma(r) = 1 - s$. By assumption either r or $1-r$ is invertible. Let r be invertible, then $\sigma(r^{-1})$ is an inverse element of s, for from $rr^{-1} = r^{-1}r = 1$ it follows that $\sigma(r)\sigma(r^{-1}) = s\sigma(r^{-1}) = \sigma(r^{-1}) \cdot s = \sigma(1) = 1 \in S$. If $1-r$ is invertible then $\sigma((1-r)^{-1})$ is an inverse element of $1-s$. □

7.1.4 *Examples of local rings*

(1) The power series ring $K[[x]]$ over a field K is local, for the non-invertible elements are precisely those with constant term $= 0$ and the set of these elements is additively closed.

(2) Localizations of commutative rings at prime ideals are local. We give briefly the definition of *localization*: Let R be a commutative ring and let $P \neq R$ be a prime ideal in R, where P is thus defined by the property

$$\forall a, b \in R[ab \in P \Rightarrow (a \in P \vee b \in P)]$$

which is equivalent to

$$\forall a, b \in R[(a \notin P \wedge b \notin P) \Rightarrow ab \notin P].$$

Let now

$$\Gamma = \{(r, a) | r \in R \wedge a \in R \backslash P\}.$$

In Γ an equivalence relation $\sim$ is introduced

$$(r_1, a_1) \sim (r_2, a_2) :\Leftrightarrow \exists a \in R \backslash P[r_1 a_2 a = r_2 a_1 a].$$

The equivalence class with the representative (r, a) is denoted by r/a. Let $R_{(P)}$ be the set of the equivalence classes, i.e.

$$R_{(P)} = \left\{\frac{r}{a} \middle| r \in R \wedge a \in R \backslash P\right\}.$$

Then by the definitions

$$\frac{r_1}{a_1} + \frac{r_2}{a_2} := \frac{r_1 a_2 + r_2 a_1}{a_1 a_2}, \qquad \frac{r_1}{a_1}\frac{r_2}{a_2} := \frac{r_1 r_2}{a_1 a_2}.$$

$R_{(P)}$ becomes a ring, as is easily verified. The zero resp. unit element of $R_{(P)}$ is the element $(0/1)$ resp. $(1/1)$ with $0 =$ zero element and $1 =$ unit element of R. The mapping

$$\varphi: R \ni r \mapsto \frac{r}{1} \in R_{(P)}$$

is a ring homomorphism and $\text{Im}(\varphi)$ is often identified with R (e.g. $\mathbb{Z}$ is considered as a subring of $\mathbb{Q}$). In $R_{(P)}$ precisely the elements of the form r/a with $r \in P$ are non-invertible, as is immediately verifiable. The set of these elements is however additively closed and consequently $R_{(P)}$ is local. As an exercise the reader may carry through the proofs in detail, in particular the demonstration of the independence of the definition of representatives.

In an integral domain R, 0 is a prime ideal and $R_{(0)}$ is the *quotient field* of R. $\mathbb{Z}$ constitutes an example of this with $\mathbb{Z}_{(0)} = \mathbb{Q}$.

If R is a principal ideal ring and $P = (p)$ then $R_{(p)}$ is written instead of $R_{(P)}$. Note $\mathbb{Q}_p \neq \mathbb{Z}_{(p)}$!

7.2 LOCAL ENDOMORPHISM RINGS

Conditions are now to be given so that the endomorphism ring of a module is local. A necessary condition for this is that the module is directly indecomposable. This condition is however not sufficient in general, as the example $\mathbb{Z}_{\mathbb{Z}}$ shows. Hence we have to set down additional properties which ensure that the endomorphism ring is local.

We begin therefore by considering ring-theoretic properties which are of interest in this connection.

7.2.1 *Definition.* Let R be a ring and let $r \in R$.

(1) r is called *nilpotent* $:\Leftrightarrow \exists n \in \mathbb{N}[r^n = 0]$.

(2) r is called *idempotent* $:\Leftrightarrow r^2 = r$.

7.2.2 COROLLARY

(1) *If r is nilpotent, then r is not invertible and $1 - r$ is invertible.*

(2) *If r is idempotent, then $1 - r$ is also idempotent.*

(3) *If r is idempotent and invertible then $r = 1$.*

Proof. (1) Suppose $rs = 1$. Let n_0 be the smallest $n \in \mathbb{N}$ with $r^n = 0$. Then $r^{n_0-1} \neq 0$ and so $0 = r^{n_0}s = r^{n_0-1}rs = r^{n_0-1} \cdot 1 = r^{n_0-1} \neq 0$ ↯. Further we have

$$(1-r)(1+r+\ldots+r^{n_0-1}) = (1+r+\ldots+r^{n_0-1})(1-r) = 1.$$

(2) $(1-r)(1-r) = 1-r-r+r^2 = 1-r-r+r = 1-r$.

(3) $r^2 = r \wedge rr' = 1 \Rightarrow r = r \cdot rr' = r^2r' = rr' = 1$. □

Examples

(1) Let R be the ring of all $n \times n$ matrices with coefficients in a field (or ring). Let d_{ij} be the matrix which in its ith row and jth column has the

entry 1 and whose other entries are 0. Then we have

$$d_{ij}d_{kl} = \delta_{jk}d_{il} = \begin{cases} 0 & \text{for } j \neq k, \\ d_{il} & \text{for } j = k, \end{cases}$$

in particular:

$$d_{ij}^2 = 0 \quad \text{for } i \neq j \quad \text{i.e. } d_{ij} \text{ is nilpotent}$$

$$d_{ii}^2 = d_{ii} \qquad \text{i.e. } d_{ii} \text{ is idempotent.}$$

(2) Let G be a finite group of order n, let K be a field and let GK be the group ring. Let

$$\gamma := \sum_{g \in G} g,$$

then we have $\gamma g = \gamma$ for every $g \in G$ and consequently $\gamma^2 = \gamma n$.

If the characteristic $\chi(K)$ of K is a divisor of n then it follows that $\gamma^2 = \gamma n = 0$, i.e. γ is nilpotent. If $\chi(K)$ is not a divisor of n then we have

$$\left(\gamma\frac{1}{n}\right)^2 = \gamma^2\frac{1}{n^2} = \gamma\frac{n}{n^2} = \gamma\frac{1}{n},$$

and consequently $\gamma\frac{1}{n}$ is idempotent.

In the following lemma some decomposition properties of rings are listed, these are also needed later on other occasions.

7.2.3 LEMMA. *Let R be a ring and let*

$$R_R = \bigoplus_{i \in I} A_i$$

be a direct decomposition of R into right ideals A_i, $i \in I$. Then we have:

(a) *The subset*

$$I_0 = \{i \mid i \in I \wedge A_i \neq 0\}$$

is finite; consequently

$$R = \bigoplus_{i \in I_0} A_i.$$

(b) *There exist elements $e_i \in A_i$ for $i \in I_0$ so that for $i, j \in I_0$ we have:*

(1) $\quad A_i = e_i R, \quad i \in I_0,$

(2) $\quad 1 = \sum_{i \in I_0} e_i,$

$$(3) \qquad e_ie_j = \begin{cases} e_i & \text{for } i = j \\ 0 & \text{for } i \neq j \end{cases}, \quad (i, j \in I_0),$$

i.e., $\{e_i | i \in I_0\}$ is a set of orthogonal idempotents.

(c) *If the A_i, $i \in I_0$ are two-sided ideals, then the elements e_i, $i \in I_0$ in (b) are from the centre of R (i.e. $e_ir = re_i$ for all $r \in R$).*

(d) *Conversely if orthogonal idempotents $e_1, \ldots, e_n \in R$ with*

$$1 = \sum_{i=1}^{n} e_i$$

are given then it follows that

$$R = \bigoplus_{i=1}^{n} e_iR,$$

and the e_iR are in fact two-sided ideals, in the case that the e_i are contained in the centre of R.

Proof. Let $1 = \sum_{i \in I} e_i$, $e_i \in A_i$, and let

$$I_0 := \{i | i \in I \wedge e_i \neq 0\}.$$

Then I_0 is finite and we have

$$1 = \sum_{i \in I_0} e_i$$

and also $e_i \neq 0$ for $i \in I_0$. Since $e_i \in A_i$ it follows also that $A_i \neq 0$ for $i \in I_0$. Let now $a_j \in A_j$ for arbitrary $j \in I$, then from

$$1 = \sum_{i \in I_0} e_i$$

by multiplication by a_j on the right we obtain

$$a_j = \sum_{i \in I_0} e_ia_j.$$

As $R_R = \bigoplus_{i \in I} A_i$ and $e_ia_j \in A_i$ there follow therefore:

(1) For $j \notin I_0$: $a_j = 0 \Rightarrow A_j = 0 \Rightarrow I_0 = \{i | i \in I \wedge A_i \neq 0\} \Rightarrow R = \bigoplus_{i \in I_0} A_i$, from which (a) is proved;

(2) For $j \in I_0$: $a_j = e_ja_j \Rightarrow A_j = e_jA_j \hookrightarrow e_jR \hookrightarrow A_j \Rightarrow A_j = e_jR$, and also $0 = e_ia_j$ for $i \neq j$. If we now restrict ourselves to $i, j \in I_0$ then we deduce for $e_j = a_j$

$$e_j = e_je_j, \qquad e_ie_j = 0 \qquad \text{for } i \neq j,$$

from which (b) is entirely proved. From $r \in R$ and $1 = \sum_{i \in I_0} e_i$ it follows that

$$r = \sum_{i \in I_0} e_i r \quad \text{and} \quad r = \sum_{i \in I_0} r e_i.$$

If the A_i are two-sided ideals, then we have $re_i \in A_i$ and as

$$\sum_{i \in I_0} e_i r = \sum_{i \in I_0} r e_i$$

the assertion $e_i r = re_i$ of (c) follows. For the proof of (d) first of all we obtain

$$R = \sum_{i=1}^{n} e_i R \quad \text{from} \quad 1 = \sum_{i=1}^{n} e_i$$

on multiplying by R on the right. Let now

$$r \in e_{i_0} R \cap \sum_{\substack{i=1 \\ i \neq i_0}}^{n} e_i R,$$

then it follows that $r = e_{i_0} r$ and

$$r = \sum_{\substack{i=1 \\ i \neq i_0}}^{n} e_i r_i,$$

thus

$$r = e_{i_0} r = \sum_{\substack{i=1 \\ i \neq i_0}}^{n} e_{i_0} e_i r_i = 0.$$

Consequently we have

$$R = \bigoplus_{i=1}^{n} e_i R.$$

If the e_i lie in the centre of R, then, as $re_i R = e_i r R \hookrightarrow e_i R$, $e_i R$ is a two-sided ideal. Thus the lemma is proved. □

7.2.4 COROLLARY. *The following are equivalent for a ring R:*
(1) R_R *is directly indecomposable.*
(2) ${}_R R$ *is directly indecomposable.*
(3) 1 *and* 0 *are the only idempotents in R.*

Proof. "(1)⇒(3)": Let e be an idempotent, then e, $1-e$ are orthogonal idempotents with $1 = e + (1-e)$. Thus it follows from 7.2.3 that

$$R = eR \oplus (1-e)R.$$

As (1) holds either $eR = 0$, thus $e = 0$ or $eR = R$. In the latter case we have

$$(1-e)R = (1-e)eR = 0,$$

thus

$$(1-e)1 = 1-e = 0.$$

"(3)⇒(1)": Assume $R_R = A \oplus B$, then by 7.2.3 there is an idempotent e with $A = eR$. From (3) it follows that $e = 1$ or $e = 0$, thus $A = R$ or $A = 0$, i.e. R_R is directly indecomposable.

Analogously we show (2)⇔(3). □

7.2.5 THEOREM. Let $S := \mathrm{End}(M_R)$, *then the following are equivalent*:

(1) M_R *is directly indecomposable.*
(2) S_S *is directly indecomposable.*
(3) ${}_SS$ *is directly indecomposable.*
(4) 0 *and* 1 *are the only idempotents in* S.

Proof. By 7.2.4 (2), (3) and (4) are equivalent.

"(1)⇒(4)": Let $e \in S$ be an idempotent, then we have

$$M = e(M) \oplus (1-e)(M),$$

since for $m \in M$ it follows that $m = e(m) + (1-e)(m)$ and if we suppose $e(m_1) = (1-e)(m_2)$ then applying e to this equation yields

$$e^2(m_1) = e(m_1) = e(1-e)(m_2) = 0.$$

From (1) it must be that $e(M) = 0$, thus $e = 0$ or $(1-e)(M) = 0$, thus $1 = e$.

"(4)⇒(1)": Assume $M_R = A \oplus B$, then

$$\eta : M \ni a + b \mapsto a \in M$$

is an endomorphism with $\eta^2 = \eta$, thus is an idempotent in S. By assumption it follows that $\eta = 0$ or $\eta = 1$. If $\eta = 0$, then it follows that $A = 0$; if $\eta = 1$, then it follows that $A = M$, i.e., M is directly indecomposable. □

7.2.6 COROLLARY. *Let* $S := \mathrm{End}(M_R)$ *be local, then* M_R *is directly indecomposable.*

Proof. By 7.2.5 it is sufficient to establish that 0 and 1 are the only idempotents in S. Let $e \in S$ be an idempotent, then $1-e$ is also an idempotent. Suppose $e \neq 0$, $e \neq 1$ then we also have $1-e \neq 0$, $1-e \neq 1$. Since e and $1-e$ are both not invertible, in the case of a local ring $1 = e + 1 - e$ must be also not invertible ↯. □

The converse of this statement holds under additional assumptions, as we shall show in two cases.

7.2.7 THEOREM. *Let* $M_R \neq 0$ *be a directly indecomposable module of finite length, then* $\mathrm{End}(M_R)$ *is local and the non-invertible elements from* $\mathrm{End}(M_R)$ *are precisely the nilpotent elements.*

Proof. Let $\varphi \in \mathrm{End}(M_R)$. Then by 6.4.2 we have

$$\exists n \in \mathbb{N}[M = \mathrm{Im}(\varphi^n) \oplus \mathrm{Ker}(\varphi^n)].$$

Since M is directly indecomposable it follows that either $\mathrm{Ker}(\varphi^n) = 0$ or $\mathrm{Im}(\varphi^n) = 0$.

Case 1. $\mathrm{Ker}(\varphi^n) = 0 \Rightarrow \mathrm{Ker}(\varphi) = 0 \Rightarrow \varphi$ is a monomorphism. Hence φ is an automorphism by 6.4.2, i.e. φ is invertible.

Case 2. $\mathrm{Im}(\varphi^n) = 0 \Rightarrow \varphi^n = 0 \Rightarrow 1 - \varphi$ invertible by 7.2.2 (1).

We have thus established: Either φ or $1 - \varphi$ is invertible; by 7.1.1 $\mathrm{End}(M_R)$ is then local. If φ is not invertible (Case 2) then φ is nilpotent. Conversely if φ is nilpotent, then by 7.2.2 φ is not invertible. □

As a special case we can deduce from this theorem the result, already known to us, that the endomorphism ring of a simple module is a skew field; for the only nilpotent endomorphism of a simple module is the zero mapping.

A further interesting case is given in the following theorem.

7.2.8 THEOREM. *Let* $Q_R \neq 0$ *be a directly indecomposable injective module, then* $\mathrm{End}(Q_R)$ *is local.*

Proof. Let $\varphi: Q \to Q$ be a monomorphism, then $\mathrm{Im}(\varphi)$ is injective, thus a direct summand in Q. Since Q is directly indecomposable, it follows that $\mathrm{Im}(\varphi) = Q$, i.e., φ is an automorphism and hence invertible in $\mathrm{End}(Q_R)$. Hence every non-invertible endomorphism of Q has a kernel different from zero.

Let now φ_1, φ_2 be two non-invertible endomorphisms of Q, then we thus have $\mathrm{Ker}(\varphi_1) \neq 0$, $\mathrm{Ker}(\varphi_2) \neq 0$. Since Q is irreducible by 6.6.2, it follows therefore that

$$0 \neq \mathrm{Ker}(\varphi_1) \cap \mathrm{Ker}(\varphi_2) \hookrightarrow \mathrm{Ker}(\varphi_1 + \varphi_2),$$

i.e. $\varphi_1 + \varphi_2$ is also not invertible. By 7.1.1 $\mathrm{End}(M_R)$ is then local. □

In view of the Krull–Remak–Schmidt Theorem that follows, it is of interest to ask which modules may be decomposed into a direct sum of submodules with local endomorphism rings. There is a positive answer to this question above all in the important cases which here follow:

(1) M is an injective module over a noetherian (or artinian) ring.
(2) M is a module of finite length.
(3) M is a semisimple module.
(4) M is a projective, semiperfect module.

Case 1 was already answered for us by 6.6.5 and 7.2.8. Case 2 is to be handled immediately below. We treat Case 3 resp. 4 in Chapter 8 resp. 11.

7.2.9 THEOREM. *Let* $M_R \neq 0$.

(a) *Let* M *be artinian or noetherian, then there are directly indecomposable submodules* $M_1, \ldots, M_n$ *of* M *with*

$$M = \bigoplus_{i=1}^{n} M_i.$$

(b) *Let* M *be of finite length (i.e. artinian and noetherian), then there are directly indecomposable submodules* $M_1, \ldots, M_n$ *of* M *with*

$$M = \bigoplus_{i=1}^{n} M_i \quad \textit{where } \mathrm{End}(M_i) \textit{ is local for } i = 1, \ldots, n.$$

Proof. (a) Let M be artinian. Let Γ be the set of the direct summands $B \neq 0$ of M. As $M \neq 0$ and $M = M \oplus 0$ we have $M \in \Gamma$, thus $\Gamma \neq \varnothing$. Let B_0 be minimal in Γ, then B_0 is directly indecomposable (since otherwise B_0 would not be minimal in Γ). Now let Λ be the set of submodules $C \hookrightarrow M$, so that finitely many directly indecomposable submodules $B_1 \neq 0, \ldots, B_l \neq 0$ exist with

$$M = B_1 \oplus \ldots \oplus B_l \oplus C.$$

Owing to the existence of B_0, $\Lambda \neq \varnothing$. Let C_0 be minimal in Λ and let

$$M = M_1 \oplus \ldots \oplus M_n \oplus C_0$$

be the corresponding decomposition. We assert that $C_0 = 0$. Otherwise, since C_0 is again artinian as a submodule of an artinian module, by the first remark, C_0 would split off a directly indecomposable direct summand $\neq 0$ in contradiction to the minimality of C_0.

Let now M be noetherian and let Γ be the set of the direct summands $A \neq M$ of M. Since $0 \in \Gamma$, we have $\Gamma \neq \varnothing$. Let A_0 be maximal in Γ and suppose we have

$$M = A_0 \oplus B_0.$$

From the maximality of A_0, it follows that B_0 is directly indecomposable and as $A_0 \neq M$ we have $B_0 \neq 0$. Let now Λ be the set of all submodules of M which are direct summands of M and are finite direct sums of directly indecomposable submodules.

As $\{0\} \in \Lambda$, we have $\Lambda \neq \varnothing$. Let

$$B_1 + \ldots + B_k = B_1 \oplus \ldots \oplus B_k$$

be a maximal element in Λ with directly indecomposable B_i. Let further

$$M = B_1 \oplus \ldots \oplus B_k \oplus C_0.$$

Suppose $C_0 \neq 0$, then by the earlier consideration the noetherian module C_0 must contain a directly indecomposable direct summand $\neq 0$. This contradicts the maximality of $B_1 \oplus \ldots \oplus B_k$. Thus $C_0 = 0$ and the proof is complete.

Remark. The "symmetry" of both proofs depends on the fact that in the first only the minimal condition and in the second only the maximal condition for direct summands is required. By Exercise 7, Chapter 6, these two conditions are however equivalent.

(b) follows from (a) 6.1.2 and 7.2.7. □

7.3 KRULL–REMAK–SCHMIDT THEOREM

We come now to the important uniqueness theorem of *Krull–Remak–Schmidt.*

7.3.1 THEOREM. *Let*

$$M_R = \bigoplus_{i \in I} M_i \quad \textit{where } \mathrm{End}(M_i) \textit{ is local for all } i \in I$$

and $M_R = \bigoplus_{j \in J} N_j$ *where* N_j *is directly indecomposable and* $N_j \neq 0$ *for all* $j \in J$. *Then a bijection* $\beta: I \to J$ *exists with* $M_i \cong N_{\beta(i)}$ *for all* $i \in I$.

We obtain the proof in several steps, which we formulate in part as lemmas.

7.3.2 LEMMA. *Let*

$$M = \bigoplus_{i \in I} M_i \quad \textit{where } \mathrm{End}(M_i) \textit{ is local for all } i \in I$$

and

$$\sigma, \tau \in \mathrm{End}(M) \quad \textit{with} \quad 1_M = \sigma + \tau.$$

Then to every $j \in I$ there exists a $U_j \hookrightarrow M$ and an isomorphism $\varphi_j : M_j \to U_j$ which is induced by σ or τ (i.e. $\varphi_j(x) = \sigma(x)$ for all $x \in M_j$ or $\varphi_j(x) = \tau(x)$ for all $x \in M_j$), so that we have

$$M = U_j \oplus \Big(\bigoplus_{\substack{i \in I \\ i \neq j}} M_i \Big).$$

Proof. Let $\pi_j : M \to M_j$ be the projections, $\iota_j : M_j \to M$ be the injections for all $j \in I$ (in the sense of Chapter 4).

From $1_M = \sigma + \tau$ it follows that

$$1_{M_j} = \pi_j 1_m \iota_j = \pi_j(\sigma + \tau)\iota_j = \pi_j \sigma \iota_j + \pi_j \tau \iota_j.$$

Since in the local ring $\mathrm{End}(M_j)$ the non-invertible elements form an ideal and 1_{M_j} is invertible, at least one of the elements $\pi_j \sigma \iota_j$, $\pi_j \tau \iota_j$ must be invertible, i.e. must be an automorphism of M_j.

Let, say, $\pi_j \sigma \iota_j$ be an automorphism. Then we define:

$$U_j := \sigma \iota_j(M_j) = \sigma(M_j),$$

$$\varphi_j : M_j \ni x \mapsto \sigma(x) \in U_j,$$

$$\iota_j' : U_j \ni y \mapsto y \in M.$$

Accordingly φ_j is an epimorphism. For $x \in M_j$ we then have

$$\iota_j' \varphi_j(x) = \varphi_j(x) = \sigma(x) = \sigma \iota_j(x) \Rightarrow \iota_j' \varphi_j = \sigma \iota_j \Rightarrow \pi_j \iota_j' \varphi_j = \pi_j \sigma \iota_j.$$

Thus we have the following commutative diagram

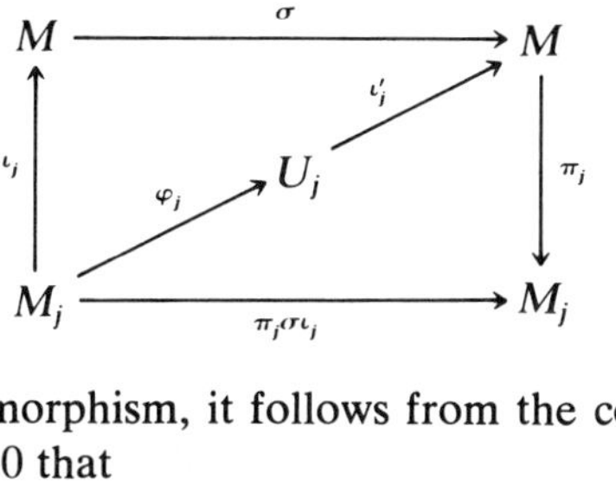

Since $\pi_j \sigma \iota_j$ is an automorphism, it follows from the commutativity of the lower triangle by 3.4.10 that

$$M = \mathrm{Im}(\iota_j' \varphi_j) \oplus \mathrm{Ker}(\pi_j) = U_j \oplus \Big(\bigoplus_{\substack{i \in I \\ i \neq j}} M_i \Big). \qquad \square$$

7.3.3 LEMMA. *Assumptions as in 7.3.2. Let further $E = \{i_1, \ldots, i_t\} \subset I$. Then there are $C_{i_j} \hookrightarrow M$, $j = 1, \ldots, t$ and isomorphisms*

$$\gamma_{i_j} : M_{i_j} \to C_{i_j},$$

which are induced either by σ or τ, so that we have:

$$M = C_{i_1} \oplus \ldots \oplus C_{i_t} \oplus \Big(\bigoplus_{\substack{i \in I \\ i \notin E}} M_i \Big).$$

Proof. The C_{i_j} are determined successively with the help of 7.3.2. For $i_1 = j$ in 7.3.2 let $C_{i_1} = U_{i_1}$, for which we then have

$$M = C_{i_1} \oplus \Big(\bigoplus_{\substack{i \in I \\ i \neq i_1}} M_i \Big).$$

As $M_{i_1} \cong C_{i_1}$, $\operatorname{End}(C_{i_1})$ is also local. In this decomposition we now exchange by 7.3.2 M_{i_2} for a C_{i_2}. Note: C_{i_2} need not be equal to U_{i_2}, since now another decomposition of M appears! After t steps (i.e. by induction) we obtain the desired result. □

7.3.4 LEMMA. *Let*

$$M = \bigoplus_{i \in I} M_i \quad \textit{where } \operatorname{End}(M_i) \textit{ is local for all } i \in I$$

and let $M = A \oplus B$ where $A \neq 0$ and directly indecomposable, $\pi' : M \to A$ the corresponding projection. Then a $k \in I$ exists so that π' induces an isomorphism of M_k onto A and $M = M_k \oplus B$ holds.

Proof. Let $\iota : A \to M$ be the inclusion and let $\pi := \iota \pi'$. As $1_M = \pi + (1_M - \pi)$ we can use 7.3.2 with $\sigma = \pi$ and $\tau = 1 - \pi$. As $A \neq 0$ there is $0 \neq a \in A$, from which we have $\pi(a) = a$. Then it follows that $(1_M - \pi)(a) = 0$. Let

$$a = \sum_{j=1}^{t} m_{i_j} \qquad \text{with } 0 \neq m_{i_j} \in M_{ij},\ i_j \in I$$

be the unique representation in $M = \bigoplus_{i \in I} M_i$.

In the sense of 7.3.3 now let the modules C_{ij} and the isomorphisms γ_{i_j} be determined. Suppose the γ_{i_j}, were all induced by $1_M - \pi$, then it would follow that

$$0 = (1_M - \pi)(a) = \sum_{j=1}^{t} (1_M - \pi)(m_{i_j})$$

with $(1_M - \pi)(m_{i_j}) = \gamma_{i_j}(m_{i_j}) \in C_{i_j}$. Because the sum of the C_{i_j} is direct, this implies $\gamma_{i_j}(m_{i_j}) = 0$, thus $m_{i_j} = 0$ and finally $a = 0$↯. Thus there is at least one i_j, so that γ_{i_j} is induced by π; let this be denoted by k. Then

$$\gamma_k : M_k \ni x \mapsto \pi(x) \in C_k$$

is thus an isomorphism. By 7.3.3 C_k is a direct summand of M; let thus $M = C_k \oplus L$. Further we have that

$$C_k = \pi(M_k) \hookrightarrow \pi(M) = A.$$

Then it follows that

$$A = M \cap A = (C_k \oplus L) \cap A = C_k \oplus (L \cap A),$$

and since A is directly indecomposable and $C_k \neq 0$ (as $M_k \neq 0$) we deduce finally that $A = C_k$.

From the commutative diagram

$$\begin{array}{ccc} M_k & \xrightarrow{\ \iota\ } & M \\ & {\scriptstyle\gamma_k}\searrow \quad \swarrow{\scriptstyle\pi'} & \\ & A = C_k & \end{array}$$

in which $\iota : M_k \to M$ is the inclusion mapping, it follows then from 3.4.10 that

$$M = \operatorname{Im}(\iota) \oplus \operatorname{Ker}(\pi') = M_k \oplus B,$$

from which the lemma is proved. □

Proof of 7.3.1. By 7.3.4 (with $A = N_j$) every N_j is isomorphic to an M_i; thus $\operatorname{End}(N_j)$ is local and the assumptions are symmetric. We now introduce into any I and J an equivalence relation and in fact let

$$i_1 \sim i_2 :\Leftrightarrow M_{i_1} \cong M_{i_2} \qquad (i_1, i_2 \in I),$$

$$j_1 \sim j_2 :\Leftrightarrow N_{j_1} \cong N_{j_2} \qquad (j_1, j_2 \in J).$$

For $i \in I$ let $\bar{\imath}$ be the equivalence class determined by i and let $\bar{I}$ be the set of all equivalence classes. Analogous notation for J.

Definition. Let $\Phi : \bar{I} \to \bar{J}$ be defined by $\Phi(i) = j$, if $M_i \cong N_j$.

Φ is a bijective mapping. Φ is defined on $\bar{I}$, since by 7.3.4 (for $A = M_i$ and $M = \bigoplus N_j$ in place of $M = \bigoplus M_i$ in 7.3.4) a $j \in J$ exists with $N_j \cong M_i$. Since the isomorphism is an equivalence relation, Φ is independent of the representative (in $\bar{I}$ and $\bar{J}$), i.e. it is in fact a mapping.

Φ is injective, for from $\Phi(\bar{\imath}_1) = \bar{\jmath}_1 = \bar{\jmath}_2 = \bar{\Phi}(\bar{\imath}_2)$ it follows $M_{i_1} \cong N_{j_i} \cong N_{j_2} \cong M_{j_2}$, thus $\bar{\imath}_1 = \bar{\imath}_2$. By 7.3.4 (with $A = N_j$) Φ is also *surjective.*

It still remains to show that for every $i \in I$ a bijection $B_{\bar{\imath}} : \bar{\imath} \to \Phi(\bar{\imath})$ exists. Then $\beta : I \ni i \mapsto \beta_{\bar{\imath}}(i) \in J$ is the desired bijection with $M_i \cong N_{\beta(i)}$. By the

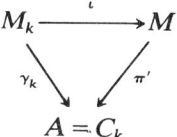

Schröder–Bernstein Theorem (to be found in any text book on set theory) it suffices to show:

There are injective mappings $\bar{\imath} \to \Phi(\bar{\imath})$ and $\Phi(\bar{\imath}) \to \bar{\imath}$.

From the symmetry of the assumptions only the existence of an injection, say $\Phi(\bar{\imath}) \to \bar{\imath}$, needs to be demonstrated.

Case 1. $\bar{\imath}$ is finite. Let the number of elements of $\bar{\imath}$ be t say. Let further $E = \{j_1, \ldots, j_s\} \subset \Phi(i)$. By 7.3.4 (with $A = N_{j_1}$) there is then an M_{i_1} with $M_{i_1} \cong N_{j_1}$, i.e. $i_1 \in \bar{\imath}$ and

$$M = M_{i_1} \oplus \Big(\bigoplus_{\substack{j \in J \\ j \neq j_1}} N_j \Big).$$

By 7.3.4 $\Big($with $A = N_{j_2}$ and $B = M_{i_1} \oplus \Big(\bigoplus_{\substack{j \in J \\ j \neq j_1, j \neq j_2}} N_j \Big)\Big)$ there is once again an M_{i_2} with $M_{i_2} \cong N_{j_2}$, i.e. $i_2 \in \bar{\imath}$ and

$$M = M_{i_1} \oplus M_{i_2} \oplus \Big(\bigoplus_{\substack{j \in J \\ j \neq j_1, j \neq j_2}} N_j \Big).$$

We obtain successively

$$M = M_{i_1} \oplus \ldots \oplus M_{i_s} \oplus \Big(\bigoplus_{\substack{j \in J \\ j \notin E}} N_j \Big) \wedge M_{i_l} \cong N_{j_l} \quad \text{for} \quad l = 1, \ldots, s.$$

Since the sum is direct, the $M_{i_1}, \ldots, M_{i_s}$ are pairwise different, thus we must have $s \leq t$. Consequently the number of the elements of $\Phi(\bar{\imath}) \leq t$ and the assertion is clear.

Case 2. $\bar{\imath}$ is infinite. Let $\pi'_j : M \to N_j$ be the projection and let for $k \in I$

$$E(k) := \{j \mid j \in J \wedge \pi'j \text{ induces an isomorphism of } M_k \text{ onto } N_j\}.$$

Assertion. $E(k)$ is finite for all $k \in I$.

Let

$$0 \neq m \in M_k \wedge m = \sum_{l=1}^{t} n_{j_l}, \qquad 0 \neq n_{i_l} \in N_{j_l} \Rightarrow \pi'_{j_l}(m) = n_{j_l};$$

in order that π'_j induces an isomorphism, we must have $\pi'_j(m) \neq 0$, i.e. $j \in \{j_1, \ldots, j_t\}$.

Assertion. $\Phi(\bar{\imath}) = \bigcup_{k \in \bar{\imath}} E(k)$.

"$\Phi(\bar{\imath}) \supset \bigcup_{k \in \bar{\imath}} E(k)$": Let $k \in \bar{\imath}$ and $j \in E(k)$.

$$\left.\begin{array}{r} k \in \bar{\imath} \Rightarrow M_k \cong M_i \\ j \in E(k) \Rightarrow M_k \cong N_j \end{array}\right\} \Rightarrow M_i \cong N_j \Rightarrow j \in \Phi(\bar{\imath}).$$

"$\Phi(i) \subset \bigcup_{k \in \bar{\imath}} E(k)$": $j \subset \Phi(i) \Rightarrow M_{i_1} \cong N_j$. By 7.3.4 there is a $k \in I$, so that π_j' induces an isomorphism of M_k onto $N_j \Rightarrow M_k \cong N_j \Rightarrow M_k \cong M_i \Rightarrow k \in \bar{\imath} \wedge j \in E(k)$. Let $\dot{\bigcup}_{k \in \bar{\imath}} E(k)$ be the disjoint union of the sets $E(k)$, then there is an injection $\Phi(\bar{\imath}) = \bigcup_{k \in \bar{\imath}} E(k) \to \dot{\bigcup}_{k \in \bar{\imath}} E(k)$. Since every $E(k)$ is finite, for every $E(k)$ there is an injection into $\mathbb{N}$. Then an injection exists

$$\dot{\bigcup_{k \in \bar{\imath}}} E(k) \to \bar{\imath} \times \mathbb{N}.$$

Since $\bar{\imath}$ is infinite, by a known result of set theory there is a bijection $\bar{\imath} \times \mathbb{N} \to \bar{\imath}$. All injections together yield an injection $\Phi(\bar{\imath}) \to \bar{\imath}$. Hence the Krull–Remak–Schmidt Theorem is proved. □

7.3.5 COROLLARY. *Let $M = \bigoplus_{i \in I} M_i$ where* End(M_i) *is local for all $i \in I$. Let $N = \bigoplus_{j \in J} N_j$ where N_j is directly indecomposable and $N_j \neq 0$ for all $j \in J$ and $M \cong N$. Then a bijection $\beta: I \to J$ exists with $M_i \cong N_{\beta(i)}$ for all $i \in I$.*

Proof. Let $\sigma: N \to M$ be an isomorphism then we have

$$M = \bigoplus_{j \in J} \sigma(N_j)$$

with directly indecomposable $\sigma(N_j)$ and by 7.3.1 (with $M = \oplus\, \sigma(N_j)$ in the place of $M = \oplus\, N_j$) it follows that $M_i \cong \sigma(N_{\beta(i)}) \cong N_{\beta(i)}$.

7.3.6 COROLLARY. *The decomposition of an injective module over a noetherian ring resp. of a module of finite length into a direct sum of directly indecomposable submodules is uniquely determined in the sense of the Krull–Remak–Schmidt Theorem.*

Proof. This follows from 6.6.5 and 7.2.8 resp. 7.2.9. □

EXERCISES

(1)

Let $\sigma: R \to S$ be a surjective ring homomorphism and let $S \neq 0$. Show: If R is local and if A is the ideal of the non-invertible elements of R, then $\sigma(A)$ is the ideal of the non-invertible elements of S.

(2)

(a) Let R be a local ring. Show that the following are equivalent for M_R:

(1) The lattice of submodules of M is totally ordered.

(2) The set of cyclic submodules of M is totally ordered.

(3) Every finitely generated submodule of M is cyclic.

(4) Every submodule of M generated by two elements is cyclic.

(b) Give an example of a ring R and a module M_R such that (3) is satisfied but not (1).

(3)

(Continuation of Exercise 11, Chapter 6.) Show that the following are equivalent for an injective module Q_R:

(1) Q satisfies the maximal condition for direct summands.

(2) Q is a direct sum of finitely many directly indecomposable submodules.

(Hint: With respect to (2)$\Rightarrow$(1) show first that every non-zero submodule of Q contains an irreducible submodule).

(4)

A module M, which satisfies the equivalent conditions of Exercise 11, Chapter 6, is called *finite-dimensional*, and the number of the directly indecomposable summands in a decomposition of $I(M)$ (Uniquely determined by the Krull–Remak–Schmidt Theorem) is then called the *dimension* of M (=dim(M)). Show:

(a) $\dim(M) = 0 \Leftrightarrow M = 0$, $\dim(M) = 1 \Leftrightarrow M$ is irreducible $\wedge\, M \neq 0$.

(b) M is finite-dimensional $\wedge\, U \hookrightarrow M \Rightarrow U$ is finite-dimensional $\wedge$ $\dim(U) \leq \dim(M)$.

(c) If M is finite-dimensional and $U \hookrightarrow M$ then $U \hookrightarrow^* M \Leftrightarrow \dim(U) = \dim(M)$.

(d) M_1, M_2 are finite-dimensional $\Rightarrow M_1 \oplus M_2$ is finite-dimensional and $\dim(M_1 \oplus M_2) = \dim(M_1) + \dim(M_2)$.

(e) If $X \hookrightarrow M$ and X and M/X are finite-dimensional then M is also finite-dimensional and $\dim(M) \leq \dim(X) + \dim(M/X)$.

(5)

(a) Exhibit a nilpotent element $\neq 0$ in $\mathbb{Z}/360\mathbb{Z}$.

(b) Exhibit seven different idempotent elements $\neq 0$ in $\mathbb{Z}/360\mathbb{Z}$.

(c) Decompose the ring $\mathbb{Z}/360\mathbb{Z}$ into a direct sum of directly indecomposable ideals.

(6)

Show:

(a) $\mathbb{Q}_{\mathbb{Z}}$ is directly indecomposable.

(b) $\mathbb{Q}_{\mathbb{Z}}$ is the sum of two proper submodules.

(c) The endomorphism ring of $\mathbb{Q}_{\mathbb{Z}}$ is ring-isomorphic to $\mathbb{Q}$ as a field.

(d) $\mathbb{Q}_{\mathbb{Z}}$ possesses a factor-module which is not directly indecomposable.

(7)

Let R be an integral domain and let K be the quotient field of R. Let V and W be K-vector spaces and let M resp. N be an R-submodule of V resp. W. Let $x_1, \ldots, x_m \in M$, $k_1, \ldots, k_m \in K$, $\sum_{i=1}^{m} x_i k_i \in M$, $\varphi \in \operatorname{Hom}_R(M, N)$. Show

$$\varphi\Big(\sum_{i=1}^{m} x_i k_i\Big) = \sum_{i=1}^{m} \varphi(x_i) k_i.$$

(Note that we may have neither $k_i \in R$ nor $x_i k_i \in M$!)

(8)

Let R be an integral domain, K the quotient field of R, $V = V_K$ an n-dimensional vector space over K, $U = U_R$ an R-submodule of $V = V_R$.

Show: There are directly indecomposable R-submodules $U_1, \ldots, U_m$ of U with $m \leq n$ and with

$$U = U_1 \oplus \ldots \oplus U_m.$$

(Hint: Let a decomposition $U = U_1 \oplus \ldots \oplus U_m$ be given and let $u_i \in U_i$, $u_i \neq 0$, then $u_1, \ldots, u_m$ are linearly independent over K.)

(9)

Let $V = V_{\mathbb{Q}}$ with a basis $x_1, \ldots, x_{m+n}$ in which $m, n \geq 2$. Let p_i, q_j be prime numbers for $1 \leq i \leq m+n$, $1 \leq j \leq m+n-1$, and let

$$A_i := \mathbb{Q}_{p_i} = \left\{\frac{z}{p_i^n} \,\middle|\, z \in \mathbb{Z} \wedge n \in \mathbb{Z}\right\}, \qquad 1 \leq i \leq m+n,$$

$$B_j := \frac{\mathbb{Z}}{q_j}, \qquad 1 \leq j \leq m+n-1,$$

$$y_j := x_j + x_{j+1}, \qquad 1 \leq j \leq m+n-1.$$

(a) Show: Let $p_1, \ldots, p_n, q_1, \ldots, q_{n-1}$ be pairwise different, then

$$U := \sum_{i=1}^{n} x_i A_i + \sum_{j=1}^{n-1} y_j B_j$$

is a directly indecomposable $\mathbb{Z}$-module.

(Hint: Suppose $U = U' \oplus U''$ with projections π' and π''. Show successively $\pi'(x_i)A_i \hookrightarrow U$, $\pi'(x_i) \in x_i A_i$, $\pi'(x_i) = 0$ or $\pi''(x_i) = 0$, U is directly indecomposable with the help of the elements $y_j \frac{1}{q_j}$.)

(b) Let p_i for $2 \leq i \leq n+m$, $i \neq n+1$ and q_j for $1 \leq j \leq n+m-1$ be pairwise different and let $p_1 = p_{n+1}$ hold. Show

$$U_1 := \sum_{i=1}^{n} x_i A_i + \sum_{j=1}^{n-1} y_j B_j$$

and

$$U_2 := \sum_{i=n+1}^{m+n} x_i A_i + \sum_{j=n+1}^{m+n-1} y_j B_j$$

are directly indecomposable.

Define $\varphi : U_1 \oplus U_2 \to U_1 \oplus U_2$ for

$$u = \sum_{i=1}^{n+m} x_i a_i + \sum_{\substack{j=1 \\ j \neq n}}^{n+m-1} y_j b_j, \; a_i \in A_i, \; b_j \in B_j$$

by

$$\varphi(u) := (q_1(a_1 + b_1) + q_{n+1}(a_{n+1} + b_{n+1}))(x_1 - x_{n+1}).$$

Show: φ is an R-homomorphism. Determine q_1, q_{n+1} so that $\varphi^2 = \varphi$ holds.

Deduce: $U_1 \oplus U_2$ may be written in two ways, different not only up to isomorphism and order, as direct sums of directly indecomposable submodules.

(10)

Let $M = M_R$. For $n \in \mathbb{N}$ let $M^n := M^{\{1,2,\ldots,n\}}$. Let

$$A_R = \bigoplus_{i \in I} A_i, \qquad B_R = \bigoplus_{j \in J} B_j.$$

Let $\mathrm{End}(A_i)$ be local for $i \in I$, and B_j be directly indecomposable for $j \in J$. Show:

(a) Let $n \in \mathbb{N}$. From $A^n \cong B^n$ it follows that $A \cong B$.

(b) Let I be finite and let S, T be non-empty sets. From $A^{(S)} \cong A^{(T)}$ it follows that S and T have the same cardinal number.

Chapter 8

Semisimple Modules and Rings

8.1 DEFINITION AND CHARACTERIZATION

There are two immediate and important generalizations of the concept of a vector space. These are:

(1) Free modules and direct summands of free modules, the projective modules with which we have already become acquainted.

(2) Modules, in which every submodule is a direct summand; these are called semisimple modules. They provide the theme for the following considerations. First some lemmas are presented.

8.1.1 LEMMA. *Let $M = M_R$ be a module, in which every submodule is a direct summand. Then every non-zero submodule contains a simple submodule.*

Proof. Let $U \hookrightarrow M$, $U \neq 0$ and U finitely generated. By 2.3.12 there is a maximal submodule $C \hookrightarrow U$. By assumption we have $M = C \oplus M_1$; hence it follows with the help of the modular law that $U = M \cap U = C \oplus (M_1 \cup U)$, thus we have $U/C \cong M_1 \cap U$. Since C is maximal in U, U/C is simple. Thus $M_1 \cap U$ is a simple submodule of U. □

8.1.2 LEMMA. *Let $M = \sum_{i \in I} M_i$ with simple submodules M_i. Further let $U \hookrightarrow M$. Then we have:*

(a) *There is $J \subset I$ so that $M = U \oplus \left(\bigoplus_{i \in J} M_i\right)$.*

(b) *There is $K \subset I$ so that $U \cong \bigoplus_{i \in K} M_i$.*

Proof. (a) Proof with the help of Zorn's Lemma. Let

$$\Gamma := \left\{ L \mid L \subset I \wedge U + \sum_{i \in L} M_i = U \oplus \left(\bigoplus_{i \in L} M_i \right) \right\}.$$

As $\bigoplus_{i \in \varnothing} M_i = 0$ we have $\varnothing \in \Gamma$, thus $\Gamma \neq \varnothing$ and Γ is ordered by $\subset$. Let Λ be a totally ordered subset in Γ. We claim that

$$L^* = \bigcup_{L \in \Lambda} L$$

is an upper bound of Λ in Γ. It is clear that L^* is an upper bound. It remains to be shown that $L^* \in \Gamma$.

Let $E \subset L^*$, E finite, then there is an $L \in \Lambda$ with $E \subset L$. Let now

$$u + \sum_{i \in E} m_i = 0, \qquad u \in U, m_i \in M_i,$$

then it follows from $E \subset L$ that: $u = m_i = 0$ for all $i \in E$. Thus we have

$$U + \sum_{i \in L^*} M_i = U \oplus \left(\bigoplus_{i \in L^*} M_i \right),$$

and consequently $L^* \in \Gamma$. By Zorn's Lemma there is then a maximal element $J \in \Gamma$. Let

$$N := U + \sum_{i \in J} M_i = U \oplus \left(\bigoplus_{i \in J} M_i \right).$$

Now consider $N + M_{i_0}$ for arbitrary $i_0 \in I$. $N + M_{i_0} = N \oplus M_{i_0}$ is not possible for then we must have

$$J \underset{\neq}{\hookrightarrow} J \cup \{i_0\} \in \Gamma$$

Thus it follows that $N \cap M_{i_0} \neq 0$. But since M_{i_0} is simple, we must have $N \cap M_{i_0} = M_{i_0}$, thus $M_{i_0} \hookrightarrow N$ holds. Then it follows that

$$M = \sum_{i \in I} M_i \hookrightarrow N \hookrightarrow M,$$

i.e. $N = M$.

(b) Let now $M = U \oplus \left(\bigoplus_{i \in J} M_{i_0} \right)$. Then (a) is applied to the submodule $\bigoplus_{i \in J} M_i$ (in the place of U in (a)). Accordingly $K \subset I$ exists with $M =$

$\left(\bigoplus_{i\in J} M_i\right)\oplus\left(\bigoplus_{i\in K} M_i\right)$. By the First Isomorphism Theorem it follows that

$$U \cong M/\bigoplus_{i\in J} M_i \cong \bigoplus_{i\in K} M_i. \qquad \square$$

We come now to the main theorem on semisimple modules.

8.1.3 THEOREM. *For a module $M = M_R$ the following conditions are equivalent*:

(1) *Every submodule of M is a sum of simple submodules.*
(2) *M is a sum of simple submodules.*
(3) *M is a direct sum of simple submodules.*
(4) *Every submodule of M is a direct summand of M.*

Proof. "(1)$\Rightarrow$(2)": (2) is a special case of (1).
"(2)$\Rightarrow$(3)": 8.1.2 (a) for $U = 0$.
"(3)$\Rightarrow$(4)": 8.1.2 (a).
"(4)$\Rightarrow$(1)": Let $U \hookrightarrow M$. Put

$$U_0 := \sum_{\substack{\text{simple } M_i \\ M_i \hookrightarrow U}} M_i.$$

Then $U_0 \hookrightarrow U$ and by (4) U_0 is a direct summand of M:

$$M = U_0 \oplus N \Rightarrow U = M \cap U = U_0 \oplus (N \cap U).$$

Case 1. $N \cap U = 0 \Rightarrow U = U_0 \Rightarrow (1)$.

Case 2. $N \cap U \neq 0 \Rightarrow$ By 8.1.1 there is a simple submodule $B \hookrightarrow N \cap U \Rightarrow B \hookrightarrow U_0$, by definition of U_0 we have $B \hookrightarrow U_0 \cap (N \cap U) = 0$ $\lightning$.
Thus only the first case can occur. $\square$

8.1.4 *Definition*

(a) A module $M = M_R$ is called *semisimple* $:\Leftrightarrow M$ satisfies the equivalent conditions of 8.1.3.

(b) A ring R is called *right* resp. *left semisimple* $:\Leftrightarrow R_R$ resp. ${}_RR$ is semisimple.

We observe that the module 0 is semisimple for

$$0 = \sum_{i\in\varnothing} M_i, \qquad \text{semisimple } M_i$$

but 0 is not a simple module, since it was assumed that, for a simple module M, $M \neq 0$.

We shall show later: R_R is semisimple$\Leftrightarrow {}_RR$ is semisimple so that with regard to a semisimple ring the statement of sidedness can be omitted.

Examples

(1) Every vector space $V = V_K$ over a skew field K is semisimple:

$$V_K = \sum_{x \in V} xK, \qquad xK \text{ is simple for } x \neq 0$$

(2) $\mathbb{Z}/n\mathbb{Z}$ with $n \neq 0$ is semisimple as a $\mathbb{Z}$-module$\Leftrightarrow n$ is square-free (i.e. n is the product of pairwise different prime numbers) or $n = \pm 1$.

Proof. Exercise for the reader. The proof follows later in a more general context.

(3) $\mathbb{Z}_{\mathbb{Z}}$ and $\mathbb{Q}_{\mathbb{Z}}$ are not semisimple since they have no simple submodules.

(4) Let $V = V_K$ be a vector space. Then we have: $\text{End}(V_K)$ is a (two-sided) semisimple ring$\Leftrightarrow \dim_K(V) < \infty$.

Proof. Later (in 8.3.1).

8.1.5 COROLLARY

(1) *Every submodule of a semisimple module is semisimple.*

(2) *Every epimorphic image of a semisimple module is semisimple.*

(3) *Every sum of semisimple modules is semisimple.*

(4) *Two decompositions of a semisimple module into a direct sum of simple modules are isomorphic in the sense of the Krull–Remak–Schmidt Theorem* (7.3.1).

Proof. (1) This follows immediately from 8.1.3.

(2) Let A be simple and let $\alpha : A \to B$ be an epimorphism, then it follows that $A/\text{Ker}(\alpha) \cong B$. If $\text{Ker}(\alpha) = 0$, then B is simple; if $\text{Ker}(\alpha) = A$, then $B = 0$. Since A is simple there are no further possibilities for $\text{Ker}(\alpha)$. The image of a sum of simple modules with respect to a homomorphism is hence a sum of simple and zero modules, of which the latter can be omitted, and therefore by 8.1.3 is again semisimple.

(3) Since by 8.1.3 every semisimple module is a sum of simple modules, a sum of semisimple modules is also again a sum of simple modules and hence by 8.1.3 again semisimple.

(4) Since the endomorphism ring of a simple module is a skew field and thus is local, the Krull–Remak–Schmidt Theorem holds in this case. □

The following theorem shows that for a semisimple module all finiteness conditions are equivalent.

8.1.6 THEOREM. *For a semisimple module $M = M_R$ the following conditions are equivalent*:

(1) *M is a sum of finitely many simple modules.*

(2) *M is a direct sum of finitely many simple modules.*

(3) *M has finite length.*
(4) *M is artinian.*
(5) *M is noetherian.*
(6) *M is finitely generated.*
(7) *M is finitely cogenerated.*

Proof. Since all statements are trivial for $M=0$ we can assume that $M \neq 0$.

"(1)⇒(2)": By 8.1.2.

"(2)⇒(3)": Let $M=\bigoplus_{i=1}^{n} M_i$, M_i simple. Then $0 \hookrightarrow M_1 \hookrightarrow M_1 \oplus M_2 \hookrightarrow \ldots \hookrightarrow \bigoplus_{i=1}^{n} M_i = M$ is a composition series because $M_1 \oplus \ldots \oplus M_i / M_1 \oplus \ldots \oplus M_{i-1} \cong M_i$ is simple.

"(3)⇒(5)", "(5)⇒(6)": By 6.1.2

"(6)⇒(1)": By 2.3.12

"(3)⇒(4)", "(4)⇒(7)": By 6.1.2

"(7)⇒(2)": Suppose that M were the direct sum of infinitely many simple submodules M_i, then a submodule of M exists of the form $M_1 \oplus M_2 \oplus \ldots$ with countably infinitely many submodules $M_1, M_2, \ldots$ Let

$$A_i := \bigoplus_{j=i}^{\infty} M_j, \qquad i \in \mathbb{N},$$

then obviously we have $\bigcap_{i=1}^{\infty} A_i = 0$ for

$$(M_1 \oplus \ldots \oplus M_n) \cap A_{n+1} = 0, \quad \text{thus} \quad (M_1 \oplus \ldots \oplus M_n) \cap \bigcap_{i=1}^{\infty} A_i = 0$$

for arbitrary $n \in \mathbb{N}$. But the intersection of any finitely many of the A_i is evidently equal to the A_i with largest i, thus unequal to 0.

Let now M_R be semisimple and let Γ denote the set of all simple submodules of M: $\Gamma = \{E \mid E \hookrightarrow M \wedge E \text{ is simple}\}$.

Then $\cong$ is an equivalence relation on Γ. Let the set of equivalence classes, which are now called isomorphism classes, be $\{\Omega_j \mid j \in J\}$, so that Ω_j is thus an isomorphism class. Therefore we then have

$$\Omega_{j_0} \cap \Omega_{j_1} = \varnothing \qquad \text{for } j_0, j_1 \in J \text{ and } j_0 \neq j_1.$$

8.1.7 *Definition* $B_j := \sum_{E \in \Omega_j} E$ is called a *homogeneous component* of M.

8.1.8 LEMMA. *Let M_R be semisimple and let B_j be a homogeneous component of M. Then we have*

(a) $U \hookrightarrow B_j \wedge U$ *is simple* $\Rightarrow U \in \Omega_j$.

(b) $M = \bigoplus_{j \in J} B_j$.

Proof. (a) This follows from 8.1.2(b); there is accordingly an $E \in \Omega_j$ with $U \cong E$, for more summands cannot appear in E since U is simple.

(b) Since M is a sum of simple submodules and every simple submodule is contained in a Ω_j, it follows that $M = \sum_{j \in J} B_j$. Suppose for $j_0 \in J$ that we had

$$D := B_{j_0} \cap \sum_{\substack{j \in J \\ j \neq j_0}} B_j \neq 0.$$

Then by 8.1.1 there is a simple submodule E of D. Since $E \hookrightarrow B_{j_0}$ it follows by (a) that $E \in \Omega_{j_0}$. Since $E \hookrightarrow \sum_{j \neq j_0} B_j$ it follows by 8.1.2(b) that a $j_1 \in J, j_1 \neq j_0$ exists with $E \in \Omega_{j_1}$. Then it would follow that $\Omega_{j_0} \cap \Omega_{j_1} \neq \varnothing$ ↯. □

If we have to determine in a concrete case whether a module is semisimple then this can be difficult and depend on very special properties. From this point of view an interesting and important example for semisimple modules (and rings) is to be considered. Let $R := GK$ the group ring of a finite group with coefficients in a field (see 4.6.2).

8.1.9 THEOREM (of Maschke). *R_R and $_RR$ are semisimple if and only if the characteristic of K is not a divisor of the order of G.*

Proof. Let the characteristic of K be not a divisor of $n := \mathrm{Ord}(G)$. Then for $0 \neq k \in K$, $nk := k + \ldots + k$ (n summands) is invertible. For the inverse of $n1$ with $1 \in K$ we write $1/n$. Let the elements of G be $g_1, \ldots, g_n$. If we consider R only as a right K-module then R is a vector space over K. For every $\varphi \in \mathrm{End}(R_K)$ a mapping $\hat{\varphi} : R \to R$ is defined by

$$\hat{\varphi}(r) := \frac{1}{n} \sum_{i=1}^{n} \varphi(rg_i)g_i^{-1}, \qquad r \in R.$$

We require to show that $\hat{\varphi} \in \mathrm{End}(R_R)$. For arbitrary $k \in K$ we have

$$\hat{\varphi}(rk) = \frac{1}{n} \sum_{i=1}^{n} \varphi(rkg_i)g_i^{-1} = \left(\frac{1}{n} \sum_{i=1}^{n} \varphi(rg_i)g_i^{-1}\right)k = \hat{\varphi}(r)k.$$

Let now $g \in G$, as $\{gg_1, \ldots, gg_n\} = \{g_1, \ldots, g_n\}$ we have

$$\hat{\varphi}(rg) = \frac{1}{n} \sum_{i=1}^{n} \varphi(rgg_i)g_i^{-1} = \frac{1}{n} \sum_{i=1}^{n} \varphi(rgg_i)(gg_i)^{-1}g = \hat{\varphi}(r)g.$$

Hence it follows that $\hat{\varphi}(rx) = \hat{\varphi}(r)x$ for arbitrary elements $r, x \in R$, i.e. $\varphi \in \operatorname{End}(R_R)$.

Let now $A \hookrightarrow R_R$, then A is also a vector subspace of R_K. Consequently a $B \hookrightarrow R_K$ exists with $R_K = A \oplus B$. Let $\pi : R_K \to R_K$ be the projection of R onto A, i.e. let $\pi(a+b) = a$ for $a \in A$, $b \in B$ hold. As $A \hookrightarrow R_R$ it follows for $a \in A$ that

$$\hat{\pi}(a) = \frac{1}{n} \sum_{i=1}^{n} \pi(ag_i)g_i^{-1} = \frac{1}{n} \sum_{i=1}^{n} ag_ig_i^{-1} = \frac{1}{n} na = a,$$

and for $r \in R$ we obtain

$$\hat{\pi}(r) = \frac{1}{n} \sum_{i=1}^{n} \pi(rg_i)g_i^{-1} \in A,$$

since $\pi(rg_i) \in A$. Therefore $\hat{\pi}$ is a projection of R_R onto A and it follows that

$$R_R = \hat{\pi}(R) \oplus (1-\hat{\pi})(R) = A \oplus (1-\hat{\pi})(R).$$

Thus R_R is semisimple (analogously for ${}_RR$; see also 8.2.1).

Let now the characteristic of K be equal to p and let p be a divisor of n. Then we show that for $r_0 := g_1 + \ldots + g_n$ the ideal r_0R is not a direct summand of R_R. For $g \in G$ we have first $r_0 g = r_0$, thus it follows that $r_0^2 = nr_0 = 0$ as well as $r_0R = r_0K$. Suppose $R_R = r_0R \oplus U$, then an idempotent e must exist with $eR = r_0R = r_0K$. But it would follow from $e = r_0k_0$ with $k_0 \in K$ that $e = e^2 = r_0^2k_0^2 = 0$, thus $r_0 = 0$ ↯. □

8.2 SEMISIMPLE RINGS

If a ring possesses a certain property on one side then it need not possess it on the other side. For example we have established that there are rings which are only one-sidedly artinian. With regard to all ring-theoretic properties which depend on the side, the question naturally arises as to whether they are in fact one-sided or whether their validity on one side implies their validity on the other side. With regard to semisimple rings the latter is the case.

8.2.1 THEOREM. *For a ring R we have:*

$$R_R \text{ is semisimple} \Leftrightarrow {}_RR \text{ is semisimple.}$$

Proof. It suffices to show: ${}_RR$ is semisimple $\Rightarrow R_R$ is semisimple for the converse implication follows analogously.

By 7.2.3 (with change of side) the semisimple ring ${}_RR$ has a decomposition

$$ {}_RR = \bigoplus_{i=1}^{n} L_i = \bigoplus_{i=1}^{n} Re_i, \qquad \text{simple } L_i \hookrightarrow {}_RR $$

with

$$ e_i \neq 0, \qquad e_ie_j = \delta_{ij}e_i, \qquad L_i = Re_i, \qquad 1 = \sum_{i=1}^{n} e_i. $$

By 7.2.3(d) the decomposition

$$ R = \bigoplus_{i=1}^{n} e_iR, $$

follows and we only have to prove that all e_iR are simple. To prove this let e be one of the e_i and let $0 \neq a = ea \in eR$. Then it follows that $aR \hookrightarrow eR$. We wish to show $aR = eR$, from which it follows at once that eR is simple.

As $ea \neq 0$ and since Re is simple

$$ \varphi\colon Re \ni re \mapsto rea = ra \in Ra $$

is an isomorphism. Let ${}_RR = Ra \oplus U$, then

$$ \psi\colon R = Ra \oplus U \ni ra + u \mapsto \varphi^{-1}(ra) = re \in R $$

is an endomorphism of ${}_RR$, which is given by right multiplication by an element $b \in R$ (for $R^{(r)} = \mathrm{End}({}_RR)$, see 3.7). Thus it follows that

$$ e = \psi(a) = ab, $$

hence $e \in aR$, i.e. $eR \hookrightarrow aR$, and so

$$ eR = aR. $$ □

8.2.2 COROLLARY

(a) *R is semisimple $\Leftrightarrow$ every right and left R-module is semisimple.*

(b) *R is semisimple $\Rightarrow R_R$ and ${}_RR$ have the same finite length.*

(c) *R is semisimple and surjective ring homomorphism $\rho\colon R \to S \Rightarrow$ S is semisimple.*

(d) *R is semisimple $\Rightarrow R_R$ and ${}_RR$ are cogenerators.*

(e) *R is semisimple $\Leftrightarrow$ every right and every left R-module is injective $\Leftrightarrow$ every right and every left R-module is projective.*

(f) *R is semisimple $\Leftrightarrow$ every simple right R-module and every simple left R-module is projective.*

Proof. (a) "$\Rightarrow$": If R_R is semisimple and if $M = M_R$, $m \in M$, then by 8.1.5 mR is semisimple as an epimorphic image of R_R. Consequently

$$M = \sum_{m \in M} mR$$

as a sum of semisimple modules is again semisimple. Analogously for the left side.

(a) "$\Leftarrow$": Special case.

(b) This is contained in the proof of 8.2.1 since, for simple Re_i, e_iR is also simple.

(c) S_S can also be considered (see also 3.2) as an R-module if we put

$$sr := s\rho(r), \qquad s \in S, r \in R$$

and thereby the ideals of S_S coincide with the submodules of S_R. Since S_R is semisimple, then S_S is also semisimple.

(d) In order to show that R_R is a cogenerator let $m \in M_R$, $m \neq 0$. Since R_R is semisimple the epimorphism

$$R \ni r \mapsto mr \in mR$$

splits, consequently mR is isomorphic to a right ideal of R; thus there is a monomorphism

$$\varphi : mR_R \to R_R.$$

Since mR is a direct summand in M_R, there is a homomorphism

$$\hat{\varphi} : M_R \to R_R \quad \text{with} \quad \hat{\varphi} \mid mR = \varphi.$$

Then it follows that $m \notin \operatorname{Ker}(\hat{\varphi})$ which was to be shown.

(e) R_R is semisimple $\Rightarrow$ every right R-module is semisimple $\Rightarrow$ every submodule is a direct summand $\Rightarrow$ every right R-module is injective, resp. projective $\Rightarrow$ every right ideal of R is a direct summand in $R_R \Rightarrow$ semisimple R_R. Similarly for the left side.

(f) "$\Rightarrow$": Clear by (e).

(f) "$\Leftarrow$": Let $\operatorname{Soc}(R_R)$ be the sum of all simple right ideals of R (detailed investigation of $\operatorname{Soc}(M_R)$ in next chapter), then it is to be shown that $R = \operatorname{Soc}(R_R)$. Suppose $R \neq \operatorname{Soc}(R_R)$, then by 2.3.11 $\operatorname{Soc}(R_R)$ is contained in a maximal right ideal A of R. Since R/A is a simple right R-module and thus projective by assumption, a homomorphism φ exists so that

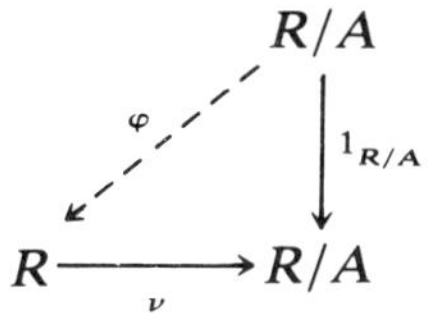

is commutative. It then follows that $\varphi \neq 0$ and

$$R = \operatorname{Im}(\varphi) \oplus A.$$

But since $\operatorname{Im}(\varphi)$ is simple, in contradiction to this we must have

$$\operatorname{Im}(\varphi) \hookrightarrow \operatorname{Soc}(R_R) \hookrightarrow A.$$

Thus in fact it follows that $R = \operatorname{Soc}(R_R)$. □

The next step in our investigation consists in decomposing a semisimple ring into a direct sum of directly indecomposable two-sided ideals. Thus let R be semisimple and let

$$R = B_1 \oplus \ldots \oplus B_m$$

be the decomposition of R_R into homogeneous components (in the sense of 8.1.8). By 7.2.3 the number of the homogeneous components, which by definition are right ideals, is finite. We wish to show that the B_j are two-sided simple ideals which mutually annihilate one another.

As a preliminary we prove first a result for an arbitrary ring R.

8.2.3 LEMMA. *Let $A \hookrightarrow R_R$ and let A be a direct summand of R_R, then the two-sided ideal RA generated by A contains all right ideals of R which are epimorphic images of A.*

Proof. Let $R_R = A \oplus B$, let $B \hookrightarrow R_R$ and let $\pi : R \to A$ be the projection. Further let $\alpha : A \to A'$ be an epimorphism, let $A' \hookrightarrow R_R$ and let $\iota' : A' \to R_R$ be the inclusion. Then it follows that $\iota'\alpha\pi \in \operatorname{Hom}_R(R_R, R_R)$. As established in 3.7, every endomorphism of R_R is by left multiplication. Thus there is a $c \in R$ with $c^{(l)} = \iota'\alpha\pi$. Then it follows from $\pi(R) = \pi(A)$ that

$$A' = \iota'\alpha\pi(R) = \iota'\alpha\pi(A) = cA \subset RA,$$

which was to be shown. □

We prove now the first part of the classical *Theorem of Wedderburn*, which Wedderburn had originally proved for algebras.

8.2.4 THEOREM. *Let $R \neq 0$ be a semisimple ring and let*

$$R_R = B_1 \oplus \ldots \oplus B_m$$

$$\textit{resp.} \quad {}_RR = C_1 \oplus \ldots \oplus C_n$$

be the decomposition of R_R resp. of ${}_RR$ into homogeneous components (8.1.8). *Then we have*:

(a) *The B_j, $j = 1, \ldots, m$, are simple two-sided ideals of R.*

(b) *$n = m$ and (with respect to an appropriate ordering)*

$$B_j = C_j,\ j = 1, \ldots, m.$$

(c) $B_iB_j = \delta_{ij}B_i,\ i, j = 1, \ldots, m.$

(d) *B_i, considered itself as a ring, is a simple ring with a unit element.*

(e) *The decomposition of R into a direct sum of simple two-sided ideals is (up to ordering) uniquely determined.*

Proof. (a) We show that: If $E \hookrightarrow B_i$ and E is simple then it follows that $RE = B_i$. For $r \in R$

$$E \ni x \mapsto rx \in rE$$

is an epimorphism. Since E is simple this is either the zero mapping, i.e. $rE = 0$, or an isomorphism, i.e. $E \cong rE$. In both cases it follows that $rE \hookrightarrow B_i$. Conversely let $E \cong E'$, then we infer from 8.2.3 that E' has the form $E' = rE$, from which it follows that $B_i \hookrightarrow RE$. Altogether this yields $RE = B_i$.

From $B_i = \sum_{E \in \Omega_i} E$ it then follows that

$$RB_i = \sum_{E \in \Omega_i} RE = \sum_{E \in \Omega_i} B_i = B_i,$$

thus B is a two-sided ideal. Let now $A \neq 0$ be a two-sided ideal contained in B_i, then A_R is semisimple and consequently there is a simple right ideal E with

$$E \hookrightarrow A_R \hookrightarrow B_i.$$

Then it follows that

$$B_i = RE \hookrightarrow RA = A \hookrightarrow B_i,$$

thus $A = B_i$, i.e. B_i is simple as a two-sided ideal.

(b) Correspondingly the $C_j,\ j = 1, \ldots, n$, are also simple two-sided ideals. Since B_iC_j is a two-sided ideal, which is contained in B_i as well as in C_j, and these are simple, we have either

$$B_iC_j = 0 \quad \text{or} \quad B_i = B_iC_j = C_j.$$

For fixed $i_0 = 1, \ldots, m$ at least one j_0 with $B_{i_0} = B_{i_0}C_{j_0} = C_{j_0}$ must exist, since otherwise it would follow that $B_{i_0}R = \sum_j B_{i_0}C_j = 0 ↯$. But there can also exist only one such j_0, for from $B_{i_0} = B_{i_0}C_{j_1} = C_{j_1}$ it would follow that $C_{j_0} = C_{j_1} ↯$. Since also correspondingly to every $C_{j_0},\ j_0 = 1, \ldots, n$, an i_0 with $B_{i_0} = C_{j_0}$ must exist, the assertion (b) follows.

(c) From $R = \bigoplus_{i=1}^{m} B_i$ it follows that $RB_j = B_j = \bigoplus_{i=1}^{m} B_iB_j$, from which (c) is obtained.

(d) According to (c) the two-sided R-ideals from B_i coincide with the two-sided B_i-ideals from B_i. Thus B_i is simple as a ring. Let $1=\sum f_i$, $f_i \in B_i$, then we have (7.2.3) $B_i = f_iR$ and the f_i are idempotents from the centre of R. For $b = f_ir \in B_i$ it then follows that $bf_i = f_ib = f_i^2r = f_ir = b$, thus f_i is the unit element of B_i.

(e) As in the proof of (b). □

8.2.5 *Definition.* The simple two-sided ideals B_i, $i = 1, \ldots, m$, in 8.2.4 are called the *blocks* of R.

8.2.6 COROLLARY. *Let R be semisimple, then we have: The number of the blocks is equal to the number of the isomorphism classes of simple right R-modules and equal to the number of isomorphism classes of simple left R-modules.*

Proof. Every simple right resp. left R-module is isomorphic to a right resp. left ideal of R (since every epimorphism of R_R onto a cyclic right R-module splits). Consequently it suffices to consider the simple right, resp. left ideals. For these the assertion follows from 8.2.4. □

8.3 THE STRUCTURE OF SIMPLE RINGS WITH A SIMPLE ONE-SIDED IDEAL

In order to elucidate entirely the structure of a semisimple ring, we now address ourselves to the investigation of the two-sided simple ideals B_j in 8.2.4. By definition the B_j are right ideals of the semisimple ring R, thus semisimple right R-modules. Therefore it follows from 8.2.4(c) that the B_j are also semisimple rings. We are here dealing with rings that are both simple and semisimple. Examples show (see Exercise 8) that not every simple ring is semisimple. Since the B_j are semisimple and it was assumed that $B_j \neq 0$, they are therefore simple rings which possess a simple right ideal.

Conversely every simple ring R, which possesses a simple right ideal E, is also semisimple, as we wish to establish immediately. Let B be the homogeneous component corresponding to E in R_R, i.e. the sum of all right ideals of R isomorphic to E, then, as a sum of simple right ideals B_R is semisimple. Further for $r \in R$ and for a simple right ideal $E' \hookrightarrow R_R$, rE' is either a right ideal isomorphic to E' or is equal to zero; thus B is a two-sided ideal $\neq 0$ in R. Since R is simple, it follows that $B = R$. Thus finally it is established that R_R is semisimple (see also 8.2.4(a)).

For rings of this sort the structure is now to be determined. It will emerge that every such ring is isomorphic to the endomorphism ring (=ring of linear transformations) of a finite-dimensional vector space V over a skew field K, which is again itself isomorphic to the ring of $n \times n$ matrices ($n = \dim_K(V)$) with coefficients in K and hence can be regarded as known.

8.3.1 THEOREM. *Let $V = {}_KV$ be a vector space over the skew field K. Then we have*

(a) *If* $1 \leq \dim_K(V) = n < \infty$ *then* $\mathrm{End}({}_KV)$ *is a simple and a semisimple ring.*

(b) *If* $\dim_K(V) = \infty$, *then* $\mathrm{End}({}_KV)$ *is neither simple nor semisimple.*

Proof. First of all we point out that we have here—with a view to the following theorem—fixed upon a left vector space $V = {}_KV$ and we wish to write the endomorphisms of V on the right of the argument: For $\varphi \in \mathrm{End}({}_KV)$ and $x \in V$ let $x\varphi$ be the image of x by φ. The result holds naturally also for the right vector spaces.

(a) Let $v_1, \ldots, v_n$ be a basis of ${}_KV$ and denote

$$V^{(i)} := \sum_{\substack{j=1 \\ j \neq i}}^{n} Kv_j, \qquad i = 1, \ldots, n;$$

$$S := \mathrm{End}({}_KV).$$

Then

$$E_i := \{\varphi \mid \varphi \in S \wedge V^{(i)} \hookrightarrow \mathrm{Ker}(\varphi)\}$$

is a simple right ideal in S and we have

$$S_S = E_1 \oplus \ldots \oplus E_n;$$

$$E_i \cong E_j \quad \text{for all} \quad i, j = 1, \ldots, n.$$

Consequently S is semisimple, and since all E_i are naturally isomorphic, S_S consists only of one homogeneous component, thus S is a simple ring. The above assertions on the E_i will not be proved here. It is a matter of simple assertions of linear algebra which are left to the reader as an exercise. The proof can also be obtained with the help of the ring of $n \times n$ matrices with coefficients in K and isomorphic to S. In this ring every row is a simple right ideal (and every column a simple left ideal), and the ring is the direct sum of its rows (resp. columns) which are all isomorphic.

(b) Let again $S := \mathrm{End}({}_KV)$.

Definition. $\varphi \in S$ is said to be of finite rank $:\Leftrightarrow \dim_K(\mathrm{Im}(\varphi)) < \infty$. Then it is easy to verify that the set of endomorphisms of finite rank is a proper two-sided ideal $A \neq 0$ in S. Thus S is not a simple ring. If S were a semisimple ring, then there would have to exist a $B \hookrightarrow S_S$ with

$$S_S = A \oplus B.$$

Since A is two-sided, it would follow that

$$BA \hookrightarrow B \cap A = 0, \quad \text{thus} \quad BA = 0.$$

Let $\beta \in B$, $\beta \neq 0$ and let $v \in V$ with $v\beta \neq 0$ and let

$$V = Kv\beta \oplus U, \qquad U \hookrightarrow {}_K V.$$

Finally for $k \in K$, $u \in U$ let the mapping α be defined by

$$\alpha : V \ni kv\beta + u \mapsto kv\beta \in V.$$

Then it follows that $\alpha \in A$ (for $\operatorname{Im}(\alpha) = Kv\beta$) and that $v\beta\alpha = v\beta \neq 0$, thus $\beta\alpha \neq 0$ in contradiction to $BA = 0$. □

Theorem 8.2.4 contains the first part of the familiar and important Wedderburn Theorem on semisimple rings. We come now to the second part of this Theorem.

8.3.2 THEOREM. *A simple ring R, which possesses a simple right ideal, is isomorphic to the endomorphism ring of a finite-dimensional vector space over a skew field.*

In particular: Let E be a simple right ideal of R and let $K := \operatorname{End}(E_R)$, then K is a skew field, $E = {}_K E$ is a left vector space of finite dimension over K and we have

$$R \cong \operatorname{End}({}_K E).$$

Proof. By Schur's Lemma (3.7.5), K is a skew field and E can be considered as a left K-module. Then E is a K-R-bimodule. For $y \in E$ we consider now the mapping

$$y_E^{(l)} : E \ni x \mapsto yx \in E,$$

i.e. the left multiplication of E by y. Then obviously we have $y_E^{(l)} \in K$. For $r \in R$ let

$$r_E^{(r)} : E \ni x \mapsto xr \in E,$$

then it follows that $r_E^{(r)} \in \operatorname{End}({}_K E)$, since for $k \in K$ we have $k(xr) = (kx)r$.

It is now to be shown that

$$\Phi : R \ni r \mapsto r_E^{(r)} \in \operatorname{End}({}_K E)$$

is a ring isomorphism.

First of all Φ is obviously a ring homomorphism. Since $\mathrm{Ker}(\Phi)$ is a two-sided ideal in R, which, as $1 \notin \mathrm{Ker}(\Phi)$, is not equal to R and since R is simple, it follows that $\mathrm{Ker}(\Phi) = 0$, thus Φ is a monomorphism. There remains to be shown: Φ is an epimorphism. As $E \neq 0$ and since RE is a two-sided ideal it follows that $RE = R$, which yields

$$(1) \qquad \Phi(R) = \Phi(RE) = \Phi(R)\Phi(E).$$

It is further to be shown that $\Phi(E)$ is a right ideal in $R'' := \mathrm{End}({}_K E)$. Let $\xi \in R''$ and let $x, y \in E$, then

$$y(x_E^{(r)}\xi) = (yx)\xi = (y_E^{(l)}x)\xi = y_E^{(l)}(x\xi) = y(x\xi) = y(x\xi)_E^{(r)}$$

hence

$$x_E^{(r)}\xi = (x\xi)_E^{(r)} \in \Phi(E)$$

and so

$$(2) \qquad \Phi(E)R'' = \Phi(E).$$

Finally as $\Phi(R) \hookrightarrow R''$ and $1_E^{(r)} = 1_E \in \Phi(R)$ we have:

$$(3) \qquad R'' = \Phi(R)R''.$$

From (1), (2), (3) it then follows that

$$R'' = \Phi(R)R'' = \Phi(R)\Phi(E)R'' = \Phi(R)\Phi(E) = \Phi(R),$$

thus in fact $\Phi(R) = R''$.

Since R is simple and $R \cong R''$, R'' must be simple. By 8.3.1 it then follows that $\dim_K(E) < \infty$, by which all is proved. □

Besides this direct proof we obtain a second proof as a corollary of the Density Theorem in the next section.

We formulate the main contents of Theorems 8.2.4 and 8.3.2 once more in a somewhat different form:

8.3.3 COROLLARY. *A semisimple ring (with unit element) is a direct sum of simple rings, which mutually annihilate one another and every one of which is isomorphic to a complete finite-dimensional matrix ring over a skew field.*

8.3.4 COROLLARY. *Let R be a simple ring with a simple right ideal E and let R be a finite-dimensional algebra over a field H. Then there exists a subfield $K_0 \hookrightarrow K := \mathrm{End}(E_R)$ with $\dim_{K_0}(K) < \infty$ and which is isomorphic to H.*

If H is algebraically closed then we have $H \cong K = \mathrm{End}(E_R)$.

Proof. For $h \in H$ let

$$h_E^{(r)}: E \ni x \mapsto xh \in E,$$

then, as $(xh)r = (xr)h$ for $r \in R$, it follows that $h_E^{(r)} \in K$ and therefore

$$\psi: H \ni h \mapsto h_E^{(r)} \in K$$

is a ring homomorphism. Let $K_0 := \operatorname{Im}(\psi)$. By assumption R_H is finite-dimensional and so also is E_H.

As $h_E^{(r)}x = xh$ for $h \in H$, $x \in H$, a basis of E_H over H is also a basis of ${}_{K_0}E$ over K_0. Hence ${}_{K_0}E$ is finite-dimensional and consequently ${}_{K_0}K$ must also be finite-dimensional (for $\dim_{K_0}(K) \cdot \dim_K(E) = \dim_{K_0}(E)$).

Since K is a finite algebraic extension field of K_0 it follows, in the case that H and thus also K_0 are algebraically closed, that $K_0 = K$, thus $H \cong K_0 = K$. □

8.4 THE DENSITY THEOREM

In our considerations so far we have mostly taken as a basic start a right R-module M_R and have written the R-homomorphisms on the left side of the arguments from M. Let $S := \operatorname{End}(M_R)$ be the endomorphism ring of M_R, then in particular M can be considered as an S-R-bimodule. This convention is in fact appropriate for many considerations, but not for all. In particular not for such considerations, as is the case in the following, in which initially we are provided with an abelian group M and the ring $T := \operatorname{End}(M_{\mathbb{Z}})$ of endomorphisms of M.

In order to show how the previously employed convention may be used in the following and to show what is the importance of the following results, we make some remarks, in which at first nothing further is assumed.

8.4.1 *Definition.* Let R resp. R° be a ring with the multiplication resp. $\circ$. R° is called the *inverse ring* to R :⇔

(1) The additive group of R is equal to the additive group of R° and

(2) $\forall r, s \in R[r \cdot s = s \circ r]$.

8.4.2 REMARKS

(a) *There is exactly one ring R° inverse to R.*

(b) $R^{\circ\circ} = R$.

(c) *R is commutative* $\Leftrightarrow R = R^\circ$.

Proof. (a) Existence of R°: Define $(R^\circ, +) := (R, +)$ together with

$$s \circ r := r \cdot s, \qquad r, s \in R,$$

then R° is an inverse ring to R.

Uniqueness: Let R^* with the multiplication $*$ be also an inverse ring to R, then it follows by definition that

$$(R^*, +) = (R, +) = (R^\circ, +).$$

Further we have $s * r = r \cdot s = s \circ r$, $r, s \in R$, thus $R^* = R^\circ$.
(b) and (c) may be left to the reader as an exercise. □

From the definition it follows further that all properties of R are carried over to R° on interchanging the sides.

8.4.3 REMARK. *Let $M = M_R$. By means of the definition*

$$r \circ m := mr \quad \text{for} \quad m \in M \quad \text{and} \quad r \in R^\circ$$

M becomes a left R°-module ${}_{R^\circ}M$. Precisely those additive subgroups of M, which are also submodules of M_R, are also the submodules of ${}_{R^\circ}M$.

Proof. For the proof of $M = {}_{R^\circ}M$ we confine ourselves to the associativity law.

$$\begin{aligned} r_1 \circ (r_2 \circ m) &= r_1 \circ (mr_2) = (mr_2)r_1 \\ &= m(r_2 r_1) = (r_2 r_1) \circ m \\ &= (r_1 \circ r_2) \circ m \quad \text{for all} \quad r_1, r_2 \in R, m \in M. \end{aligned}$$

Let $U \hookrightarrow M_R$. Then

$$r \circ U = Ur \subset U \quad \text{for all} \quad r \in R \Rightarrow U \hookrightarrow {}_{R^\circ}M.$$

In the same way it follows that: $U \hookrightarrow {}_{R^\circ}M \Rightarrow U \hookrightarrow M_R$.
All properties of M_R carry over accordingly to ${}_{R^\circ}M$ (on interchanging the sides). □

Since we have clarified the significance of the change of sides, we shall now assume that $M = {}_RM$. Further let $T := \operatorname{End}(M_{\mathbb{Z}})$ (=ring of all group endomorphisms of M) in which the endomorphisms are to be applied on the left, so that we have $M = {}_TM$. For every $r \in R$ the left multiplication

$$r^{(l)}: M \ni x \mapsto rx \in M$$

is then an element from T and the mapping

$$\psi : R \ni r \mapsto r^{(l)} \in T$$

is, as is directly verified, a ring homomorphism.

$R^{(l)} := \mathrm{Im}(\psi)$ is called the ring of left multiplications of the module $_RM$. $\mathrm{Ker}(\psi)$ is a two-sided ideal in R and consists of all $r \in R$ with $rM = 0$.

8.4.4 *Definition.* The module M is called *faithful* $:\Leftrightarrow$

$$\forall r \in R[rM = 0 \Rightarrow r = 0] \Leftrightarrow \mathrm{Ker}(\psi) = 0.$$

In the case of a faithful module we can identify R with $R^{(l)}$ so that $R \hookrightarrow T$ holds.

8.4.5 *Definition.* Let T be an arbitrary ring and let $A \subset T$ (A subset of T). Then

$$\mathrm{Cen}_T(A) := \{t \mid t \in T \wedge \forall a \in A[at = ta]\}$$

is called the *centralizer* of A in T.

As we see immediately, $\mathrm{Cen}_T(A)$ is a unitary subring of T and $\mathrm{Cen}_T(T)$ is the centre of T.

8.4.6 LEMMA. *Let* $M = {}_RM$, $T := \mathrm{End}(M_{\mathbb{Z}})$, $S := \mathrm{End}({}_RM)$ (*all applied on the left*), *then*:

(a) $S = R' := \mathrm{Cen}_T(R^{(l)})$.

(b) $R^{(l)} \subset R'' := \mathrm{Cen}_T(\mathrm{Cen}_T(R^{(l)}))$.

(c) $R' = R''' := \mathrm{Cen}_T(\mathrm{Cen}_T(\mathrm{Cen}_T(R^{(l)})))$.

Proof. (a) $S \hookrightarrow \mathrm{Cen}_T(R^{(l)})$: Let $\sigma \in S$, then we have for all $r \in R$, $x \in M$: $\sigma(rx) = r(\sigma x)$, thus $\sigma r^{(l)} = r^{(l)}\sigma \Rightarrow \sigma \in \mathrm{Cen}_T(R^{(l)})$. $\mathrm{Cen}_T(R^{(l)}) \hookrightarrow S$. Let $\tau \in \mathrm{Cen}_T(R^{(l)}) \Rightarrow \tau r^{(l)} = r^{(l)}\tau$ for all $r \in R \Rightarrow \tau(rx) = r(\tau x) \Rightarrow \tau \in S$.

(b) and (c) follow by the definition of centralizer. □

On account of this situation the interesting question arises as to the assumptions under which $R^{(l)} = R''$ holds and as to the relationships which exist in case $R^{(l)} \neq R''$ between $R^{(l)}$ and R''. We observe moreover that $R^{(l)}$ and R'' evidently depend on $M = {}_RM$ which is not apparent from the notation.

8.4.7 *Examples*

(1) If $M = {}_RM \neq 0$ is a free R-module then $_RM$ is a faithful R-module and we have $R^{(l)} = R''$. Prove as an exercise.

(2) Let ${}_RM = {}_{\mathbb{Z}}\mathbb{Q}$, then we have $\mathbb{Z} \cong \mathbb{Z}^{(l)}$ and $S = \mathbb{Z}' = \mathbb{Z}'' \cong \mathbb{Q}$. Prove as an exercise.

(3) Let $V = V_K$ be an infinite-dimensional vector space. Let R be the subring of $T := \mathrm{End}(V_K)$ which is generated by the identity mapping of V and by all linear mappings of V_K of finite rank. We claim:

(a) $V = {}_RV$ is a simple R-module.

(b) $R' = \mathrm{End}({}_RV) = K^{(r)} (\cong K)$.

(c) $R^{(l)} = R \neq R'' = \mathrm{End}(V_{R'})$.

(d) For any finitely many elements $v_1, \ldots, v_t \in V$ and $\sigma \in R''$ there is an $r \in R$ with $\sigma v_i = r v_i$, $i = 1, \ldots, t$.

Proof of (a), (b), (c) is an exercise for the reader; (d) is a special case of the Density Theorem to follow.

8.4.8 *Definition.* Let R and S be rings and let as well ${}_RM$ and ${}_SM$ be modules with the same additive group. ${}_RM$ is called *dense* in ${}_SM$:$\Leftrightarrow$for any finitely many elements $x_1, \ldots, x_t \in M$ and $s \in S$ there is an $r \in R$ with $s x_i = r x_i$, $i = 1, \ldots, t$.

8.4.9 THEOREM. *Every semisimple module ${}_RM$ is dense in ${}_{R''}M$.*

Proof. The proof follows in three steps.

(1) First let $N = {}_RN$ be an arbitrary module with U a direct summand of N, thus $N = U \oplus N_1$. Let now $R'' = R''_N$ be the double centralizer of R with respect to N. We claim that $R''U = U$, i.e. U is an R''-submodule of ${}_{R''}N$. For the proof let π be the projection of N onto U, and η the inclusion of U in N, then it follows that $\eta\pi \in R' = \mathrm{Hom}_R(N, N)$ and $\mathrm{Im}(\eta\pi) = U$. For $r'' \in R''$ and $u \in U$ we obtain therefore

$$r''u = r''\eta\pi(u) = \eta\pi r''(u) = \eta\pi(r''u) \in U,$$

which was to be shown.

Let now N be semisimple and let $x \in N$, then Rx is a direct summand in N and it follows (for $U = Rx$) that $Rx = R''Rx$. Since, by 8.4.6(b), we also have $R''Rx = R''x$ we deduce that $Rx = R''x$. Thus to every $x \in N$ and $r'' \in R''$ there is an $r_0 \in R$ with $r_0 x = r'' x$.

(2) Let now $M = {}_RM$ be semisimple and let $N := \coprod_{i=1}^{n} M_i$ with $M_i = M$ for $i = 1, \ldots, n$. Then we have (see Chapter 4)

$$N = \coprod_{i=1}^{n} M_i = \bigoplus_{i=1}^{n} M_i' \quad \text{with} \quad M_i' \cong M_i = M.$$

Consequently N is a direct sum of the semisimple modules M_i' and therefore (by 8.1.5) is itself again semisimple. Let now the double centralizer of R with regard to M resp. N be denoted by R''_M resp. R''_N.

Assertion. For $r'' \in R''_M$ and $(x_1, \ldots, x_n) \in N$ by means of the definition

$$\hat{r}''(x_1 \ldots x_n) := (r''x_1 \ldots r''x_n)$$

N becomes an R''_M-module and the mapping

$$\hat{r}'' : N \ni (x_1 \ldots x_n) \mapsto (r''x_1 \ldots r''x_n) \in N$$

is an element of R''_N. (Indeed $R''_M \ni r'' \mapsto \hat{r}'' \in R''_N$ is a ring monomorphism.) The module property is clear. It remains to be shown that $\hat{r}'' \in R''_N$. Let

$$\pi_i : N \to M_i \text{ resp. } \eta_i : M_i \to N$$

be the projection resp. the inclusion (see Chapter 4), then we have

$$r''\pi_i(x_1 \ldots x_n) = r''x_i = \pi_i\hat{r}''(x_1 \ldots x_n),$$

$$\hat{r}''\eta_i x_i = \hat{r}''(0 \ldots 0 x_i 0 \ldots 0) = \eta_i r'' x_i,$$

i.e.

$$r''\pi_i = \pi_i\hat{r}'' \quad \text{and} \quad \hat{r}''\eta_i = \eta_i r''.$$

As $\sum_{i=1}^{n} \eta_i\pi_i = 1_N$ we have for arbitrary $\varphi \in \mathrm{Hom}_R(N, N)$

$$\varphi = 1_N\varphi 1_N = \sum_{i=1}^{n}\sum_{j=1}^{n} \eta_i\pi_i\varphi\eta_j\pi_j,$$

in which (as $M_i = M$) $\pi_i\varphi\eta_j \in \mathrm{Hom}_R(M, M)$.

Hence it follows that

$$\hat{r}''\varphi = \hat{r}''\sum_i\sum_j \eta_i\pi_i\varphi\eta_j\pi_j = \sum_i\sum_j \eta_i r''(\pi_i\varphi\eta_j)\pi_j$$

$$= \sum_i\sum_j \eta_i(\pi_i\varphi\eta_j)r''\pi_j = \left(\sum_i\sum_j \eta_i\pi_i\varphi\eta_j\pi_j\right)\hat{r}'' = \varphi\hat{r}'',$$

which was to be shown.

(3) Let now $x_1, \ldots, x_n \in N$ and $r'' \in R''_M$ be given. By (1) applied to

$$N := \coprod_{i=1}^{n} M_i \quad \text{with} \quad M_i = M \quad \text{for} \quad i = 1, \ldots, n$$

and with $x = (x_1, \ldots, x_n) \in N$ and $\hat{r}'' \in R''_N$ ($\hat{r}''$, in the sense of (2) corresponding to r'') there is an $r_0 \in R$ with

$$r_0 x = (r_0 x_1 \ldots r_0 x_n) = \hat{r}'' x = (r''x_1 \ldots r''x_n);$$

thus we have

$$r_0 x_i = r'' x_i, \qquad i = 1, \ldots, n. \qquad \square$$

8.4.10 COROLLARY. *Let $_RM$ be simple and let M be finite-dimensional over $K = \mathrm{End}(_RM)$, then we have $R^{(l)} = R''$.*

Proof. Let $x_1, \ldots, x_n$ be a basis of $_KM$, then to every $\sigma \in R''$ there is an $r \in R$ with $\sigma x_i = r x_i$, $i = 1, \ldots, n$. Since σ and $r^{(l)}$ are linear mappings, it follows that $\sigma = r^{(l)}$, thus $R'' \hookrightarrow R^{(l)}$. Since, on the other hand, $R^{(l)} \hookrightarrow R''$ it follows $R^{(l)} = R''$. $\square$

8.4.11 COROLLARY. *Let $_RM$ be simple and $_RR$ be artinian. Then M is finite-dimensional over K and we have $R^{(l)} = R''$.*

Proof. By 8.4.10 we have only to show that $_KM$ is finite-dimensional. Suppose that that were not the case, then there would exist a countably infinite set of linearly independent elements in $_KM$:

$$x_1, x_2, x_3, \ldots.$$

Let

$$A_n = \{a \mid a \in R \wedge a x_1 = \ldots = a x_n = 0\},$$

then A_n is a left ideal in R. Since an $a_n \in A$ with $a_n x_{n+1} \neq 0$ exists (from 8.4.9), $A_n \supsetneqq A_{n+1}$ holds, and we would obtain the infinite chain of left ideals

$$A_1 \supsetneqq A_2 \supsetneqq A_3 \supsetneqq \ldots$$

which contradicts the fact that $_RR$ is artinian. $\square$

As a corollary from 8.4.9 we prove once more the structure theorem for simple rings (8.3.2).

8.4.12 COROLLARY. *Let R be a simple ring with a simple left ideal. Then R is isomorphic to the endomorphism ring of a finite-dimensional vector space over a skew field.*

Proof. As established at the beginning of 8.3, R is semisimple and hence by 8.1.6 (two-sided) artinian. Let $_RM$ be a simple left ideal in R. Then

$$\psi : R \to R_M^{(l)}$$

is an isomorphism, since $\mathrm{Ker}(\psi)$ as a two-sided ideal in a simple ring must be equal to 0, for $1 \notin \mathrm{Ker}(\psi)$. By 8.4.11 the assertion follows. $\square$

EXERCISES

(1)

(a) Let p be a prime number and let $n \in \mathbb{N}$. Which is the largest semisimple $\mathbb{Z}$-submodule of $\mathbb{Z}/p^n\mathbb{Z}$?

(b) Which is the smallest $\mathbb{Z}$-submodule U of $\mathbb{Z}/p^n\mathbb{Z}$, so that $(\mathbb{Z}/p^n\mathbb{Z})/U$ is semisimple?

(c) Give an example of a module M and a $U \hookrightarrow M$ so that M is not semisimple but M/U and U are semisimple.

(2)

Let R be a ring and let $Jk(R)$ denote the number of isomorphism classes of simple right R-modules. (In the class of all simple right R-modules $\cong$ is an equivalence relation; the isomorphism classes are the equivalence classes with respect to $\cong$).

(a) For every $n \in \mathbb{N}$ give an example of a ring R with $Jk(R) = n$.

(b) Give an example of a ring R with $Jk(R) = \infty$.

(c) Does the case $Jk(R) = 0$ occur?

(3)

Let e be an idempotent element of a ring R. Show:

(a) $\mathrm{End}(eR_R) \cong eRe$.

(b) Let R be simple and let eRe be a skew field, then eR is a simple right ideal of R.

(4)

Let R_i, $i = 1, \ldots, n$ be rings and let $R := \prod_{i=1}^{n} R_i$ with componentwise addition and multiplication.

(a) Show: R is a semisimple ring $\Leftrightarrow \forall i = 1, \ldots, n$. [$R_i$ is semisiple.]

(b) Does (a) hold also for infinite products?

(5)

(a) Let V_K be a vector space of countably infinite dimension. Show: The ideal of all endomorphisms of V_K of finite rank is the only proper two-sided ideal $\neq 0$ in $\mathrm{End}(V_K)$.

(b) Does (a) also hold if the dimension of V is greater than countably infinite?

(6)

Let $M = M_R$ be semisimple and let $S := \mathrm{End}(M_R)$. Show: ${}_SM$ is semisimple.

(7)

Let M_R be a semisimple R-module with only finitely many homogeneous components:

$$M_R = \bigoplus_{j=1}^{n} B_j.$$

Show:

(a) $S := \mathrm{End}(M_R) = \bigoplus_{j=1}^{n} S_j$, where the S_j are two-sided ideals in S and $S_j \cong \mathrm{End}(B_{jR})$.

(b) If M_R is finitely generated, then S is semisimple.

(c) If M_R is finitely generated and all simple submodules are isomorphic, then S is simple and semisimple.

(d) If M_R is not finitely generated, then S is neither simple nor semisimple.

(8)

Let $K := \mathbb{R}(x)$ be the field of rational functions in x with real coefficients, and let for $k \in K$

$$k' := \frac{\mathrm{d}}{\mathrm{d}x}(k)$$

be the usual derivative. Further let $R := K[y]$ be the additive group of all polynomials in y with coefficients in K. Define in $K[y]$ a (non-commutative) multiplication by induction over $n = 0, 1, 2, \ldots$ for fixed $m = 0, 1, 2, \ldots$

$$(ay^0)(by^m) := aby^m, \qquad a, b \in K,$$

$$(ay^n)(by^m) := ay^{n-1}(by^{m+1} + b'y^m) \qquad \text{for } n > 0,$$

and further require that the associative and distributive laws hold. Show:

(a) R is a simple ring. (Hint: If a polynomial of degree n with $n \geq 1$ lies in a two-sided ideal then so also does a polynomial of degree $n-1$.)

(b) R contains no simple right or left ideal, thus R is not semisimple.

(9)

Prove the assertions in 8.4.7.

(10)

Show for a module M_R:

(a) M is semisimple $\Leftrightarrow M$ has no large proper submodule.

(b) Let M be finitely generated. Then we have: M is semisimple $\Leftrightarrow M$ has no large maximal submodule.

(c) Construct a non-semisimple module which possesses no large maximal submodule.

Chapter 9

Radical and Socle

In the historical development of the theory of rings it had already been early established that in every finite-dimensional algebra A a two-sided nilpotent ideal B exists such that A/B is a semisimple algebra. (B is called nilpotent if $B^n = 0$ for some natural number n.)

This result yields three avenues for the investigation of A:

(1) the investigation of the semi-simple algebra A/B (for which the theory of semisimple algebras is at our disposal);

(2) the investigation of the nilpotent ideal B;

(3) the investigation of the relation between A/B and A, which is given by the epimorphism $A \to A/B$; in particular the question arises as to whether properties of A/B can be "lifted" to A.

Since the formulation of these questions was very fruitful for the investigation of algebras, the desire arises of having at our disposal an object corresponding to B in an arbitrary ring or module. We cannot enter here into the interesting historical development of this question. It would lead in any case to current concepts of the radical which are to be developed in this paragraph. The radical of a module M_R, denoted by $\text{Rad}(M_R)$, is accordingly the intersection of all maximal submodules of M_R or is equal to the sum of all small submodules of M_R. In consequence we then have $\text{Rad}(M/\text{Rad}(M)) = 0$ and $\text{Rad}(M)$ is contained in every submodule $U \hookrightarrow M$ with $\text{Rad}(M/U) = 0$. The three possibilities, listed above, have also to be reconsidered if $M/\text{Rad}(M)$ is in general no longer semisimple.

The concept dual to that of the radical is the socle. The socle of the module M_R, denoted by $\text{Soc}(M_R)$, is the sum of all minimal (=simple) submodules of M_R and therefore is the largest semisimple submodule of M_R. It is equal to the intersection of all large submodules of M_R.

9.1 DEFINITION OF RADICAL AND SOCLE

9.1.1 THEOREM. *Let $M = M_R$ be given. Then we have*

(a)

$$\sum_{A \overset{\circ}{\hookrightarrow} M} A = \bigcap_{\substack{B \hookrightarrow M \\ \text{maximal } B}} B = \bigcap_{\substack{\text{semisimple } N_R \\ \varphi \in \operatorname{Hom}_R(M,N)}} \operatorname{Ker}(\varphi).$$

(b)

$$\bigcap_{A \overset{*}{\hookrightarrow} M} A = \sum_{\substack{B \hookrightarrow M \\ \text{minimal } B \\ (=\text{simple } B)}} B = \sum_{\substack{\text{semisimple } N_R \\ \varphi \in \operatorname{Hom}_R(N,M)}} \operatorname{Im}(\varphi).$$

Proof. (a) In the order written the submodules of M, for which the equality is to be shown, are denoted by U_1, U_2, U_3.

"$U_2 \hookrightarrow U_1$": Let $a \in U_2$. Suppose aR were not a small submodule of M, then there would be by 5.1.4 a maximal submodule C of M with $a \notin C$, thus $a \notin U$. Consequently aR is small and hence $a \in aR \hookrightarrow U_1$.

"$U_3 \hookrightarrow U_2$": Let B be maximal in M and let $\nu_B : M \to M/B$ be the natural epimorphism onto the simple module M/B. Then $\operatorname{Ker}(\nu_B) = B$ and it follows that

$$U_3 \subset \bigcap_{\substack{B \hookrightarrow M \\ \text{maximal } B}} \operatorname{Ker}(\nu_B) = \bigcap_{\substack{B \hookrightarrow M \\ \text{maximal } B}} B = U_2.$$

"$U_1 \hookrightarrow U_3$": By 5.1.3(c) we have $A \overset{\circ}{\hookrightarrow} M \Rightarrow \varphi(A) \overset{\circ}{\hookrightarrow} N$ for every homomorphism $\varphi : M \to N$. If N is semisimple, then 0 is the unique small submodule of N, then we must have $\varphi(A) = 0$, i.e. $A \hookrightarrow \operatorname{Ker}(\varphi)$ holds. Consequently we have $U_1 \hookrightarrow U_3$.

(b) Let the submodules again be denoted in order by U_1, U_2, U_3.

"$U_2 \hookrightarrow U_1$": If B is a simple submodule of M and $A \overset{*}{\hookrightarrow} M$. Then $A \cap B \neq 0$ so $A \cap B = B$, $B \hookrightarrow A$ and hence $U_2 \hookrightarrow U_1$.

"$U_3 \hookrightarrow U_2$": Since the image of a semisimple module under a homomorphism is again semisimple and likewise so also the sum of semisimple modules (8.1.5), U_3 is a semisimple submodule of M, thus is the sum of simple submodules of M. Consequently we have $U_3 \hookrightarrow U_2$, since U_2 is the sum of all simple submodules of M.

"$U_1 \hookrightarrow U_3$": We claim that U_1 is semisimple. Let $C \hookrightarrow U_1$ and let C' be inco of C in M, then we have $C + C' = C \oplus C' \overset{*}{\hookrightarrow} M$ (5.2.5), thus $U_1 \hookrightarrow C + C'$. By the modular law (note $C \hookrightarrow U_1$) it follows that $U_1 = C \oplus (C' \cap U_1)$, thus U_1 is semisimple. Let $\iota : U_1 \to M$ be the inclusion, then it follows that $U_1 = \operatorname{Im}(\iota) \hookrightarrow U_3$. □

9.1.2 *Definition*

(1) The submodule of M defined in 9.1.1(a) is called the *radical* of M and is denoted by $\operatorname{Rad}(M)$.

(2) The submodule of M defined in 9.1.1(b) is called the *socle* of M and is denoted by $\operatorname{Soc}(M)$.

9.1.3 COROLLARY

(a) *For* $m \in M_R$ *we have*: $mR \overset{\circ}{\hookrightarrow} M \Leftrightarrow m \in \operatorname{Rad}(M)$.

(b) $\operatorname{Soc}(M)$ *is the largest semisimple submodule of* M.

Proof. (a) $mR \overset{\circ}{\hookrightarrow} M \Rightarrow m \in mR \hookrightarrow \operatorname{Rad}(M)$ by 9.1.1. The converse, $m \in \operatorname{Rad}(M) \Rightarrow mR \overset{\circ}{\hookrightarrow} M$, was shown in the proof of 9.1.1(a) with regard to "$U_2 \hookrightarrow U_1$".

(b) By definition $\operatorname{Soc}(M)$ is semisimple as the sum of simple submodules. Let C be a semisimple submodule of M, then C is contained in $\operatorname{Soc}(M)$ being the image of the inclusion $\iota: C \to M$, thus $\operatorname{Soc}(M)$ is the largest semisimple submodule of M. □

We come now to the main theorem on the radical and socle.

9.1.4 THEOREM

(a) $\varphi \in \operatorname{Hom}_R(M, N) \Rightarrow \varphi(\operatorname{Rad}(M)) \hookrightarrow \operatorname{Rad}(N) \wedge \varphi(\operatorname{Soc}(M)) \hookrightarrow \operatorname{Soc}(N)$

(b) $\operatorname{Rad}(M/\operatorname{Rad}(M)) = 0 \wedge \forall C \hookrightarrow M[\operatorname{Rad}(M/C) = 0 \Rightarrow \operatorname{Rad}(M) \hookrightarrow C]$ i.e. $\operatorname{Rad}(M)$ *is the smallest submodule of* M *with* $\operatorname{Rad}(M/C) = 0$.

(c) $\operatorname{Soc}(\operatorname{Soc}(M)) = \operatorname{Soc}(M) \wedge \forall C \hookrightarrow M\ [\operatorname{Soc}(C) = C \Rightarrow C \hookrightarrow \operatorname{Soc}(M)]$ *i.e.* $\operatorname{Soc}(M)$ *is the largest submodule which coincides with its socle.*

Proof. (a) From $\operatorname{Rad}(M) = \sum_{A \overset{\circ}{\hookrightarrow} M} A$ it follows that $\varphi(\operatorname{Rad}(M)) = \sum_{A \overset{\circ}{\hookrightarrow} M} \varphi(A)$. As shown in 5.1.3 we have $\varphi(A) \overset{\circ}{\hookrightarrow} N$, thus it follows that $\varphi(\operatorname{Rad}(M)) \hookrightarrow \operatorname{Rad}(N)$. Since the image of a semisimple module is again semisimple, we have also $\varphi(\operatorname{Soc}(M)) \hookrightarrow \operatorname{Soc}(N)$.

(b) *Assertion.* The maximal submodules Δ of M/C are obtained as images of the maximal submodules $B \hookrightarrow M$ with $C \hookrightarrow B$ by $\nu: M \to M/C$.

Proof. See 3.1.13 or directly as follows. $\nu\nu^{-1}(\Delta) = \Delta \cap \operatorname{Im}(\nu) = \Delta$. Let $B := \nu^{-1}(\Delta)$. Then $\nu(B) = \Delta \wedge C \hookrightarrow B \hookrightarrow M$. Since Δ is maximal $(M/C)/\Delta = (M/C)/(B/C) \cong M/B$ simple, thus B is maximal in M.

Assertion. If $(B_i | i \in I)$ is a family of submodules of $M \wedge \forall i \in I[C \hookrightarrow B_i]$, then we have

$$\bigcap_{i\in I} (B_i/C) = \Big(\bigcap_{i\in I} B_i\Big)\Big/C.$$

Proof. It is clear that $(\bigcap B_i)/C \hookrightarrow \bigcap(B_i/C)$. Let now $v + C \in \bigcap(B_i/C)$. Then for every i there is a $b_i \in B_i$ with $v + C = b_i + C$, so $v = b_i + c_i \in B_i + C = B_i$ for all $i \in I$ hence $v + C \in (\bigcap B_i)/C$.

We now apply the two statements established above.

$$\mathrm{Rad}(M/\mathrm{Rad}(M)) = \bigcap_{\max \Delta \text{ in } M/\mathrm{Rad}(M)} \Delta = \bigcap_{\substack{\max B \hookrightarrow M\\ \mathrm{Rad}(M)\hookrightarrow B}} (B/\mathrm{Rad}(M))$$

$$= \Big(\bigcap_{\substack{\max B \hookrightarrow M\\ \mathrm{Rad}(M)\hookrightarrow B}} B\Big)\Big/\mathrm{Rad}(M) = \Big(\bigcap_{\max B \hookrightarrow M} B\Big)\Big/\mathrm{Rad}(M)$$

$$= \mathrm{Rad}(M)/\mathrm{Rad}(M) = 0.$$

Let now $C \hookrightarrow M \wedge \mathrm{Rad}(M/C) = 0$, then it follows for the mapping $\nu: M \to M/C$ by (a) that

$$\nu(\mathrm{Rad}(M)) \hookrightarrow \mathrm{Rad}(M/C) = 0$$

and consequently

$$\mathrm{Rad}(M) \hookrightarrow \mathrm{Ker}(\nu) = C.$$

(c) A semisimple module coincides with its socle. Since $\mathrm{Soc}(M)$ is the largest semisimple submodule of M, it is hence clear that $\mathrm{Soc}(\mathrm{Soc}(M)) = \mathrm{Soc}(M)$. Let $\mathrm{Soc}(C) = C$, then C is semisimple and it follows that $C \hookrightarrow \mathrm{Soc}(M)$. □

The properties (a), (b), (c) of this theorem can be formulated functorially and motivate the definition of *preradical* ((a)), *radical* ((a) and (b)) and *socle* ((a) and (c)) in *categories.*

9.1.5 Corollaries

(a) *Epimorphism* $\varphi: M \to N \wedge \mathrm{Ker}(\varphi) \overset{\circ}{\hookrightarrow} M \Rightarrow \varphi(\mathrm{Rad}(M)) = \mathrm{Rad}(N) \wedge \mathrm{Rad}(M) = \varphi^{-1}(\mathrm{Rad}(N)).$

Monomorphism $\varphi: M \to N \wedge \mathrm{Im}(\varphi) \overset{*}{\hookrightarrow} N \Rightarrow \varphi(\mathrm{Soc}(M)) = \mathrm{Soc}(N) \wedge \mathrm{Soc}(M) = \varphi^{-1}(\mathrm{Soc}(N)).$

(b) $C \hookrightarrow M \Rightarrow \mathrm{Rad}(C) \hookrightarrow \mathrm{Rad}(M) \wedge \mathrm{Soc}(C) \hookrightarrow \mathrm{Soc}(M).$

(c) $M = \bigoplus_{i\in I} M_i \Rightarrow \mathrm{Rad}(M) = \bigoplus_{i\in I} \mathrm{Rad}(M_i) \wedge \mathrm{Soc}(M) = \bigoplus_{i\in I} \mathrm{Soc}(M).$

(d) $M = \bigoplus_{i\in I} M_i \Rightarrow M/\mathrm{Rad}(M) \cong \bigoplus_{i\in I} (M_i/\mathrm{Rad}(M_i)).$

Proof. (a) $\varphi(\text{Rad}(M)) \hookrightarrow \text{Rad}(N)$ holds by 9.1.4. Now let $U \overset{\circ}{\hookrightarrow} N$ and for $A \hookrightarrow M$ let $A + \varphi^{-1}(U) = M$.

Since φ is an epimorphism it then follows that $\varphi(A) + U = N$, thus $\varphi(A) = N$ and consequently

$$A + \text{Ker}(\varphi) = M.$$

As $\text{Ker}(\varphi) \overset{\circ}{\hookrightarrow} M$ we obtain $A = M$, i.e. $\varphi^{-1}(U) \overset{\circ}{\hookrightarrow} M \Rightarrow \varphi^{-1}(U) \hookrightarrow \text{Rad}(M) \Rightarrow \varphi(\varphi^{-1}(U)) = U \hookrightarrow \varphi(\text{Rad}(M))$, thus $\text{Rad}(N) \hookrightarrow \varphi(\text{Rad}(M))$, which was to be shown.

From $\varphi(\text{Rad}(M)) = \text{Rad}(N)$ it follows finally as $\text{Ker}(\varphi) \hookrightarrow \text{Rad}(M)$ that

$$\text{Rad}(M) = \text{Rad}(M) + \text{Ker}(\varphi) = \varphi^{-1}\varphi(\text{Rad}(M)) = \varphi^{-1}(\text{Rad}(N)).$$

For the socle we have on the other hand by 9.1.4 $\varphi(\text{Soc}(M)) \hookrightarrow \text{Soc}(N)$. Let now $E \hookrightarrow N$ be simple, then as $\text{Im}(\varphi) \overset{*}{\hookrightarrow} N$ we have: $E \hookrightarrow \text{Im}(\varphi) \Rightarrow \varphi^{-1}(E) \hookrightarrow \text{Soc}(M) \Rightarrow \varphi\varphi^{-1}(E) = E \hookrightarrow \varphi(\text{Soc}(M)) \Rightarrow \text{Soc}(N) \hookrightarrow \varphi(\text{Soc}(M))$. From $\varphi(\text{Soc}(M)) = \text{Soc}(N)$ it follows finally that

$$\text{Soc}(M) = \varphi^{-1}\varphi(\text{Soc}(M)) = \varphi^{-1}(\text{Soc}(N)).$$

(b) Let $\iota: C \to M$ be the inclusion, then it follows by 9.1.4 that

$$\text{Rad}(C) = \iota(\text{Rad}(C)) \hookrightarrow \text{Rad}(M) \wedge \text{Soc}(C) = \iota(\text{Soc}(C)) \hookrightarrow \text{Soc}(M).$$

(c) $\text{Rad}(M_i) \hookrightarrow \text{Rad}(M)$ from (b) hence

$$\sum_{i \in I} \text{Rad}(M_i) = \bigoplus_{i \in I} \text{Rad}(M_i) \hookrightarrow \text{Rad}(M).$$

Let now $m = \sum m_i \in \text{Rad}(M)$ and let $\pi_i: M \to M_i$ be the ith projection. Then $\pi_i(m) = m_i \in \text{Rad}(M_i)$ from 9.1.4 and so $m \in \bigoplus \text{Rad}(M_i)$. Hence $\text{Rad}(M) \hookrightarrow \bigoplus \text{Rad}(M_i)$ whence Assertion. Analogously for the socle.

(d) We exhibit explicitly an isomorphism

$$\varphi: M/\text{Rad}(M) \to \bigoplus_{i \in I} (M_i/\text{Rad}(M_i)).$$

Let $\sum m_i \in \bigoplus M_i$ with $m_i \in M_i$ be an arbitrary element from M, then let

$$\varphi((\textstyle\sum m_i) + \text{Rad}(M)) := \sum (m_i + \text{Rad}(M_i)) \in \bigoplus_{i \in I} (M_i/\text{Rad}(M_i)).$$

"φ is a mapping": Let $(\sum m_i) + \text{Rad}(M) = (\sum m_i') + \text{Rad}(M)$ with $m_i, m_i' \in M_i$, then it follows that $\sum(m_i - m_i') \in \text{Rad}(M)$ hence, by (c), $m_i - m_i' \in \text{Rad}(M_i)$ and therefore it follows that $m_i + \text{Rad}(M_i) = m_i' + \text{Rad}(M_i)$, thus

$$\sum(m_i + \text{Rad}(M_i)) = \sum(m_i' + \text{Rad}(M_i)).$$

"φ is a monomorphism": Let

$$\varphi((\sum m_i)+\mathrm{Rad}(M))=\sum(m_i+\mathrm{Rad}(M_i))=0,$$

then it follows that $m_i \in \mathrm{Rad}(M_i)$ for all occurring m_i. As $\mathrm{Rad}(M_i) \hookrightarrow \mathrm{Rad}(M)$ we deduce therefore that

$$(\sum m_i)+\mathrm{Rad}(M)=\mathrm{Rad}(M),$$

thus $\mathrm{Ker}(\varphi)=0$.

"φ is an epimorphism": Clear. □

Examples

(1) $\mathrm{Rad}(\mathbb{Z}_\mathbb{Z})=0$, since by 5.1.2 0 is the only small ideal in $\mathbb{Z}$. $\mathrm{Soc}(\mathbb{Z}_\mathbb{Z})=0$ for $\mathbb{Z}$ has no simple ideals.

(2) $\mathrm{Rad}(\mathbb{Q}_\mathbb{Z})=\mathbb{Q}$, since for every $q \in \mathbb{Q}$, $q\mathbb{Z}$ is small in $\mathbb{Q}$ (see 5.1.2). This is equivalent to saying that $\mathbb{Q}$ has no maximal submodules.

(3) Let $n \in \mathbb{Z}$, $n>1$ with the unique decomposition into powers of prime numbers

$$n=p_1^{m_1}\dots p_k^{m_k},\ p_i \neq p_j \quad \text{for} \quad i\neq j,\ m_i>0.$$

The maximal ideals of $\mathbb{Z}$ are the prime ideals generated by prime numbers. The maximal ideals which contain $n\mathbb{Z}$ are then the ideals $p_i\mathbb{Z}$, $i=1,\dots,k$, and we have

$$\bigcap_{i=1}^{k} p_i\mathbb{Z}=p_1\dots p_k\mathbb{Z}.$$

Hence we have

$$\mathrm{Rad}(\mathbb{Z}/n\mathbb{Z})=\left(\bigcap_{i=1}^{k} p_i\mathbb{Z}\right)\Big/ n\mathbb{Z}=p_1\dots p_k\mathbb{Z}/n\mathbb{Z}.$$

Therefore it follows that

$$\mathrm{Rad}(\mathbb{Z}/n\mathbb{Z})=0 \Leftrightarrow n=p_1\dots p_k.$$

Likewise for $n=0$ and $n=1$ we have $\mathrm{Rad}(\mathbb{Z}/n\mathbb{Z})=0$.

We now wish to determine $\mathrm{Soc}(\mathbb{Z}/n\mathbb{Z})$. This is equal to 0 for $n=0$ and $n=1$. Now again let $n>1$ with the decomposition into powers of primes as given above. First of all we establish: $\mathbb{Z}/n\mathbb{Z}$ is a simple $\mathbb{Z}$-module, if and only if n is a prime number. If namely $n=p$ is a prime number, then $\mathbb{Z}/p\mathbb{Z}$ (as a ring) is a field and hence is simple as a $\mathbb{Z}$-module. If n has at least one proper divisor q, then $q\mathbb{Z}/n\mathbb{Z}$ is a proper submodule $\neq 0$ of $\mathbb{Z}/n\mathbb{Z}$. As

$$\frac{n}{p_i}\mathbb{Z}/n\mathbb{Z}\cong\mathbb{Z}/p_i\mathbb{Z} \qquad (i=1,\dots,k)$$

the modules $\frac{n}{p_i}\mathbb{Z}/n\mathbb{Z}$ are simple submodules of $\mathbb{Z}/n\mathbb{Z}$. Then

$$\sum_{i=1}^{k}\frac{n}{p_i}\mathbb{Z}/n\mathbb{Z}=\left(\sum_{i=1}^{k}\frac{n}{p_i}\mathbb{Z}\right)\Big/n\mathbb{Z}=\frac{n}{p_1\ldots p_k}\mathbb{Z}/n\mathbb{Z}\hookrightarrow\operatorname{Soc}(\mathbb{Z}/n\mathbb{Z}).$$

On the other hand let $q\mathbb{Z}/n\mathbb{Z}$ with $n=qn_1$ be a simple submodule of $\mathbb{Z}/n\mathbb{Z}$. Since

$$q\mathbb{Z}/n\mathbb{Z}\cong\mathbb{Z}/n_1\mathbb{Z}$$

n_1 must then be one of the prime numbers $p_1,\ldots,p_k$, say p_i; thus $q=\frac{n}{p_i}$, and it follows that

$$\operatorname{Soc}(\mathbb{Z}/n\mathbb{Z})=\frac{n}{p_1\ldots p_k}\mathbb{Z}/n\mathbb{Z}.$$

We point out the following special cases:

$$\operatorname{Rad}(\mathbb{Z}/p_1\ldots p_k\mathbb{Z})=0,\qquad \operatorname{Soc}(\mathbb{Z}/p_1\ldots p_k\mathbb{Z})=\mathbb{Z}/p_1\ldots p_k\mathbb{Z},$$

$$\operatorname{Rad}(\mathbb{Z}/p^n\mathbb{Z})=p\mathbb{Z}/p^n\mathbb{Z}\cong\mathbb{Z}/p^{n-1}\mathbb{Z},\qquad \operatorname{Soc}(\mathbb{Z}/p^n\mathbb{Z})=p^{n-1}\mathbb{Z}/p^n\mathbb{Z}\cong\mathbb{Z}/p\mathbb{Z}.$$

9.2 FURTHER PROPERTIES OF THE RADICAL

We collect together several other properties of the radical in the following theorem.

9.2.1 THEOREM. *Let* $M=M_R$, *then we have*

(a) *M is semisimple* $\Rightarrow\operatorname{Rad}(M)=0$.

(b) $M\operatorname{Rad}(R_R)\hookrightarrow\operatorname{Rad}(M)$.

(c) *M is finitely generated* $\Rightarrow\operatorname{Rad}(M)\overset{\circ}{\hookrightarrow}M$, in particular $\operatorname{Rad}(R_R)\overset{\circ}{\hookrightarrow}R_R$.

(d) *M is finitely generated* $\wedge\, A\hookrightarrow\operatorname{Rad}(R_R)$ $(\Leftrightarrow A\overset{\circ}{\hookrightarrow}R_R)\Rightarrow MA\overset{\circ}{\hookrightarrow}M$ (*Nakayama's Lemma*).

(e) *M is finitely generated* $\wedge\, M\neq 0\Rightarrow\operatorname{Rad}(M)\neq M$.

(f) $\operatorname{Rad}(R_R)$ *is a two-sided ideal of* R.

(g) *For every projective module* P_R *we have*: $\operatorname{Rad}(P)=P\operatorname{Rad}(R_R)$.

(h) $C\hookrightarrow M\Rightarrow C+\operatorname{Rad}(M)/C\hookrightarrow\operatorname{Rad}(M/C)$.

Proof. (a) M is semisimple $\Rightarrow$ every submodule is a direct summand $\Rightarrow 0$ is the only small submodule $\Rightarrow\operatorname{Rad}(M)=0$.

(b) Let $m\in M$, then $\varphi_m\colon R_R\ni r\to mr\in M_R$ is a homomorphism. By 9.1.4 we have

$$m\operatorname{Rad}(R_R)=\varphi_m(\operatorname{Rad}(R_R))\hookrightarrow\operatorname{Rad}(M)$$

$$\Rightarrow\sum_{m\in M}m\operatorname{Rad}(R_R)=M\operatorname{Rad}(R_R)\hookrightarrow\operatorname{Rad}(M).$$

(c) Let $\mathrm{Rad}(M)+C=M$. Suppose $C\neq M$. Then, since M is finitely generated, C is contained (2.3.11) in a maximal submodule $B\hookrightarrow M$; hence $M=\mathrm{Rad}(M)+C\hookrightarrow B\lightning$. Thus we have $C=M$ and so $\mathrm{Rad}(M)\overset{\circ}{\hookrightarrow} M$.

(d) $MA\hookrightarrow M\ \mathrm{Rad}(R_R)\hookrightarrow \mathrm{Rad}(M)\overset{\circ}{\hookrightarrow} M\Rightarrow MA\overset{\circ}{\hookrightarrow} M$.

(e) Since $M\neq 0\wedge \mathrm{Rad}(M)\overset{\circ}{\hookrightarrow} M$ we have $\mathrm{Rad}(M)\neq M$, since from $\mathrm{Rad}(M)=M$ we should have $\mathrm{Rad}(M)+0=M$, thus $0=M$ would follow.

(f) This follows from (b) with $M_R=R_R$.

(g) Let (y_i, φ_i) be a "projective basis" in the sense of the Dual Basis Lemma (5.4.2). For $u\in \mathrm{Rad}(P)$ it then follows that $\varphi_i(u)\in \mathrm{Rad}(R_R)$ (by 9.1.4) and hence we have

$$u=\sum y_i\varphi_i(u)\in P\ \mathrm{Rad}(R_R),$$

thus $\mathrm{Rad}(P)\hookrightarrow P\ \mathrm{Rad}(R_R)$. Since by (b) the reverse inclusion also holds, the assertion follows.

(h) Let $\nu: M\to M/C$ be the natural epimorphism, then we have $C+\mathrm{Rad}(M)/C=\nu(\mathrm{Rad}(M))\hookrightarrow \mathrm{Rad}(M/C)$. □

We point out meantime that we need

(f) in order to prove in 9.3 that $\mathrm{Rad}(R_R)=\mathrm{Rad}({}_RR)$.

We now wish to show: If M is artinian then $M/\mathrm{Rad}(M)$ is semisimple. We deduce this from the following more general theorem.

9.2.2 THEOREM

(a) *Every submodule of M has an* adco *in M and* $\mathrm{Rad}(M)=0\Leftrightarrow M$ *is semisimple.*

(b) *M is artinian and* $\mathrm{Rad}(M)=0\Leftrightarrow M$ *is semisimple and M is finitely generated.*

Proof. (a) "$\Rightarrow$": Let $C\hookrightarrow M\wedge C$ adco of C in M. Then $M=C+C^{\cdot}\wedge C\cap C^{\cdot}\hookrightarrow \mathrm{Rad}(M)=0$. Then (by 5.2.4(a)) $M=C\oplus C^{\cdot}\Rightarrow M$ is semisimple.

(a) "$\Leftarrow$": Clear.

(b) "$\Rightarrow$": M is artinian$\Rightarrow$every submodule has an adco. By (a) it then follows that M is semisimple. Since M is semisimple and artinian, M is finitely generated (8.1.6).

(b) "$\Leftarrow$": Since M is semisimple and finitely generated, M is artinian (8.1.6). $\mathrm{Rad}(M)=0$ is clear. □

9.2.3 COROLLARY. *M is artinian$\Rightarrow M/\mathrm{Rad}(M)$ is semisimple. Special case: R_R is artinian$\Rightarrow R/\mathrm{Rad}(R_R)$ is semisimple.*

Proof. M is artinian$\Rightarrow M/\mathrm{Rad}(M)$ is artinian. Since $\mathrm{Rad}(M/\mathrm{Rad}(M))=0$ by 9.1.4(b), it follows by 9.2.2 that $M/\mathrm{Rad}(M)$ is semisimple. □

We remark further in the special case of $M_R = R_R$ being artinian that first of all $R/\mathrm{Rad}(R_R)$ is semisimple as a right R-module. Since $\mathrm{Rad}(R_R)$ is, by 9.2.1(f), a two-sided ideal, $\bar{R} := R/\mathrm{Rad}(R_R)$ is also as a ring right-sided semisimple. As we have earlier shown (8.2.1), ${}_{\bar{R}}\bar{R}$ is then also semisimple and consequently also ${}_R\bar{R}$. Hence by 9.1.4(b) we must have

$$\mathrm{Rad}({}_RR) \hookrightarrow \mathrm{Rad}(R_R).$$

From the basic symmetry the reverse inclusion also holds and equality then follows. This equality is proved in the next section for arbitrary rings.

9.3 THE RADICAL OF A RING

The main result of this section is the equation

$$\mathrm{Rad}(R_R) = \mathrm{Rad}({}_RR).$$

We lead up to the proof by means of a lemma.

9.3.1 LEMMA. *The following statements are equivalent for* $A \hookrightarrow R_R$.

(1) $A \overset{\circ}{\hookrightarrow} R_R$.

(2) $A \hookrightarrow \mathrm{Rad}(R_R)$.

(3) $\forall a \in A$ [$1-a$ *has a right inverse in* R].

(4) $\forall a \in A$ [$1-a$ *has an inverse in* R].

Proof. "(1)$\Rightarrow$(2)": By definition of the radical.

"(2)$\Rightarrow$(1)": By 9.2.1(c) we have $\mathrm{Rad}(R_R) \overset{\circ}{\hookrightarrow} R_R$, thus $A \overset{\circ}{\hookrightarrow} R_R$.

"(1)$\Rightarrow$(3)": For arbitrary $r \in R$ we have $ar + (1-a)r = r \Rightarrow A + (1-a)R = R \Rightarrow (1-a)R = R$ (since $A \overset{\circ}{\hookrightarrow} R_R$)$\Rightarrow$(3).

"(3)$\Rightarrow$(4)": Let $(1-a)r = 1$; then $r = 1 + ar = 1 - (-ar)$. Since $-ar \in A$, there exists $s \in R$ with $rs = (1-(-ar))s = 1$. Thus r has $1-a$ as left inverse and s as right inverse which then must coincide and it follows that $1 = rs = r(1-a)$, i.e. r is an inverse of $(1-a)$.

"(4)$\Rightarrow$(1)": Let $A + B = R_R$. Then $1 = a + b$ with $a \in A$, $b \in B$; i.e. $b = 1-a$ and so there exists r with $br = (1-a)r = 1$ hence $B = R$, i.e. $A \overset{\circ}{\hookrightarrow} R_R$. □

Remarks

(a) In the literature a right ideal with property (3) is also called *quasi-regular.*

(b) Evidently this lemma holds also "on the left side" i.e. if we interchange the right and left sides.

9.3.2 THEOREM. $\mathrm{Rad}(R_R) = \mathrm{Rad}({}_RR)$.

Proof. We apply the lemma to $A = \mathrm{Rad}(R_R)$. For it (4) then also holds. Since A is a two-sided ideal (9.2.1(f)) A is also a left ideal, and so (4) of the "left-sided" version of the lemma holds; thus it follows that

$$\mathrm{Rad}(R_R) \hookrightarrow \mathrm{Rad}({}_R R).$$

On the basis of symmetry the reverse inclusion also holds and the equality follows. □

9.3.3 *Definition.* $\mathrm{Rad}(R) := \mathrm{Rad}(R_R) = \mathrm{Rad}({}_R R)$.

In general $R/\mathrm{Rad}(R)$ is not semisimple; e.g. we have for $R = \mathbb{Z}$ since $\mathrm{Rad}(\mathbb{Z}) = 0$: $\mathbb{Z}/\mathrm{Rad}(\mathbb{Z}) = \mathbb{Z}/0 \cong \mathbb{Z}$ and $\mathbb{Z}$ is not semisimple. If however the case arises that $R/\mathrm{Rad}(R)$ is semisimple, then interesting statements can be made.

9.3.4 THEOREM. *If R is a ring such that $R/\mathrm{Rad}(R)$ is semisimple then we have:*

(a) *Every simple right resp. left R-module is isomorphic to a submodule of* $(R/\mathrm{Rad}(R))_R$ *resp.* ${}_R(R/\mathrm{Rad}(R))$.

(b) *The number of the blocks of $R/\mathrm{Rad}(R)$ is finite and equal to the number of the isomorphism classes of simple right R-modules and equal to the number of isomorphism classes of simple left R-modules.*

Proof. (a) Since every cyclic right R-module $M_R = mR$ is an epimorphic image of R_R, it follows that $M \cong R/A$ with $A \hookrightarrow R_R$. If now M_R is simple, then A must be maximal. Consequently $\mathrm{Rad}(R) \hookrightarrow A$ then holds and we obtain

$$M \cong R/A \cong (R/\mathrm{Rad}(R))/(A/\mathrm{Rad}(R)).$$

Since $\bar{R} := R/\mathrm{Rad}(R)$ is semisimple, $\bar{A} := A/\mathrm{Rad}(R)$ is a direct summand, thus

$$\bar{R}_R = \bar{A} \oplus \bar{B},$$

from which $M_R \cong \bar{B}_R$ follows. Analogously for the left side.

(b) As we know, the R-submodules of $\bar{R}_R$ coincide with the right ideals of $\bar{R}$, and two R-submodules are isomorphic if and only if they are isomorphic as right ideals of $\bar{R}$. The assertion then follows from 7.2.3 and 8.2.6. □

We had established in 9.2.1, that for an arbitrary module we have

$$M\,\mathrm{Rad}(R) \hookrightarrow \mathrm{Rad}(M).$$

Here we give a condition sufficient to ensure $M\,\mathrm{Rad}(R) = \mathrm{Rad}(M)$.

9.3.5 THEOREM. *If $R/\mathrm{Rad}(R)$ is a semisimple ring then we have for every module M_R:*

(1) $\mathrm{Rad}(M) = M\,\mathrm{Rad}(R)$.

(2) $\mathrm{Soc}(M) = l_M(\mathrm{Rad}(R)) := \{m \mid m \in M \wedge m\,\mathrm{Rad}(R) = 0\}$.

Proof. (1) Since $(M/M\,\mathrm{Rad}(R))\,\mathrm{Rad}(R) = 0$, $M/M\,\mathrm{Rad}(R)$ can be considered as an $R/\mathrm{Rad}(R)$-module, in which the R-submodules and the $R/\mathrm{Rad}(R)$ submodules are the same. As a module over the semisimple ring $R/\mathrm{Rad}\,R$, by 8.2.2 $M/M\,\mathrm{Rad}(R)$ is semisimple, thus we have by 9.2.1 (a) $\mathrm{Rad}(M/M\,\mathrm{Rad}(R)) = 0$. By 9.1.4(b) it follows therefore that $\mathrm{Rad}(M) \hookrightarrow M\,\mathrm{Rad}(R)$ and then 9.2.1 (b) implies (1).

(2) First of all from 9.2.1 (a) and (b) it follows that $\mathrm{Soc}(M) \hookrightarrow l_M(\mathrm{Rad}(R))$. On the other hand $l_M(\mathrm{Rad}(R))$ is semisimple as an $R/\mathrm{Rad}(R)$ module and hence also as an R-module. Thus we have also $l_M(\mathrm{Rad}(R)) \hookrightarrow \mathrm{Soc}(M)$. □

9.3.6 *Definition.* A right, left or two-sided ideal A of a ring R is called a *nil ideal* $:\Leftrightarrow \forall a \in A\, \exists n \in \mathbb{N}[a^n = 0]$, resp. *nilpotent ideal* $:\Leftrightarrow \exists n \in \mathbb{N}[A^n = 0]$.

9.3.7 COROLLARY

(a) *Every one-sided or two-sided nilpotent ideal is a nil ideal.*

(b) *The sum of two nilpotent right, left or two-sided ideals is again nilpotent.*

(c) *If R_R is noetherian then every two-sided nil ideal is nilpotent.*

Proof. (a) Clear.

(b) Let $A \hookrightarrow R_R$, $B \hookrightarrow R_R$ and $A^m = 0$, $B^n = 0$. We assert that $(A+B)^{m+n} = 0$. Let $a_i \in A$, $b_i \in B$, $i = 1, \ldots, m+n$, then by the Binomial Theorem

$$\prod_{i=1}^{m+n} (a_i + b_i)$$

is a sum of products of $m+n$ factors of which either at least m factors are from A or at least n factors are from B. Since A and B are right ideals the assertion follows.

(c) Let N be a two-sided nil ideal of R. Since R_R is noetherian, among the nilpotent right ideals contained in N there is a maximal one; let A be one such and suppose we have $A^n = 0$. By (b) A is indeed the largest nilpotent right ideal contained in N. Since for $x \in R$ xA is also a nilpotent right ideal contained in N, A is in fact a two-sided ideal. If for an element $b \in N$ we have: $(bR)^k \hookrightarrow A$, then it follows that $(bR)^{kn} = 0$, thus $bR \hookrightarrow A$.

We claim that $A = N$. Suppose $A \neq N$, then let $b \in N \backslash A$ (set-complement to A in N) be chosen so that

$$\boldsymbol{r}_R(b, A) := \{r | r \in R \wedge br \in A\}$$

is maximal. For an arbitrary $x \in R$ we then have $xb \in N$ as well as

$$\boldsymbol{r}_R(b, A) \hookrightarrow \boldsymbol{r}_R(xb, A),$$

since N and A are two-sided ideals. Consequently for $xb \notin A$ we must have

$$\boldsymbol{r}_R(b, A) = \boldsymbol{r}_R(xb, A).$$

For $xb \notin A$ let $(xb)^k \in A$ and $(xb)^{k-1} \notin A$ (k exists, since xb is nilpotent!), then it follows that

$$\boldsymbol{r}_R(b, A) = \boldsymbol{r}_R((xb)^{k-1}, A),$$

thus $bxb \in A$ and consequently $(bR)^2 \hookrightarrow A$, in which the two-sidedness of A for $xb \in A$ is used. As established at the beginning, it follows that $bR \hookrightarrow A$, thus $b \in A$ ↯. □

We now investigate the relation between the recently introduced concepts and the concepts of the radical.

9.3.8 THEOREM. *Every (one-sided or two-sided) nil ideal is contained in* $\mathrm{Rad}(R)$.

Proof. Let A be a nil right ideal and let $a \in A$, $a^n = 0$, then we have

$$(1 + a + \ldots + a^{n-1})(1 - a) = (1 - a)(1 + a + \ldots + a^{n-1})$$
$$= 1 - a^n = 1,$$

i.e. $1 - a$ has an inverse element. By Lemma 9.3.1 it follows that $A \hookrightarrow \mathrm{Rad}(R)$. □

We consider now the radical of an artinian ring.

9.3.9 THEOREM. R_R *is artinian* $\Rightarrow \mathrm{Rad}(R)$ *is nilpotent.*

Proof. For brevity let $U := \mathrm{Rad}(R)$. Since R_R is artinian, the chain

$$R \hookleftarrow U \hookleftarrow U^2 \hookleftarrow \ldots$$

is stationary, i.e. there is an $n \in \mathbb{N}$ with $U^n = U^{n+i}$ ($i \in \mathbb{N}$). It is to be shown that $U^n = 0$. Suppose $U^n \neq 0$. Then the set of right ideals

$$\Gamma := \{A | A \hookrightarrow R_R \wedge AU^n \neq 0\}$$

is not empty, since $U \in \Gamma$. By assumption there is a minimal $A_0 \in \Gamma$. Then there exists $a_0 \in A_0$ with $a_0 U^n \neq 0$, thus also $a_0 R U^n \neq 0$, and from the minimality of A_0 it follows that $a_0 R = A_0$. As $U^n = U^{n+1}$ and $RU = U$ we deduce further that

$$a_0 R U^n = a_0 R U U^n = a_0 U \cdot U^n,$$

so that indeed $a_0 U = a_0 R = A_0$ holds. Since R_R is finitely generated and $U = \mathrm{Rad}(R)$ it follows on the other hand by Nakayama's Lemma (9.2.1) that: $a_0 U = a_0 R U \overset{\circ}{\hookrightarrow} a_0 R$, thus $a_0 U \neq a_0 R$ ↯. □

9.3.10 COROLLARIES

(a) *R_R is artinian* $\Rightarrow \mathrm{Rad}(R)$ *is the largest nilpotent right, left or two-sided ideal of R.*

(b) *R is commutative and artinian* $\Rightarrow \mathrm{Rad}(R)$ *is the set of all nilpotent elements of R.*

(c) *R_R is artinian* $\Rightarrow$ *for every right R-module M_R* resp. *for every left R-module ${}_RM$ we have*

$$\mathrm{Rad}(M) = M\,\mathrm{Rad}(R) \overset{\circ}{\hookrightarrow} M \text{ resp. } \mathrm{Rad}(M) = \mathrm{Rad}(R)M \overset{\circ}{\hookrightarrow} M.$$

Proof. (a): $\mathrm{Rad}(R)$ is nilpotent and every nilpotent ideal is contained in it.

(b): Since $\mathrm{Rad}(R)$ is nilpotent, every one of its elements is nilpotent. Let now $a \in R$, $a^n = 0$. Then it follows that since R is commutative

$$(aR)^n = a^n R^n = a^n R = 0R = 0,$$

thus aR is nilpotent and consequently $a \in aR \hookrightarrow \mathrm{Rad}(R)$.

(c) By 9.2.3 and 9.3.5 we have $\mathrm{Rad}(M) = M\,\mathrm{Rad}(R)$ resp. $\mathrm{Rad}(M) = \mathrm{Rad}(R)M$. Since by 9.3.9 $\mathrm{Rad}(R)$ is nilpotent, there is an $n \in \mathbb{N}$ with $(\mathrm{Rad}(R))^n = 0$. Let now for $U \hookrightarrow M_R$

$$M = U + M\,\mathrm{Rad}(R),$$

then by substituting the equality for M $(n-1)$times into $M\,\mathrm{Rad}(R)$ it follows that on the right side of the equality we have

$$M = U + M(\mathrm{Rad}(R))^n = U,$$

thus $M\,\mathrm{Rad}(R) \overset{\circ}{\hookrightarrow} M$ holds. This equally holds for left R-modules. □

9.3.11 THEOREM. *Let $R/\mathrm{Rad}(R)$ be semisimple and let* $\mathrm{Rad}(R)$ *be nilpotent. Then the following are equivalent for a module M_R:*

(1) *M_R is artinian.*

(2) *M_R is noetherian.*

(3) *M_R has finite length.*

(Analogously for left R-modules.)

Proof. Since $(1)\wedge(2)\Leftrightarrow(3)$ it suffices to show that $(1)\Leftrightarrow(2)$. Put $U := \mathrm{Rad}(R)$; then we define

$$e(M) := \mathrm{Min}\{i \mid i \in \mathbb{N} \wedge MU^i = 0\},$$

then this $e(M)$ exists, since there is an n with $U^n = 0$, thus also $MU^n = 0$. We now prove $(1)\Leftrightarrow(2)$ by means of induction over $e(M)$ for all modules $M_R \neq 0$.

Beginning: $e(M) = 1$, i.e. $MU = 0$. Then by putting

$$m(r+U) := mr, \qquad r \in R,\ m \in M$$

M becomes an $\bar{R} := R/U$-module, in which the R- and $\bar{R}$-submodules coincide. Since $\bar{R}$ is semisimple, M is semisimple (8.2.2(a)) and $(1)\Leftrightarrow(2)$ holds by 8.1.6.

Now let the assertion be satisfied for all M with $e(M) \leq k$ and suppose $e(M) = k+1$. Then it follows that $e(MU^k) = 1$. As $(M/MU^k)U^k = 0$ we have further $e(M/MU^k) \leq k$.

Let now M be artinian resp. noetherian, then by 6.1.2 MU^k and M/MU^k are both artinian resp. noetherian. Then by the induction assumption both are noetherian resp. artinian, and by 6.1.2 M is noetherian resp. artinian. □

9.3.12 COROLLARY

(a) *Let R_R be artinian and let M_R be artinian* resp. *noetherian, then M_R is also noetherian* resp. *artinian.*

(b) *If R_R is artinian, then R_R is noetherian.*

(c) *If R_R is artinian and ${}_RR$ is noetherian then ${}_RR$ is artinian.*

Proof. (a) By 9.2.3 $R/\mathrm{Rad}(R)$ is semisimple and by 9.3.9 $\mathrm{Rad}(R)$ is nilpotent. The assertion then follows from 9.3.11.

(b) Special case of (a) for $R_R = M_R$.

(c) By 9.3.11 for ${}_RR = {}_RM$. □

9.4 CHARACTERIZATIONS OF FINITELY GENERATED AND FINITELY COGENERATED MODULES

We have already become acquainted earlier with finitely generated and finitely cogenerated modules and in particular we have used them for the characterization of noetherian and artinian modules (in Chapter 6). We are now in a position to present further characterizations.

9.4.1 THEOREM. *M_R is finitely generated if and only if we have:*

(a) $\mathrm{Rad}(M)$ *is small in* M; *and*

(b) $M/\mathrm{Rad}(M)$ *is finitely generated.*

Proof. First let M_R be finitely generated. Then (a) holds by 9.2.1(c). As with M every epimorphic image of M is finitely generated, thus (b) also holds. Let us now assume (a) and (b). Thus let $\bar{x}_i = x_i + \text{Rad}(M)$, $i = 1, \ldots, n$, be a generating set of $M/\text{Rad}(M)$. Then it follows that

$$x_1R + \ldots + x_nR + \text{Rad}(M) = M,$$

since $\text{Rad}(M) \overset{\circ}{\hookrightarrow} M$ we deduce that

$$x_1R + \ldots + x_nR = M,$$

thus M is finitely generated. □

9.4.2 COROLLARY. *A module M_R is noetherian if and only if for every $U \hookrightarrow M$ we have:*
(a) $\text{Rad}(U) \overset{\circ}{\hookrightarrow} U$; *and*
(b) $U/\text{Rad}(U)$ *is finitely generated.*

Proof. This follows by 6.1.2 and 9.4.1. □

We now consider finitely cogenerated modules.

9.4.3 THEOREM. *For a module $M_R \neq 0$ the following conditions are equivalent:*
(1) *M is finitely cogenerated.*
(2) (a) $\text{Soc}(M)$ *is large in M and* (b) $\text{Soc}(M)$ *is finitely cogenerated.*
(3) *For an injective hull $I(M)$ of M we have*

$$I(M) = Q_1 \oplus \ldots \oplus Q_n,$$

where every Q_i is an injective hull of a simple R-module.

Proof. "(1)⇒(2)": (a) With the help of Zorn's Lemma we show that every submodule $U \hookrightarrow M$, $U \neq 0$ contains a simple submodule E, so that $U \cap \text{Soc}(M) \neq 0$ then also holds. Let

$$\Gamma := \{U_i \mid i \in I\}$$

be the set of all submodules $U_i \neq 0$ of U. As $U \in \Gamma$, $\Gamma \neq \varnothing$. In Γ we define an ordering by

$$U_i \leq U_j :\Leftrightarrow U_j \hookrightarrow U_i$$

(reverse inclusion). Let

$$\Lambda = \{A_j \mid j \in J\}$$

be a totally ordered subset of Γ. We then show that

$$D := \bigcap_{j \in J} A_j$$

is an upper bound of Λ in Γ. If we suppose $D = 0$, then by (1) the intersection of finitely many of the A_j must already be equal to zero. Since Λ is totally ordered, under these finitely many A_j there is a largest element (with respect to the reverse inclusion), and this must then be already equal to zero: contradiction to $U_i \neq 0$! Thus $D \neq 0$ and consequently $D \in \Gamma$. By Zorn's Lemma there is now a maximal element U_0 in Γ and this U_0 is obviously a simple submodule of U.

(b) By definition of "finitely cogenerated" as well as M every submodule of M is finitely cogenerated, thus also $\mathrm{Rad}(M)$.

"(2)$\Rightarrow$(1)": From

$$\bigcap_{i \in I} A_i = 0 \qquad \text{with} \quad A_i \hookrightarrow M$$

it follows that $\bigcap_{i \in I} \mathrm{Soc}(A_i) = 0$.

As

$$\mathrm{Soc}(A_i) \hookrightarrow \mathrm{Soc}(M)$$

and since $\mathrm{Soc}(M)$ is finitely cogenerated, there is a finite subset $I_0 \subset I$ with

$$\bigcap_{i \in I_0} \mathrm{Soc}(A_i) = 0.$$

For an arbitrary submodule $A \hookrightarrow M$ we have by the definition of the socle

$$\mathrm{Soc}(A) = A \cap \mathrm{Soc}(M).$$

Therefore it follows that

$$0 = \bigcap_{i \in I_0} \mathrm{Soc}(A_i) = \bigcap_{i \in I_0} (A_i \cap \mathrm{Soc}(M)) = \Big(\bigcap_{i \in I_0} A_i\Big) \cap \mathrm{Soc}(M).$$

Since by assumption $\mathrm{Soc}(M)$ is large in M, we obtain finally

$$\bigcap_{i \in I_0} A_i = 0.$$

(2)$\Rightarrow$(1) is therefore proved.

"(2)$\Rightarrow$(3)": Let $I(M)$ be an injective hull of M with $M \hookrightarrow I(M)$ and let $M \neq 0$. As $\mathrm{Soc}(M) \overset{*}{\hookrightarrow} M$ it follows that $\mathrm{Soc}(M) \neq 0$. Let

$$\mathrm{Soc}(M) = E_1 \oplus \ldots \oplus E_n$$

with simple modules E_i and let $Q_i \hookrightarrow I(M)$ be the injective hull of E_i. Then by 5.1.7 we have

$$\sum_{i=1}^{n} Q_i = \bigoplus_{i=1}^{n} Q_i$$

(as sums in $I(M)$), as well as

$$\operatorname{Soc}(M) \hookrightarrow \bigoplus_{i=1}^{n} Q_i.$$

As a finite direct sum of injective modules $\bigoplus_{i=1}^{n} Q_i$ is injective and consequently is a direct summand in $I(M)$. As $\operatorname{Soc}(M) \overset{*}{\hookrightarrow} M$ and $M \overset{*}{\hookrightarrow} I(M)$ it follows that $\operatorname{Soc}(M) \overset{*}{\hookrightarrow} I(M)$, thus also

$$\bigoplus_{i=1}^{n} Q_i \overset{*}{\hookrightarrow} I(M).$$

From the last two statements we deduce that

$$\bigoplus_{i=1}^{n} Q_i = I(M)$$

which was to be shown.

"(3)⇒(2)": Without loss it can again be supposed that

$$M \hookrightarrow I(M) = \bigoplus_{i=1}^{n} Q_i$$

and that as well as $E_i \overset{*}{\hookrightarrow} Q_i$, E_i is simple. As $E_i \overset{*}{\hookrightarrow} Q_i$ E_i is the only simple submodule of Q_i. Hence by 9.1.5 we have

$$\operatorname{Soc}(I(M)) = \bigoplus_{i=1}^{n} \operatorname{Soc}(Q_i) = \bigoplus_{i=1}^{n} E_i.$$

As $M \overset{*}{\hookrightarrow} I(M)$ we have $E_i \hookrightarrow M$ for $i = 1, \ldots, n$, thus

$$\operatorname{Soc}(M) = \bigoplus_{i=1}^{n} E_i.$$

By 8.1.6 $\operatorname{Soc}(M)$ is finitely cogenerated, i.e. (2)(b) is satisfied. As

$$\operatorname{Soc}(M) = \operatorname{Soc}(I(M)) \overset{*}{\hookrightarrow} I(M)$$

we also have $\operatorname{Soc}(M) \overset{*}{\hookrightarrow} M$, i.e. (2)(a) is also satisfied. □

9.4.4 COROLLARY. *A module M_R is artinian if and only if for every factor-module M/U we have:*

(a) $\operatorname{Soc}(M/U) \overset{*}{\hookrightarrow} M/U$; *and*

(b) $\operatorname{Soc}(M/U)$ *is finitely cogenerated.*

Proof. This follows from 6.1.2 and 9.4.3. □

9.5 ON THE CHARACTERIZATION OF ARTINIAN AND NOETHERIAN RINGS

In Chapter 6 the following Theorem (6.6.4) was stated but was there proved only in part.

9.5.1 THEOREM

(a) *The following conditions are equivalent*:

(1) R_R *is noetherian.*

(2) *Every injective module* Q_R *is a direct sum of directly indecomposable (injective) submodules.*

(b) *The following conditions are equivalent*:

(1) R_R *is artinian.*

(2) *Every injective module* Q_R *is a direct sum of injective hulls of simple R-modules.*

The implication (1)⇒(2) was proved in 6.6.5, from which it now suffices, by 9.3.12, only to assume in (b) that R_R is artinian (and not additionally, as in Chapter 6, that R_R is noetherian). The lemma for proving the converse is now available.

Proof of (b). "(2)⇒(1)": In view of 9.9.4 it suffices to show that every factor module $M = R/A$ of R_R satisfies condition (3) in 9.4.3. Let $I(R/A)$ be an injective hull of R/A with $R/A \hookrightarrow I(R/A)$. By assumption we have

$$I(R/A) = \bigoplus_{i \in I} Q_i,$$

where the Q_i are the injective hulls of simple R-modules. Since R/A is cyclic, R/A is already contained in a finite subsum:

$$R/A \hookrightarrow \bigoplus_{i \in I_0} Q_i, \qquad \text{finite } I_0.$$

From $R/A \overset{*}{\hookrightarrow} I(R/A)$ it then follows that $I = I_0$, i.e. $I(R/A) = \bigoplus_{i \in I_0} Q_i$, which was to be shown.

Proof of (a). "(2)⇒(1)": The proof is established by showing that condition (3) in 6.5.1 is satisfied. Let

$$M = \bigoplus_{i=1}^{\infty} Q_i$$

be a direct sum of injective hulls Q_i of simple R-modules $E_i \hookrightarrow Q_i$. Let $I(M)$ be the injective hull of M with $M \hookrightarrow I(M)$. We prove that $M = I(M)$.

As $M \overset{*}{\hookrightarrow} I(M)$ we have $\text{Soc}(M) = \text{Soc}(I(M))$. Further we have

$$\text{Soc}(M) = \bigoplus_{i=1}^{\infty} \text{Soc}(Q_i) = \bigoplus_{i=1}^{\infty} E_i.$$

We now use the assumption $I(M) = \bigoplus_{j \in J} D_j$ where the D_j are directly indecomposable injective modules. Let

$$J_1 = \{j \mid j \in J \wedge \text{Soc}(D_j) \neq 0\},$$

then we have

$$\text{Soc}(I(M)) = \bigoplus_{j \in J_1} \text{Soc}(D_j).$$

If $\text{Soc}(D_j) \neq 0$ then by 6.6.3 $F_j := \text{Soc}(D_j)$ is simple and D_j is the injective hull of F_j. Consequently we have

$$\text{Soc}(I(M)) = \bigoplus_{i=1}^{\infty} E_i = \bigoplus_{j \in J_1} F_j,$$

and by the Krull–Remak–Schmidt Theorem these two decompositions are isomorphic (in the sense of 7.3.1). If $E_i \cong F_j$ then by 5.6.3 it follows that $Q_i \cong D_j$ and by consideration of the bijection in 7.3.1 we obtain

$$M = \bigoplus_{i=1}^{\infty} Q_i \cong \bigoplus_{j \in J_1} D_j.$$

As

$$I(M) = \Big(\bigoplus_{j \in J_1} D_j\Big) \oplus \Big(\bigoplus_{j \in J \setminus J_1} D_j\Big),$$

M is therefore isomorphic to a direct summand of the injective module $I(M)$ and is itself thereby injective which was to be shown. □

9.6 THE RADICAL OF THE ENDOMORPHISM RING OF AN INJECTIVE OR PROJECTIVE MODULE

For certain considerations it is of interest to know the radical of the endomorphism ring of an injective or projective module. We wish to concern ourselves here with this issue. As an application it is then to be shown that for a projective module $P \neq 0$ we always have $\text{Rad}(P) \neq P$, which also indicates that P always contains a maximal submodule.

9.6.1 THEOREM

(a) *Let Q_R be injective and let $S := \text{End}(Q_R)$, then we have for $\alpha \in S$:*

$$S\alpha \overset{\circ}{\hookrightarrow} {}_SS \Leftrightarrow \alpha \in \text{Rad}(S) \Leftrightarrow \text{Ker}(\alpha) \overset{*}{\hookrightarrow} Q_R.$$

(b) *Let P_R be projective and let $S := \mathrm{End}(P_R)$, then we have for $\alpha \in S$:*

$$\alpha S \overset{\circ}{\hookrightarrow} S_S \Leftrightarrow \alpha \in \mathrm{Rad}(S) \Leftrightarrow \mathrm{Im}(\alpha) \overset{\circ}{\hookrightarrow} P_R.$$

Proof. (a) "$S\alpha \overset{\circ}{\hookrightarrow} {}_SS \Leftrightarrow \alpha \in \mathrm{Rad}(S)$": This holds by 9.1.3.

(a) "$\alpha \in \mathrm{Rad}(S) \Rightarrow \mathrm{Ker}(\alpha) \overset{*}{\hookrightarrow} Q_R$": Let $U \hookrightarrow Q_R$ with $\mathrm{Ker}(\alpha) \cap U = 0$. Then $\alpha_0 := \alpha | U$ is a monomorphism and there exists a commutative diagram

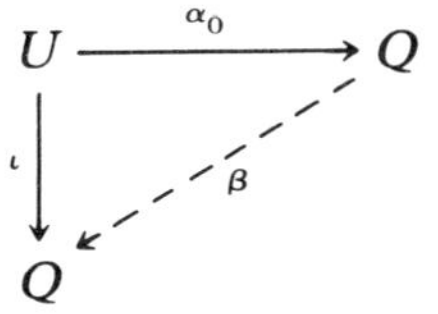

As $u = \iota(u) = \beta\alpha_0(u) = \beta\alpha(u)$, $u \in U$, we have $U \hookrightarrow \mathrm{Ker}(1 - \beta\alpha)$. Since $\alpha \in \mathrm{Rad}(S)$, it follows that $\beta\alpha \in \mathrm{Rad}(S)$. From 9.3.1 $1 - \beta\alpha$ is then invertible, thus $\mathrm{Ker}(1 - \beta\alpha) = 0$, from which $U = 0$ follows. Hence we have shown that $\mathrm{Ker}(\alpha)$ is large in Q.

(a) "$\mathrm{Ker}(\alpha) \overset{*}{\hookrightarrow} Q_R \Rightarrow S\alpha \overset{\circ}{\hookrightarrow} {}_SS$": Let $S\alpha + \Gamma = {}_SS$ with $\Gamma \hookrightarrow {}_SS$, then there are $\sigma \in S$, $\gamma \in \Gamma$ with $\sigma\alpha + \gamma = 1$. From this it follows that $\mathrm{Ker}(\alpha) \cap \mathrm{Ker}(\gamma) = 0$, and as $\mathrm{Ker}(\alpha) \overset{*}{\hookrightarrow} Q_R$ we deduce that $\mathrm{Ker}(\gamma) = 0$. Then there exists a commutative diagram

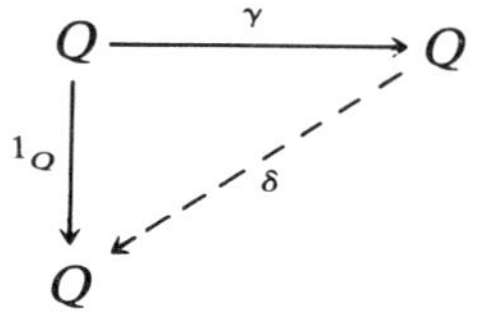

i.e., we have $1_Q = \delta\gamma$ and hence it follows that $\Gamma = S$, thus $S\alpha \overset{\circ}{\hookrightarrow} {}_SS$.

(b) "$\alpha S \overset{\circ}{\hookrightarrow} S_S \Leftrightarrow \alpha \in \mathrm{Rad}(S)$": This holds by 9.1.3.

(b) "$\alpha \in \mathrm{Rad}(S) \Rightarrow \mathrm{Im}(\alpha) \overset{\circ}{\hookrightarrow} P_R$": Let $U \hookrightarrow P_R$ with $\mathrm{Im}(\alpha) + U = P$, and let $\nu : P \to P/U$ be the natural epimorphism. Then $\nu\alpha$ is an epimorphism and we obtain the commutative diagram

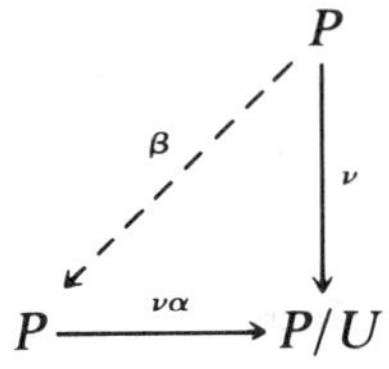

From $\nu = \nu\alpha\beta$ it follows that $\nu(1 - \alpha\beta) = 0$, thus $\mathrm{Im}(1 - \alpha\beta) \hookrightarrow U$. As $\alpha \in \mathrm{Rad}(S)$ we also have $\alpha\beta \in \mathrm{Rad}(S)$, and by 9.3.1 $1 - \alpha\beta$ is then invertible,

thus

$$P = \mathrm{Im}(1-\alpha\beta) \hookrightarrow U \hookrightarrow P,$$

i.e. $U = P$. Hence we have shown that $\mathrm{Im}(\alpha) \overset{\circ}{\hookrightarrow} P_R$.

(b) "$\mathrm{Im}(\alpha) \overset{\circ}{\hookrightarrow} P_R \Rightarrow \alpha S \overset{\circ}{\hookrightarrow} S_S$": Let $\alpha S + \Gamma = S_S$ with $\Gamma \hookrightarrow S_S$, then there are $\sigma \in S$, $\gamma \in \Gamma$ with $\alpha\sigma + \gamma = 1$. From this it follows that $\mathrm{Im}(\alpha) + \mathrm{Im}(\gamma) = P$, thus $\mathrm{Im}(\gamma) = P$ as $\mathrm{Im}(\alpha) \overset{\circ}{\hookrightarrow} P_R$. Hence γ is an epimorphism and consequently there exists a δ so that the diagram

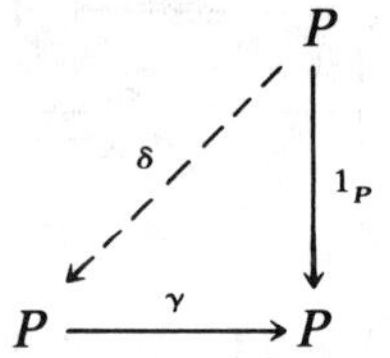

is commutative, thus we have $1_P = \gamma\delta$. It then follows that $\Gamma = S$, which was to be shown. □

9.6.2 COROLLARY. *Let Q_R be injective and let $S := \mathrm{End}(Q_R)$. Then to every $\alpha \in S$ there is a $\gamma \in S$ with $\alpha\gamma\alpha - \alpha \in \mathrm{Rad}(S)$.*

Proof. Let $\alpha \in S$ and let U be an inco of $\mathrm{Ker}(\alpha)$ in Q. By 5.2.5 we then have $\mathrm{Ker}(\alpha) + U \overset{*}{\hookrightarrow} Q$. As $\mathrm{Ker}(\alpha) \cap U = 0$, $\alpha_0 := \alpha | U$ is a monomorphism. Hence there exists a $\gamma \in S$ so that the diagram

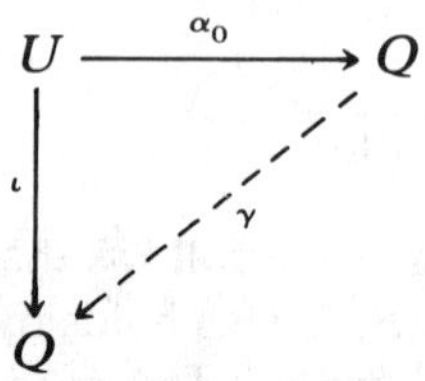

is commutative (ι = the inclusion mapping). For $u \in U$ it then follows that

$$\gamma\alpha(u) = \gamma\alpha_0(u) = u.$$

Hence we have

$$\mathrm{Ker}(\alpha) + U \hookrightarrow \mathrm{Ker}(\alpha\gamma\alpha - \alpha)$$

and as

$$\mathrm{Ker}(\alpha) + U \overset{*}{\hookrightarrow} Q$$

it also follows that

$$\mathrm{Ker}(\alpha\gamma\alpha - \alpha) \overset{*}{\hookrightarrow} Q.$$

From 9.6.1 it then follows that

$$\alpha\gamma\alpha - \alpha \in \operatorname{Rad}(S). \qquad \square$$

This result says that $S/\operatorname{Rad}(S)$ is a regular ring. Regular rings are introduced in the next paragraph and investigated in detail.

The Dual Basis Lemma and 9.6.1(b) also yield an interesting result on the radical of a projective module.

9.6.3 THEOREM. For every projective module $P \neq 0$

$$\operatorname{Rad}(P) \neq P.$$

Proof. We consider generally: If $p \in P_R$ and $\varphi \in P^* = \operatorname{Hom}_R(P_R, R_R)$, then $p\varphi$ can be considered as an element from $S = \operatorname{End}(P_R)$; namely let for $x \in P$

$$(p\varphi)(x) := p\varphi(x),$$

then from

$$\begin{aligned}(p\varphi)(x_1r_1 + x_2r_2) &= p\varphi(x_1r_1 + x_2r_2)\\ &= p(\varphi(x_1)r_1 + \varphi(x_2)r_2) = (p\varphi(x_1))r_1 + (p\varphi(x_2))r_2\\ &= (p\varphi)(x_1)r_1 + (p\varphi)(x_2)r_2\end{aligned}$$

this is in fact an element of S. Let now $p \in \operatorname{Rad}(P)$, then $pR \overset{\circ}{\hookrightarrow} R_R$, and consequently we also have $\operatorname{Im}(p\varphi) = p\varphi(P) \overset{\circ}{\hookrightarrow} P_R$ (as $p\varphi(P) \hookrightarrow pR$). By 9.6.1(b) it follows that $p\varphi S \overset{\circ}{\hookrightarrow} S_S$. Let

$$x = \sum_{\varphi_i(x) \neq 0} p_i\varphi_i(x)$$

be a representation of x in the sense of the Dual Basis Lemma 5.4.2. If we now suppose $x \neq 0$ and let (after a change of indices) $i = 1, \ldots, n$ be the indices with $\varphi_i(x) \neq 0$, then it follows that

$$1_P(x) = x = \sum_{i=1}^{n} p_i\varphi_i(x) = \left(\sum_{i=1}^{n} p_i\varphi_i\right)(x)$$

$$\Rightarrow \quad \left(1_P - \sum_{i=1}^{n} p_i\varphi_i\right)(x) = 0$$

in the sense of the earlier interpretation of the $p_i\varphi_i$ as elements of S. If we now suppose $\operatorname{Rad}(P) = P$, then we have $\operatorname{Im}(p_i\varphi_i) \overset{\circ}{\hookrightarrow} P_R$, thus $p_i\varphi_i S \overset{\circ}{\hookrightarrow} S_S$, thus $p_i\varphi_i \in \operatorname{Rad}(S)$ and finally

$$\sum_{i=1}^{n} p_i\varphi_i \in \operatorname{Rad}(S).$$

By 9.3.1

$$1_P - \sum_{i=1}^{n} p_i\varphi_i$$

is then an invertible element in S; let $\sigma \in S$ be the inverse element, then it follows that

$$x = 1_P(x) = \sigma\left(1 - \sum_{i=1}^{n} p_i\varphi_i\right)(x) = \sigma(0) = 0 \quad ↯.$$

The supposition $0 \neq x \in P$ was thus false, and under the assumption $\mathrm{Rad}(P) = P$ we have necessarily $P = 0$. □

As we have already remarked at the beginning, it follows from $\mathrm{Rad}(P) \neq P$ that P has at least one maximal submodule.

9.6.4 COROLLARY. *If P is projective and we have $P = P_1 \oplus P_2$ with $P_2 \hookrightarrow \mathrm{Rad}(P)$ then it follows that $P_2 = 0$.*

Proof. Let $\pi : P \to P_2$ be the projection of P onto P_2, then from $P_2 \hookrightarrow \mathrm{Rad}(P)$ it follows by 9.1.4 that $P_2 \hookrightarrow \pi(\mathrm{Rad}(P)) \hookrightarrow \mathrm{Rad}(P_2)$, thus $P_2 = \mathrm{Rad}(P_2)$. Since P_2 is projective, it follows from 9.6.3 that $P_2 = 0$. □

9.7 GOOD RINGS

As we have seen in 9.2.1(b) we always have $M\,\mathrm{Rad}(R) \hookrightarrow \mathrm{Rad}(M)$. The question arises as to when equality holds. By no means is this the case for an arbitrary ring and module; e.g. $\mathrm{Rad}(\mathbb{Z}) = 0$ but there are, as we know, $\mathbb{Z}$-modules with non-zero radical, as say $\mathbb{Z}/4\mathbb{Z}$ or $\mathbb{Q}_{\mathbb{Z}}(\mathrm{Rad}(\mathbb{Q}_{\mathbb{Z}}) = \mathbb{Q}_{\mathbb{Z}}!)$.

Additionally the following theorem gives certain information.

9.7.1 THEOREM. *Let $\mathcal{M}_R$ be the category of unitary right R-modules, and let $\bar{R} := R/\mathrm{Rad}(R)$, then the following are equivalent:*

(1) $\forall M \in \mathcal{M}_R\ [M\,\mathrm{Rad}(R) = \mathrm{Rad}(M)]$.
(2) $\forall M \in \mathcal{M}_R\ [M\,\mathrm{Rad}(R) = 0 \Rightarrow \mathrm{Rad}(M) = 0]$.
(3) $\forall \Omega \in \mathcal{M}_{\bar{R}}\ [\mathrm{Rad}(\Omega) = 0]$.
(4) $\forall M, N \in \mathcal{M}_R \forall \varphi \in \mathrm{Hom}_R(M, N)[\varphi(\mathrm{Rad}(M)) = \mathrm{Rad}(\varphi(M))]$.
(5) $\forall M \in \mathcal{M}_R \forall U \hookrightarrow M[\mathrm{Rad}(M) + U/U = \mathrm{Rad}(M/U)]$.
(6) $\forall M \in \mathcal{M}_R \forall U \hookrightarrow M[\mathrm{Rad}(M) = 0 \Rightarrow \mathrm{Rad}(M/U) = 0]$.

Proof. We prove (1)⇒(2)⇒(3)⇒(1) and (1)⇒(4)⇒(5)⇒(6)⇒(1).

"(1)⇒(2)": Special case.

"(2)⇒(3)": Let $\Omega \in M_{\bar{R}}$, then Ω can be made into a right R-module by means of the following definition:

$$\omega r := \omega \bar{r}, \qquad \omega \in \Omega, \qquad \bar{r} = r + \mathrm{Rad}(R) \in \bar{R}.$$

For Ω considered as a right R-module we then obviously have $\Omega\,\mathrm{Rad}(R) = 0$. By (2) it follows that $\mathrm{Rad}(\Omega_R) = 0$. But since by the definition of Ω_R the R- and $\bar{R}$-submodules of Ω coincide, it follows also that $\mathrm{Rad}(\Omega_{\bar{R}}) = 0$.

"(3)⇒(1)": As $(M/M\,\mathrm{Rad}(R))\,\mathrm{Rad}(R) = 0$, $M/M\,\mathrm{Rad}(R)$ can be made into an $\bar{R}$-module by the following definition

$$\bar{m}\bar{r} = (m + M\,\mathrm{Rad}(R))(r + \mathrm{Rad}(R)) := \bar{m}r = mr + M\,\mathrm{Rad}(R),$$

in which the $\bar{R}$- and R-submodules of $M/M\,\mathrm{Rad}(R)$ again coincide. It then follows from $\mathrm{Rad}((M/M\,\mathrm{Rad}(R))_{\bar{R}}) = 0$ that also $\mathrm{Rad}(M/M\,\mathrm{Rad}(R))_R) = 0$, and hence from 9.1.4(b) we have $\mathrm{Rad}(M) \hookrightarrow M\,\mathrm{Rad}(R)$, thus from 9.2.1(b) it follows that $\mathrm{Rad}(M) = M\,\mathrm{Rad}(R)$.

"(1)⇒(4)": From $M\,\mathrm{Rad}(R) = \mathrm{Rad}(M) \wedge \varphi(M)\,\mathrm{Rad}(R) = \mathrm{Rad}(\varphi M))$ it follows that $\varphi(\mathrm{Rad}(M)) = \varphi(M\,\mathrm{Rad}(R)) = \varphi(M)\,\mathrm{Rad}(R) = \mathrm{Rad}(\varphi(M))$.

"(4)⇒(5)": Special case $\varphi = \nu : M \to M/U$.

"(5)⇒(6)": Special case for $\mathrm{Rad}(M) = 0$.

"(6)⇒(1)": By 9.1.5(a) (1) is preserved under isomorphisms of modules. Since every module is an epimorphic image of a free module, it suffices to prove (1) for modules of the form F/U, where F is a free module and $U \hookrightarrow F$. By 9.2.1(g) we have $\mathrm{Rad}(F) = F\,\mathrm{Rad}(R)$. Hence we have $\mathrm{Rad}(F/F\,\mathrm{Rad}(R)) = 0$, thus by (6) we also have

$$\mathrm{Rad}(F/F\,\mathrm{Rad}(R))/(F\,\mathrm{Rad}(R) + U/F\,\mathrm{Rad}(R)) = 0.$$

Since

$$\begin{aligned}(F/F\,\mathrm{Rad}(R))/(F\,\mathrm{Rad}(R) + U/F\,\mathrm{Rad}(R)) &\cong F/(F\,\mathrm{Rad}(R) + U)\\ &\cong (F/U)/(F\,\mathrm{Rad}(R) + U/U)\end{aligned}$$

it then follows that

$$\mathrm{Rad}((F/U)/(F\,\mathrm{Rad}(R) + U/U) = 0,$$

thus by 9.1.4(b)

$$\mathrm{Rad}(F/U) \hookrightarrow F\,\mathrm{Rad}(R) + U/U = (F/U)\,\mathrm{Rad}(R).$$

By reference to 9.2.1(b) it follows therefore that $\mathrm{Rad}(F/U) = (F/U)\mathrm{Rad}(R)$ which was to be shown. □

9.7.2 *Definition.* Let a ring, which satisfies the conditions of Theorem 9.7.1, be called a *right good ring.* Correspondingly let a *left good ring* be defined. Let a two-sided good ring be called a *good ring.*

9.7.3 COROLLARIES

(a) *A ring R, for which $\bar{R} := R/\mathrm{Rad}(R)$ is semisimple, is by* (3) *a good ring.*

(b) *By* 9.2.3 *every (one-sided) artinian ring is consequently a good ring.*

(c) *If R is right good ring then by* 9.1.5(b) *and* 9.7.1(1) *we have for an arbitrary module M_R:*

$$M = \sum_{i \in I} M_i \Rightarrow \mathrm{Rad}(M) = \sum_{i \in I} \mathrm{Rad}(M_i).$$

Finally we remark that there are good rings for which $R/\mathrm{Rad}(R)$ is not semisimple; e.g. this is the case if $R/\mathrm{Rad}(R)$ is commutative and regular (see Chapter 10, Exercise 18), but is not semisimple.

EXERCISES

(1)

(a) Show that for a ring R the following statements are equivalent:
   (1) For every right R-module $\mathrm{Rad}(M) \overset{\circ}{\hookrightarrow} M$.
   (2) There is no right R-module $M \neq 0$ with $\mathrm{Rad}(M) = M$.

(b) Show that for a ring R the following statements are equivalent:
   (1) For every right R-module $\mathrm{Soc}(M) \overset{*}{\hookrightarrow} M$.
   (2) For every cyclic right R-module M $\mathrm{Soc}(M) \overset{*}{\hookrightarrow} M$.
   (3) There is no right R-module $M \neq 0$ with $\mathrm{Soc}(M) = 0$.

(2)

(a) Let $\mathrm{Soc}(M) \hookrightarrow B_R \hookrightarrow M_R \wedge a \in M \wedge a \notin B$. Show: Then there exists $C \overset{*}{\hookrightarrow} M$ with $B \hookrightarrow C \wedge a \notin C$.

(b) Show: $\mathrm{Soc}(M) \hookrightarrow B_R \hookrightarrow M_R \Rightarrow B = \bigcap_{B \hookrightarrow C \overset{*}{\hookrightarrow} M} C.$

(c) Show: $\mathrm{Soc}(M) \hookrightarrow A \hookrightarrow M \wedge \mathrm{Soc}(M/A) \overset{*}{\hookrightarrow} M/A \Rightarrow A \overset{*}{\hookrightarrow} M.$

(3)

Show: R_R is a cogenerator if and only if the injective hull of every finitely cogenerated right R-module is projective.

(**4**)

(a) Determine $\mathrm{Rad}(R)$, $\mathrm{Soc}({}_RR)$, $\mathrm{Soc}(R_R)$ and find out whether $\mathrm{Soc}(R_R) = \mathrm{Soc}({}_RR)$ holds for the following rings:

$$R := \left\{ \begin{pmatrix} q & r \\ 0 & s \end{pmatrix} \middle| q \in \mathbb{Q} \wedge r, s \in \mathbb{R} \right\}$$

$$R := \left\{ \begin{pmatrix} z & a \\ 0 & b \end{pmatrix} \middle| z \in \mathbb{Z} \wedge a, b \in \mathbb{Q} \right\}.$$

In the above let $\mathbb{R}$ be the field of real numbers.

(b) Assumptions as in Exercise 6 of Chapter 6. Show

$$\mathrm{Rad}(R) = \left\{ \begin{pmatrix} a & m \\ 0 & b \end{pmatrix} \middle| a \in \mathrm{Rad}(A),\ m \in M,\ b \in \mathrm{Rad}(B) \right\}.$$

(Hint: Determine the right-invertible elements in R.)

(**5**)

Show: The following statements are equivalent for M_R (Compare 9.2.2(b)):

(1) M is finitely cogenerated and $\mathrm{Rad}(M) = 0$.

(2) M is finitely generated and semisimple.

(**6**)

Let Δ be a complete lattice (see 3.1). Let the smallest element be denoted by 0 and the largest by M. For $A, B \in \Delta$ with $A \leqslant B$ let

$$[A, B] := \{L \in \Delta | A \leqslant L \leqslant B\};$$

under the lattice structure induced from Δ this is again a complete lattice.

Definitions

(a) A family $\Gamma = (A_i | i \in I)$ of elements from Δ is said to be *directed upwards* if to any two elements A_i, A_j from Γ there exists an element A_k from Γ with $A_i \leqslant A_k$ and $A_j \leqslant A_k$.

(b) An element $A \in \Delta$ is called *compact*, if in every directed family $(A_i | i \in I)$ with $A \leqslant \bigcup_{i \in I} A_i$ there exists an A_j with $A \leqslant A_j$.

(c) Δ is called *compactly generated*, if every element from Δ is a union of compact elements.

(d) Δ is called *modular* $\Leftrightarrow \forall A, B, C \in \Delta [B \leqslant A \Rightarrow A \cap (B \cup C) = B \cup (A \cap C)]$.

(e) $A \in \Delta$ is called *small* in $\Delta :\Leftrightarrow \forall B \in \Delta \backslash \{M\} [A \cup B \neq M]$.

(f) $\mathrm{Rad}(\Delta) := \bigcap_{\text{max. } B \text{ in } \Delta \backslash \{M\}} B.$

Show:

(1) $\bigcup_{A \text{ small in } \Delta} A \leqslant \text{Rad}(\Delta)$.

(2) A is compact and $A \leqslant \text{Rad}(\Delta) \Rightarrow A$ is small in Δ.

(3) Δ is compactly generated $\Rightarrow \bigcup_{A \text{ small in } \Delta} A = \text{Rad}(\Delta)$.

(4) For $A \in \Delta$ we have $A \cup \text{Rad}(\Delta) \leqslant \text{Rad}([A, M])$.

(5) If $A \leqslant \text{Rad}(\Delta)$, then we have $\text{Rad}(\Delta) = \text{Rad}([A, M])$.

(6) If Δ is compact $\text{Rad}(\Delta)$ is small in Δ.

(7) Let Δ be modular and $A \in \Delta$ then we have $\text{Rad}([0\ A]) \leqslant \text{Rad}(\Delta)$.

(8) In the lattice Δ of submodules of a module M what does it mean if $A \in \Delta$ is compact? (Observe: $\bigcup$ is then $+$).

(7)

For a module M_R we define:

(a) M is semiartinian $:\Leftrightarrow \forall U \subsetneq M[\text{Soc}(M/U) \neq 0]$.

(b) $\text{Sa}(M) := \sum_{\substack{U \hookrightarrow M \\ \text{semiartinian } U}} U$.

Show:

(1) M is semiartinian $\Rightarrow \forall U \hookrightarrow M[M/U$ is semiartinian$]$.

(2) For arbitrary M $\text{Sa}(M)$ is semiartinian.

(3) $\forall M, N \in \mathcal{M}_R \forall \varphi \in \text{Hom}_R(M, N)[\varphi(\text{Sa}(M)) \hookrightarrow \text{Sa}(N)]$.

(4) $\text{Sa}(\text{Sa}(M)) = \text{Sa}(M)$.

(5) $\text{Sa}(M/\text{Sa}(M)) = 0$.

(6) M is semiartinian $\Rightarrow \text{Soc}(M) \overset{*}{\hookrightarrow} M$.

(7) M is semiartinian $\Rightarrow \forall U \hookrightarrow M[\text{Soc}(M/U) \overset{*}{\hookrightarrow} M/U]$.

(8) Let $U \hookrightarrow M \Rightarrow M$ is semiartinian $\Leftrightarrow M/U$ is semiartinian $\wedge$ U is semiartinian.

(9) A is semiartinian and M is noetherian $\Leftrightarrow M$ is artinian and M is noetherian.

(10) M is semiartinian and R_R is noetherian $\Rightarrow M$ is the sum of its artinian submodules.

(8)

Definition. (a) M is seminoetherian $:\Leftrightarrow \forall U \hookrightarrow M,\ U \neq 0[\text{Rad}(U) \neq U]$.

(b) $\text{Snr}(M) := \sum_{\substack{U \hookrightarrow M \\ \text{Rad}(U) = U}} U$.

Consider whether the properties dual to those given in problem 7 hold and consider respectively under which additional assumptions they hold.

(9)

Definition. Coatomic:

$$M: \Leftrightarrow \forall U \overset{\neq}{\hookrightarrow} M \exists A \hookrightarrow M[U \hookrightarrow A \wedge A \text{ maximal in } M].$$

Show:

(a) If A is semisimple or finitely generated then M is coatomic.

(b) There is a coatomic $\mathbb{Z}$-module M which is neither semisimple nor finitely generated.

(c) M is semisimple$\Leftrightarrow M$ is coatomic and every maximal submodule of M is a direct summand in M.

(d) $U \hookrightarrow \mathrm{Rad}(M)$ and U is coatomic$\Rightarrow U \overset{\circ}{\hookrightarrow} M$.

(e) M is coatomic$\Rightarrow \mathrm{Rad}(M) \overset{\circ}{\hookrightarrow} M$.

(f) There is a module M with $\mathrm{Rad}(M)=0$ but M is not coatomic.

(10)

Let $M = M_{\mathbb{Z}}$ be an abelian group and let

$$T(M) := \{m \in M \mid \exists z \neq 0[mz = 0]\}$$

be the torsion subgroup. Show:

(a) $\mathrm{Soc}(M) \overset{*}{\hookrightarrow} M \Leftrightarrow T(M) = M$; $\mathrm{Soc}(M) = 0 \Leftrightarrow T(M) = 0$.

(b) $U \overset{*}{\hookrightarrow} M \Leftrightarrow \mathrm{Soc}(M) \hookrightarrow U \hookrightarrow M \wedge T(M/U) = M/U$.

(c) M is semisimple$\Leftrightarrow T(M) = M$ and $\mathrm{Rad}(M) = 0$.

(11)

Show that for a ring R the following statements are equivalent:

(1) For every family $(M_i \mid i \in I)$ of right R-modules we have:

$$\mathrm{Soc}\left(\prod_{i \in I} M_i\right) = \prod_{i \in I} \mathrm{Soc}(M_i).$$

(2) Every product of semisimple right R-modules is again semisimple.

(3) Every radical-free right R-module (i.e. with $\mathrm{Rad}(M)=0$) is semisimple.

(4) $R/\mathrm{Rad}(R)$ is semisimple.

(12)

For a right R-module M show:

(a) $U \hookrightarrow M \wedge U$ is a direct summand in $M \Rightarrow \mathrm{Rad}(M/U) = (\mathrm{Rad}(M) + U)/U$, $\mathrm{Soc}(M/U) = (\mathrm{Soc}(M) + U)/U$.

(b) $\forall U \hookrightarrow M[\mathrm{Rad}(U) = U \cap \mathrm{Rad}(M)] \Leftrightarrow \mathrm{Rad}(M) = 0$.

(c) $\forall U \hookrightarrow M[\mathrm{Soc}(M/U) = (\mathrm{Soc}(M) + U)/U] \Leftrightarrow M$ semisimple.

(13)

(a) Show that for a left ideal $U \hookrightarrow {}_R R$ the following statements are equivalent:

(1) $\left(\prod_{i\in I} M_i\right)U = \prod_{i\in I}(M_i U)$ for every family $(M_i \mid i \in I)$ of right R-modules.

(2) ${}_R U$ is finitely generated.

(b) If R is right good, then the radical in $\mathcal{M}_R$ is permutable with direct products if and only if ${}_R\mathrm{Rad}(R)$ is finitely generated.

(c) If R is commutative and noetherian then the radical is permutable with direct products.

(14)

For a commutative ring R we have $\mathrm{Rad}(M_R) = \bigcap\{MA \mid A$ maximal ideal in $R\}$. A corresponding result will be shown more generally for rings in which every maximal right ideal is two-sided.

Definition: For a right R-module M_R let

$$D(M) := \bigcap\{MA \mid A \text{ maximal right ideal in } R\}.$$

Show:

(a) $D(M)$ is a submodule of M_R and for every homomorphism $f: M \to N$ we have $f(D(M)) \subset D(N)$ (i.e. D is a preradical in $\mathcal{M}_R$).

(b) $D(R_R) = R \Leftrightarrow$ no maximal right ideal is two-sided.

(c) $D(M) = \mathrm{Rad}(M)$ for all $M \in \mathcal{M}_R \Leftrightarrow$ every maximal right ideal is two-sided.

(15)

Let $M = M_{\mathbb{Z}}$ be an abelian group.

Show: There is an abelian group N with $\mathrm{Rad}(N) = M$. (Hint: Choose an injective extension $M \hookrightarrow Q$ and consider $\mathrm{Soc}(Q/M)$.)

(16)

Notations as in Chapter 5, Exercise 27. Show:

(a) S is local.

(b) The socle of S_S has length $n+1$.

(17)

For every R-module M we define an ascending sequence of submodules $M_i (i = 0, 1, 2, \ldots)$ by

$$M_0 := 0 \qquad \text{and} \qquad M_{i+1}/M_i := \mathrm{Soc}(M/M_i)$$

(more precisely: let $\nu: M \to M/M_i$ be the natural epimorphism, then let $M_{i+1} := \nu^{-1}(\mathrm{Soc}(M/M_i))$. Show:

(a) If M is artinian, then we have for every $i \geq 0$:

(1) M_{i+1} has finite length.

(2) If $B \hookrightarrow M_{i+1}$ and if the length of B is $\leq i$ then it follows that $B \hookrightarrow M_i$.

(b) If M is artinian and a self-generator then M is also noetherian. (M is called a self-generator, if for every submodule U of M we have:

$$U = \sum_{f \in \mathrm{Hom}_R(M, U)} \mathrm{Im}(f).$$

Hint: With regard to (b) show that the set $\{A \mid A \hookrightarrow M \wedge M/A \text{ noetherian}\}$ has a smallest element A_0 and apply (a) with i = length of M/A_0.)

(18)

For every R-module M we define a descending sequence of submodules $M^i (i = 0, 1, 2, \ldots)$ by

$$M^\circ := M \qquad \text{and} \qquad M^{i+1} := \mathrm{Rad}(M^i).$$

Show:

(a) If $R/\mathrm{Rad}(R)$ is semisimple and if M_R is noetherian then we have for $i \geq 0$:

(1) M/M^{i+1} has finite length,

(2) if $M^{i+1} \hookrightarrow B \hookrightarrow M$ and if the length of M/B is $\leq i$ then it follows that $M^i \hookrightarrow B$.

(b) If $R/\mathrm{Rad}(R)$ is semisimple and if M is a noetherian selfcogenerator, then M is also artinian.

(c) Question: In (b) can we omit the assumption "$R/\mathrm{Rad}(R)$ semisimple"? (M is called a selfcogenerator if for every submodule U of M we have: $0 = \bigcap_{f \in \mathrm{Hom}_R(M/U, M)} \mathrm{Ker}(f)$.)

Chapter 10

The Tensor Product, Flat Modules and Regular Rings

The significance of the tensor product depends above all on the two following facts:

(1) The tensor product has an important factorization property, namely every tensorial mapping can be factorized over the tensor product and the tensor product is uniquely determined up to isomorphism by this property.

(2) The tensor product is a functor (10.3.1) and in fact is an adjoint functor to the functor Hom (10.3.4).

10.1 DEFINITION AND FACTORIZATION PROPERTY

The tensor product links a module A_S and a module $_SU$ into a new module

$$A \underset{S}{\otimes} U,$$

which, in general, is a $\mathbb{Z}$-module, under suitable assumptions however it can also be a module over other rings.

In order to define $A \underset{S}{\otimes} U$ let

$$A \times U = \{(a, u) \mid a \in A \wedge u \in U\}$$

be the product set of A and U and let $F = F(A \times U, \mathbb{Z})$ denote the free right $\mathbb{Z}$-module (or left module—the side for $\mathbb{Z}$ plays no role) with the basis $A \times U$ (see 4.4). We again denote the basis elements of F by (a, u). Finally let K be the submodule of F (as a $\mathbb{Z}$-module) generated by the set $D_1 \cup D_2 \cup T$ with

$$D_1 = \{(a + a', u) - (a, u) - (a', u) \mid a, a' \in A \wedge u \in U\},$$

$$D_2 = \{(a, u+u') - (a, u) - (a, u') \mid a \in A \wedge u, u' \in U\},$$

$$T = \{(as, u) - (a, su) \mid a \in A \wedge u \in U \wedge s \in S\}.$$

10.1.1 *Definition.* The factor module F/K is called the *tensor product* of A_S and ${}_SU$ over S, notationally

$$A \underset{S}{\otimes} U := F/K.$$

The image of the element $(a, u) \in F$ under the natural epimorphism $F \to F/K$ is called the *tensor product* of a and u and is denoted by $a \otimes u$:

$$a \otimes u := (a, u) + K.$$

If it is clear from the relationship that we have a tensor product over S, then we write only $A \otimes U$.

For the tensor product the following operational rules hold.

10.1.2 *Operational Rules*

$$(1) \qquad (a+a') \otimes u = a \otimes u + a' \otimes u,$$

$$(2) \qquad a \otimes (u+u') = a \otimes u + a \otimes u',$$

$$(3) \qquad as \otimes u = a \otimes su,$$

$$(4) \qquad 0 \otimes u = a \otimes 0 = 0,$$

$$(5) \qquad -(a \otimes u) = (-a) \otimes u = a \otimes (-u),$$

$$(6) \qquad (a \otimes u)z = (az) \otimes u = a \otimes (uz), \qquad z \in \mathbb{Z}.$$

Proof. (1), (2), (3) by definition of K.

$$(4) \qquad 0 \otimes u + 0 \otimes u = (0+0) \otimes u = 0 \otimes u \Rightarrow 0 \otimes u = 0;$$

analogously it follows that $a \otimes 0 = 0$.

$$(5) \qquad a \otimes u + (-a) \otimes u = (a-a) \otimes u = 0 \otimes u = 0 \Rightarrow (-a) \otimes u = -(a \otimes u);$$

analogously for $a \otimes (-u)$.

$$(6) \quad z > 0 : (a \otimes u)z = \underbrace{a \otimes u + \ldots + a \otimes u}_{z \text{ summands}} = (a + \ldots + a) \otimes u = (az) \otimes u;$$

$$z = 0 : (a \otimes u)0 = 0 = 0 \otimes u = (a0) \otimes u;$$

$$z < 0 : (a \otimes u)(-z) = (a(-z)) \otimes u = (-az) \otimes u = -((az) \otimes u)$$

by (5). Since we also have

$$(a \otimes u)(-z)+(a \otimes u)z=(a \otimes u)(-z+z)=(a \otimes u)0=0$$

the uniquely determined element negative to $-((az)\otimes u)$ is on the one hand $az \otimes u$ and on the other hand $(a \otimes u)z$, thus we have $(az)\otimes u = (a \otimes u)z$; analogously for u. □

10.1.3 *Remarks*

(1) The free right $\mathbb{Z}$-module F can also be considered as a free left $\mathbb{Z}$-module; the side is of no significance, and in the following the side for F and $A \underset{S}{\otimes} U$ is chosen which is the more convenient for the purpose under consideration.

(2) By Rule (6) every element $t \in A \underset{S}{\otimes} U$ can be written as a finite sum of the form

$$t=\sum a_i \otimes u_i.$$

(3) The representation $t=\sum a_i \otimes u_i$ is not uniquely determined in general, and indeed not even if it is a representation of "shortest length".

(4) The tensor product of two modules different from zero can be zero.

Example for (3) and (4). Let $A=(\mathbb{Z}/2\mathbb{Z})_{\mathbb{Z}}$, $U={}_{\mathbb{Z}}(\mathbb{Z}/3\mathbb{Z})$, then we have for arbitrary $a \in A$, $u \in U$ in $A \underset{\mathbb{Z}}{\otimes} U$:

$$\begin{aligned} 0\otimes 0=0=a \otimes 0-0 \otimes u &= a \otimes (3u)-(2a)\otimes u \\ &= 3(a \otimes u)-2(a \otimes u)=a \otimes u, \end{aligned}$$

thus

$$A \otimes U=0.$$

10.1.4 *Definition.* Let A_S, ${}_SU$, $M_{\mathbb{Z}}$ be given.

(1) A mapping of $\varphi: A \times U \to M$ is called *biadditive* :⇔

$$\begin{aligned} \forall a, a' \in A \forall u, u' \in U[\varphi(a+a', u)=\varphi(a, u)+\varphi(a', u) \wedge \\ \varphi(a, u+u')=\varphi(a, u)+\varphi(a, u')]. \end{aligned}$$

(2) A biadditive mapping φ is called an *S-tensorial mapping* :⇔

$$\forall a \in A \forall u \in U \forall s \in S[\varphi(as, u)=\varphi(a, su)].$$

10.1.5 COROLLARY. *Let $\nu: F \to F/K = A \underset{S}{\otimes} U$ be the natural epimorphism and let τ be its restriction onto the basis $A \times U$ of F,*

$$\tau := \nu | A \times U, \qquad i.e. \quad \tau(a, u)=a \otimes u,$$

then we have: For every $\mathbb{Z}$-homomorphism $\lambda : A \underset{S}{\otimes} U \to M$ the mapping

$$\varphi := \lambda\tau : A \times U \to M$$

is an S-tensorial mapping. In particular τ is an S-tensorial mapping.

Proof. We have

$$\lambda\tau(a+a', u) = \lambda((a+a') \otimes u) = \lambda(a \otimes u + a' \otimes u) = \lambda(a \otimes u) + \lambda(a' \otimes u) = \lambda\tau(a, u) + \lambda\tau(a', u).$$

The other properties follow analogously. □

In the following it is important that we can express the image of an element under λ by $\varphi := \lambda\tau$:

(10.1.6) $\quad \lambda(\sum a_i \otimes u_i) = \sum \lambda(a_i \otimes u_i) = \sum \lambda\tau(a_i, u_i) = \sum \varphi(a_i, u_i).$

Now let $\mathrm{Tens}(A \times U, M)$ denote the set of S-tensorial mappings of $A \times U$ into M; then by the definition

$$(\varphi_1 + \varphi_2)(a, u) := \varphi_1(a, u) + \varphi_2(a, u), \qquad (-\varphi)(a, u) := -\varphi(a, u)$$

this set obviously becomes a $\mathbb{Z}$-module and the mapping

$$\Phi : \mathrm{Hom}_{\mathbb{Z}}(A \underset{S}{\otimes} U, M) \ni \lambda \mapsto \varphi := \lambda\tau \in \mathrm{Tens}(A \times U, M)$$

is a $\mathbb{Z}$-homomorphism.

10.1.7 THEOREM. *Φ is an isomorphism.*

Proof. Injectivity of Φ: This follows from 10.1.6. Surjectivity of Φ: Given $\varphi \in \mathrm{Tens}(A \times U, M)$ we seek a $\lambda \in \mathrm{Hom}_{\mathbb{Z}}(A \underset{S}{\otimes} U, M)$ with $\varphi = \lambda\tau$.

First of all φ is extended to $\hat{\varphi} \in \mathrm{Hom}_{\mathbb{Z}}(F, M)$ by the definition

$$\hat{\varphi}(\sum (a_i, u_i)z_i) := \sum \varphi(a_i, u_i)z_i$$

Since φ is S-tensorial it follows that $K \hookrightarrow \mathrm{Ker}(\hat{\varphi})$. Consequently $\hat{\varphi}$ can be factorized over $F/K = A \underset{S}{\otimes} U$ (3.4.7 special case); the factorizing mapping, which is again a $\mathbb{Z}$-homomorphism, we call λ and therefore we have $\varphi = \lambda\tau$. □

For later applications we summarize 10.1.6 and 10.1.7 in the following statements:

10.1.8 COROLLARY. *For every S-tensorial mapping* $\varphi : A \times U \to M$ *there is exactly one* $\mathbb{Z}$*-homomorphism* $\lambda : A \otimes_S U \to M$ *with* $\varphi = \lambda\tau$, *such that*

$$\lambda(\sum a_i \otimes u_i) = \sum \varphi(a_i, u_i)$$

also holds.

Finally it is to be shown that the tensor product $A \otimes_S U$ is uniquely determined by 10.1.8 up to isomorphism.

More precisely: Given $C_{\mathbb{Z}}$, let $\gamma : A \times U \to C$ be an S-tensorial mapping so that for every $\mathbb{Z}$-module M and every S-tensorial mapping $\varphi : A \times U \to M$ there exists exactly one $\mathbb{Z}$-homomorphism

$$\eta : C \to M$$

with $\varphi = \eta\gamma$, then we have $A \otimes_S U \cong C$ as $\mathbb{Z}$-modules.

In the proof we can certainly make do with weaker assumptions as the following theorem shows.

10.1.9 THEOREM. *Let* $\gamma : A \times U \to C$ *be an S-tensorial mapping with the following properties:*

(1) *There exists a* $\mathbb{Z}$*-homomorphism*

$$\sigma : C \to A \otimes_S U$$

with $\tau = \sigma\gamma$ *(i.e. factorization of* τ *over* γ *is possible).*

(2) *The equation* $\gamma = \eta\gamma$ *with* $\eta \in \mathrm{Hom}_{\mathbb{Z}}(C, C)$ *is only satisfied for* $\eta = 1_C$ *(i.e. factorization of* γ *over* γ *is unique).*

Then we have

$$C \cong A \otimes_S U \qquad \text{as } \mathbb{Z}\text{-modules.}$$

Proof. By 10.1.8 there is a $\rho : A \otimes_S U \to C$ with $\gamma = \rho\tau$ and by assumption we have $\tau = \sigma\gamma$. From the two equations together it follows that:

$$\tau = \sigma\rho\tau, \qquad \gamma = \rho\sigma\gamma.$$

By 10.1.8 and assumption (2) it then follows that

$$\sigma\rho = 1_{A \otimes U}, \qquad \rho\sigma = 1_C,$$

thus $C \cong A \otimes_S U$. □

Nevertheless with regard to the tensor product only the equation $\gamma = \rho\tau$ and the uniqueness of the factorization $\tau = \sigma\rho\tau$ will be used.

10.2 FURTHER PROPERTIES OF THE TENSOR PRODUCT

10.2.1 THE TENSOR PRODUCT OF HOMOMORPHISMS

Let S-modules A_S, B_S as well as ${}_SU$, ${}_SV$ and S-homomorphisms

$$\alpha : A \to B, \qquad \mu : U \to V$$

be given. Then we consider the mapping

$$\varphi : A \times U \ni (a, u) \mapsto \alpha(a) \otimes \mu(u) \in B \underset{S}{\otimes} V.$$

As is immediately verified, this is an S-tensorial mapping of $A \times U$ into $B \underset{S}{\otimes} V$, in which thus $\varphi(a, u) = \alpha(a) \otimes \mu(u)$ holds. The $\mathbb{Z}$-homomorphism of $A \underset{S}{\otimes} U$ into $B \underset{S}{\otimes} V$, which exists in the sense of 10.1.8, is to be denoted by $\alpha \otimes \mu$; thus we have:

$$\alpha \otimes \mu : A \underset{S}{\otimes} U \ni \sum a_i \otimes u_i \mapsto \sum \alpha(a_i) \otimes \mu(u_i) \in B \underset{S}{\otimes} V,$$

i.e. we apply α and μ to the respective components.

Definition. $\alpha \otimes \mu$ is called the *tensor product of the homomorphisms* α and μ.

The following properties of this tensor product of homomorphisms are immediately clear:

(1) $1_A \otimes 1_U = 1_{A \underset{S}{\otimes} U}$.

(2) Besides α and μ let the homomorphisms $\beta : B_S \to C_S$, $\nu : {}_SV \to {}_SW$ be given, then we have $(\beta\alpha) \otimes (\nu\mu) = (\beta \otimes \nu)(\alpha \otimes \mu)$.

(3) Let α and μ be isomorphisms, then $\alpha \otimes \mu$ is an isomorphism and $(\alpha \otimes \mu)^{-1} = \alpha^{-1} \otimes \mu^{-1}$ holds.

10.2.2 MODULE PROPERTIES OF THE TENSOR PRODUCT

Now let R be also a ring and let ${}_RA_S$ be a bimodule. It is to be established that according to the definition

$$r(\sum a_i \otimes u_i) := \sum (ra_i) \otimes u_i, \qquad r \in R,$$

$A \underset{S}{\otimes} U$ is then a left R-module. For this purpose we consider for fixed $r \in R$ the mapping

$$A \ni a \mapsto ra \in A.$$

Since ${}_RA_S$ is a bimodule, this is evidently an S-homomorphism which is to be denoted by $r^{\cdot}$. By 10.2.1 $r^{\cdot} \otimes 1_U$ is then a homomorphism with

$$r^{\cdot} \otimes 1_U : A \underset{S}{\otimes} U \ni \sum a_i \otimes u_i \mapsto \sum (ra_i) \otimes u_i \in A \underset{S}{\otimes} U,$$

so that the definition above in fact makes $A \underset{S}{\otimes} U$ into a left R-module. ($\sum (ra_i) \otimes u_i$ is uniquely determined by r and $\sum a_i \otimes u_i$, independently of the representation of $\sum a_i \otimes u_i$!)

If ${}_SU_T$ is a bimodule, then by the definition

$$(\sum a_i \otimes u_i)t := \sum a_i \otimes (u_i t), \qquad t \in T$$

$A \underset{S}{\otimes} U$ becomes a right T-module and in the case ${}_RA_S$, ${}_SU_T$, we have $A \underset{S}{\otimes} U$ as an R-T-bimodule.

Let homomorphisms

$$\alpha : {}_RA_S \to {}_RB_S, \qquad \mu : {}_SU \to {}_SV$$

be given, then $\alpha \otimes \mu$ is an R-homomorphism

$$\alpha \otimes \mu : {}_R(A \underset{S}{\otimes} U) \to {}_R(B \underset{S}{\otimes} V),$$

for

$$\begin{aligned}(\alpha \otimes \mu)(r \sum a_i \otimes u_i) &= \sum \alpha(ra_i) \otimes \mu(u_i)\\ &= \sum r\alpha(a_i) \otimes \mu(u_i) = r \sum \alpha(a_i) \otimes \mu(u_i)\\ &= r(\alpha \otimes \mu)(\sum a_i \otimes u_i).\end{aligned}$$

Correspondingly for

$$\alpha : {}_RA_S \to {}_RB_S, \qquad \mu : {}_SU_T \to {}_SV_T$$

$\alpha \otimes \mu$ is an R-T-bimodule homomorphism of ${}_R(A \underset{S}{\otimes} U)_T$ into ${}_R(B \underset{S}{\otimes} V)_T$.

If R is a subring of the centre of S (for commutative S e.g. $R = S$) then by definition

$$ra := ar, \qquad ur := ru, \qquad r \in R, \quad a \in A, \quad u \in U$$

A_S and $_SU$ become R-S resp. S-R bimodules and $A \underset{S}{\otimes} U$ is a two-sided R-bimodule. We then have

$$r(\sum a_i \otimes u_i) = \sum (ra_i) \otimes u_i = \sum (a_i r) \otimes u_i$$
$$= \sum a_i \otimes (ru_i) = \sum a_i \otimes (u_i r) = (\sum a_i \otimes u_i)r.$$

The special case of S commutative and $R = S$ is of particular interest.

Now let $_RA_S$, $_SU$ and $_RM$ be given, as well as an S-tensorial mapping

$$\varphi : A \times U \to M \quad \text{with} \quad \varphi(ra, u) = r\varphi(a, u), \quad r \in R.$$

We consider $A \underset{S}{\otimes} U$ as a left R-module and show that the $\mathbb{Z}$-homomorphism λ which exists in the sense of 10.1.8 is also an R-homomorphism:

$$\lambda(r \sum a_i \otimes u_i) = \lambda(\sum (ra_i) \otimes u_i) = \sum \varphi(ra_i, u_i)$$
$$= \sum r\varphi(a_i, u_i) = r\lambda(\sum a_i \otimes u_i).$$

A corresponding statement holds also in the case $_RA_S$, $_SU_T$, $_RM_T$.

Finally we wish to establish that the mapping

$$\lambda : A \underset{S}{\otimes} S \ni \sum a_i \otimes s_i \mapsto \sum a_i s_i \in A$$

is an S-isomorphism of the right S-modules $A \underset{S}{\otimes} S$ and A_S. Since

$$\varphi : A \times S \ni (a, s) \mapsto as \in A$$

is S-tensorial and $\varphi(a, ss_1) = \varphi(a, s)s_1$ holds, λ is an S-homomorphism and indeed is obviously an epimorphism. Let $\sum a_i \otimes s_i \in \text{Ker}(\lambda)$, thus $\sum a_i s_i = 0$, then it follows that

$$\sum a_i \otimes s_i = \sum (a_i s_i \otimes 1) = (\sum a_i s_i) \otimes 1 = 0 \otimes 1 = 0,$$

i.e. λ is also a monomorphism, thus an isomorphism. Analogously we also have $_S(S \underset{S}{\otimes} U) \cong {_SU}$.

10.2.3 ASSOCIATIVITY OF THE TENSOR PRODUCT

We have to show here that the tensor product is associative up to isomorphism. Let modules A_R, $_RM_S$, $_SU$ be given, then we assert:

$$(A \underset{R}{\otimes} M) \underset{S}{\otimes} U \cong A \underset{R}{\otimes} (M \underset{S}{\otimes} U),$$

and this isomorphism is obtained from

$$(*) \qquad \sum (a_i \otimes m_i) \otimes u_i \mapsto \sum a_i \otimes (m_i \otimes u_i).$$

For the proof we consider the mapping

$$\varphi_u: A \times M \ni (a, m) \mapsto a \otimes (m \otimes u) \in A \underset{R}{\otimes} (M \underset{S}{\otimes} U)$$

with respect to a fixed $u \in U$.

As we see immediately, this is an R-tensorial mapping, so that by 10.1.8 the homomorphism

$$\lambda_u : A \underset{R}{\otimes} M \ni \sum a_i \otimes m_i \mapsto \sum a_i \otimes (m_i \otimes u) \in A \underset{R}{\otimes} (M \underset{S}{\otimes} U)$$

exists. Consequently the element $\sum a_i \otimes (m_i \otimes u)$ is uniquely determined by $\sum a_i \otimes m_i$ and u (independently of the representation of $\sum a_i \otimes m_i$). Consequently the mapping

$$(A \underset{R}{\otimes} M) \times U \ni (\sum a_i \otimes m_i, u) \mapsto \sum a_i \otimes (m_i \otimes u) \in A \underset{R}{\otimes} (M \underset{S}{\otimes} U)$$

is an S-tensorial mapping. By 10.1.8(*) is then a homomorphism ρ. Similarly a corresponding homomorphism σ exists in the reverse direction, and hence we have obviously

$$\sigma\rho = 1_{(A \otimes M) \otimes U}, \qquad \rho\sigma = 1_{A \otimes (M \otimes U)};$$

thus ρ and σ are isomorphisms.

On the basis of the associativity of the tensor product we can omit brackets in many tensor products if we are not concerned about isomorphisms.

10.2.4 COMMUTABILITY OF THE TENSOR PRODUCT WITH THE DIRECT SUM

Let now modules A_S, ${}_SU$ with

$$A = \bigoplus_{i \in I} A_i, \qquad U = \bigoplus_{j \in J} U_j$$

be given. Let M_{ij} denote the subgroup of $A \underset{S}{\otimes} U$ which is generated by the elements $a_i \otimes u_j$, $a_i \in A_i$, $u_j \in U_j$. Then we have

(1) $$A \underset{S}{\otimes} U = \bigoplus_{i \in I, j \in J} M_{ij}, \qquad M_{ij} \cong A_i \underset{S}{\otimes} U_j$$

and consequently

(2) $$\left(\bigoplus_{i \in I} A_i\right) \underset{S}{\otimes} \left(\bigoplus_{j \in J} U_j\right) \cong \bigoplus_{i \in I, j \in J} (A_i \underset{S}{\otimes} U_j).$$

Proof of (1). By definition of M_{ij} we have first of all $A \underset{S}{\otimes} U = \sum_{i \in I, j \in J} M_{ij}$. Let

$$\iota_i : A_i \to A, \qquad \iota_j' : U_j \to U$$

be the inclusion mappings and let also

$$\pi_i : A \to A_i, \qquad \pi_j' : U \to U_j$$

be the projections with respect to the direct sums. Then we have

$$\pi_i \iota_i = 1_{A_i}, \qquad \pi_j' \iota_j' = 1_{U_j},$$

and consequently

$$1_{A_i \otimes U_j} = \pi_i \iota_i \otimes \pi_j' \iota_j' = (\pi_i \otimes \pi_j')(\iota_i \otimes \iota_j').$$

Hence $\iota_i \otimes \iota_j'$ is a monomorphism with $\mathrm{Im}(\iota_i \otimes \iota_j') \hookrightarrow M_{ij}$. By definition of M_{ij} and $\iota_i \otimes \iota_j'$ we even have $\mathrm{Im}(\iota_i \otimes \iota_j') = M_{ij}$, i.e. $\iota_i \otimes \iota_j'$ induces (by restriction of the codomains) an isomorphism ω_{ij} between $A_i \underset{S}{\otimes} U_j$ and M_{ij}. This means that we do not have to differentiate between the elements $a_i \otimes u_j \in A_i \underset{S}{\otimes} U$ with $a_i \in A_i$, $u_j \in U_j$ and the elements $a_i \otimes u_j \in A_i \underset{S}{\otimes} U_j$.

Note: The first $a_i \otimes u_j$ is regarded as an element from $A \underset{S}{\otimes} U$, the second $a_i \otimes u_j$ is regarded as an element from $A_i \underset{S}{\otimes} U_j$.

Since ω_{ij} is an isomorphism, it follows that $\pi_i \otimes \pi_j' \,|\, M_{ij}$ is also an isomorphism. Hence we have

$$\omega_{ij}(\pi_i \otimes \pi_j') \,|\, M_{ij} = 1_{M_{ij}}$$

and consequently $\omega_{ij}(\pi_i \otimes \pi_j')$ is the projection of $A \underset{S}{\otimes} U$ on M_{ij}. Therefore we obtain finally $A \underset{S}{\otimes} U = \oplus M_{ij}$. □

10.2.5 THE TENSOR PRODUCT OF FREE MODULES

Now let A_S be a free S-module with basis $\{x_l \mid l \in L\}$, so that in consequence, $A = \bigoplus_{l \in L} x_l S$ holds.

PROPOSITION. *Every element of* $A \underset{S}{\otimes} U$ *is representable as a finite sum*

$$\sum x_l \otimes u_l, \qquad u_l \in U$$

in which the $u_l \neq 0$ *are uniquely determined.*

Proof. By the use of 10.2.4 $x_l S \cong S$ and $x_l S \otimes_S U \cong S \otimes_S U \cong U$. The proof can also be inferred directly from 10.1.8. By the distributive law for the tensor product (10.1.2) it is clear that every element from $A \otimes_S U$ can be written as a finite sum $\sum x_l \otimes u_l$. The uniqueness remains to be shown. Let

$$a = \sum x_l s_l, \qquad s_l \in S$$

be the representation of $a \in A$ in terms of a basis and let $k \in L$ be fixed.

$$(a, u) \mapsto \begin{cases} s_k u, & \text{if } x_k \text{ appears in the basis representation of } a, \\ 0, & \text{otherwise,} \end{cases}$$

is obviously an S-tensorial mapping $A \times U \to U$. Consequently there exists a homomorphism $A \otimes_S U \to U$ for which the following holds:

$$\sum x_l \otimes u_l \mapsto \begin{cases} u_k, & \text{if } x_k \text{ appears in the sum } \sum x_l \otimes u_l, \\ 0, & \text{otherwise,} \end{cases}$$

Since the image with regard to a homomorphism (independently of the representation $\sum x_l \otimes u_l$) is uniquely determined, the uniqueness of the $u_k \neq 0$ follows. □

If A and U are vector spaces over the same field of dimension m and n then the tensor product is a vector space of dimension mn over this field. More generally we have the following.

PROPOSITION. *Let S be a commutative ring, let A_S be a free S-module with a basis $x_1, \ldots, x_m$ and let ${}_SU$ be a free S-module with a basis $z_1, \ldots, z_n$, then $A \otimes_S U$ is a free S-module with the basis*

$$\{x_i \otimes z_j \mid i = 1, \ldots, m; j = 1, \ldots, n\}.$$

Proof. This follows from 10.2.4 or from the preceding proposition. Accordingly the $u_l \neq 0$ in $\sum x_l \otimes u_l$ are uniquely determined thus also the coefficients $\neq 0$ in the representation

$$u_l = \sum s_{lk} z_k$$

of u_l in terms of a basis. Then in the representation $\sum x_l \otimes u_l = \sum (x_l \otimes z_k) s_{lk}$ the $s_{lk} \neq 0$ are uniquely determined. □

10.3 FUNCTORIAL PROPERTIES OF THE TENSOR PRODUCT

Let M_S resp. ${}_S\mathsf{M}$ denote the categories of the (unitary) right resp. left S-modules and A the category of $\mathbb{Z}$-modules, i.e. of abelian groups (definition see Chapter 1).

10.3.1 THEOREM. *The tensor product is a functor of $\mathsf{M}_S \times {}_S\mathsf{M}$ into A which is covariant in both arguments.*

Proof. It is to be shown that the conditions of 1.3.4 are satisfied. First of all it is clear that

$$\mathrm{Obj}(\mathsf{M}_S)\times\mathrm{Obj}({}_S\mathsf{M}) \ni (A, U) \mapsto A \underset{S}{\otimes} U \in \mathsf{A}$$

and

$$\mathrm{Hom}_S(A, B)\times\mathrm{Hom}_S(U, V) \ni (\alpha, \mu) \mapsto \alpha \otimes \mu \in \mathrm{Hom}_{\mathbb{Z}}(A \underset{S}{\otimes} U, B \underset{S}{\otimes} V)$$

are mappings with the proper codomains. Further we have, as shown in 10.2.1,

$$1_A \otimes 1_U = 1_{A\underset{S}{\otimes}U}, \qquad \beta\alpha \otimes \nu\mu = (\beta \otimes \nu)(\alpha \otimes \mu).$$

Hence the theorem is proved. □

In addition we may observe that the tensor product can be considered as a covariant functor of the form

$$\underset{S}{\otimes} : {}_R\mathsf{M}_S \times {}_S\mathsf{M}_T \to {}_R\mathsf{M}_T.$$

We direct our attention now to the proof of the fact that the tensor product and Hom for a suitable fixed argument are adjoint functors in the other argument. We deduce this as a special case from the following general theorem. In order to understand the formulation of this theorem we have first to recall some earlier statements.

Let the modules X_S, ${}_SU_T$, Y_T be given. If we apply the homomorphisms from $\mathrm{Hom}_T(U, Y)$ on the left of the elements of U, i.e. let $\mu(u)$ be the image of $u \in U$ under $\mu \in \mathrm{Hom}_T(U, Y)$, then by the following prescription $\mathrm{Hom}_T(U, Y)$ becomes a right S-module

$$(\mu s)(u) := \mu(su), \qquad u \in U, \quad s \in S, \quad \mu \in \mathrm{Hom}_T(U, Y).$$

In this sense then $\mathrm{Hom}_T(U, Y)$ is to be considered as a right S-module in $\mathrm{Hom}_S(X, \mathrm{Hom}_T(U, Y))$. Further, with regard to ${}_SU_T$, $X \underset{S}{\otimes} U$ is a right T-module which appears as such in $\mathrm{Hom}_T(X \underset{S}{\otimes} U, Y)$.

Now let a homomorphism $\Phi_{(X,U,Y)}$ of $\mathrm{Hom}_T(X \underset{S}{\otimes} U, Y)$ into $\mathrm{Hom}_S(X, \mathrm{Hom}_T(U, Y))$ (as additive groups) be given. For this purpose for every $\rho \in \mathrm{Hom}_T(X \underset{S}{\otimes} U, Y)$ there must be made explicit an image $\rho^* \in \mathrm{Hom}_S(X, \mathrm{Hom}_T(U, Y))$. For $x \in X$ we must then have $\rho^*(x) \in \mathrm{Hom}_T(U, Y)$. The application of $\rho^*(x)$ on $u \in U$ is to be written in the form $\rho^*(x)(u)$. We now define:

$$\rho^*(x)(u) := \rho(x \otimes u), \qquad x \in X, \quad u \in U, \quad x \otimes u \in X \underset{S}{\otimes} U,$$

$$\rho \in \mathrm{Hom}_T(X \underset{S}{\otimes} U, Y).$$

By this means ρ^* is evidently uniquely defined for every $x \in X$ and $u \in U$. If we now consider for $x_1, x_2 \in X$, $u_1, u_2 \in U$, $s_1, s_2 \in S$, $t_1, t_2 \in T$

$$\begin{aligned}\rho^*(x_1s_1 + x_2s_2)(u_1t_1 + u_2t_2) &= \rho((x_1s_1 + x_2s_2) \otimes (u_1t_1 + u_2t_2)) \\ &= \rho(x_1 \otimes s_1u_1)t_1 + \rho(x_1 \otimes s_1u_2)t_2 + \rho(x_2 \otimes s_2u_1)t_1 + \rho(x_2 \otimes s_2u_2)t_2 \\ &= \rho^*(x_1)(s_1u_1)t_1 + \rho^*(x_1)(s_1u_2)t_2 + \rho^*(x_2)(s_2u_1)t_1 + \rho^*(x_2)(s_2u_2)t_2,\end{aligned}$$

from which it follows that $\rho^* \in \mathrm{Hom}_S(X_S, \mathrm{Hom}_T(U, Y))$. Let now $\rho_1, \rho_2 \in \mathrm{Hom}_T(X \underset{S}{\otimes} U, Y)$, then evidently we have

$$\begin{aligned}(\rho_1 + \rho_2)^*(x)(u) &= (\rho_1 + \rho_2)(x \otimes u) \\ &= \rho_1(x \otimes u) + \rho_2(x \otimes u) = \rho_1^*(x)(u) + \rho_2^*(x)(u),\end{aligned}$$

thus

$$(\rho_1 + \rho_2)^* = \rho_1^* + \rho_2^*.$$

Altogether

(10.3.2) $$\Phi_{(X,U,Y)}: \mathrm{Hom}_T(X \underset{S}{\otimes} U, Y) \ni \rho \mapsto \rho^* \in \mathrm{Hom}_S(X, \mathrm{Hom}_T(U, Y))$$

with

$$\rho^*(x)(u) := \rho(x \otimes u), \qquad x \in X, \quad u \in U$$

is a homomorphism of the additive groups.

10.3.3 THEOREM. (1) *For every triple* X_S, ${}_SU_T$, Y_T, $\Phi_{(X,U,Y)}$ *is an isomorphism.*

(2) *Let* $\xi : X'_S \to X_S$, $\mu : {}_SU'_T \to {}_SU_T$, $\eta : Y_T \to Y'_T$, *then the following diagram is commutative*:

$$\begin{array}{ccc}
\mathrm{Hom}_T(X \underset{S}{\otimes} U, Y) & \xrightarrow{\Phi_{(X,U,Y)}} & \mathrm{Hom}_S(X, \mathrm{Hom}_T(U, Y)) \\
\Big\downarrow {\scriptstyle \mathrm{Hom}(\xi\otimes\mu,\eta)} & & \Big\downarrow {\scriptstyle \mathrm{Hom}(\xi,\mathrm{Hom}(\mu,\eta))} \\
\mathrm{Hom}_T(X' \underset{S}{\otimes} U', Y') & \xrightarrow{\Phi_{(X',U',Y')}} & \mathrm{Hom}_S(X', \mathrm{Hom}_T(U', Y')).
\end{array}$$

Proof. (1) Put $\Phi := \Phi_{(X,U,Y)}$. Φ is a monomorphism for $\rho^* = 0$ signifies that $\rho^*(x)(u) = \rho(x \otimes u) = 0$ for all $x \in X$, $u \in U$, thus $\rho = 0$. Let $\sigma \in \mathrm{Hom}_S(X, \mathrm{Hom}_T(U, Y))$, then consider the S-tensorial mapping

$$X \times U \ni (x, u) \mapsto \sigma(x)(u) \in Y;$$

in addition there is a T-homomorphism

$$\rho : X \underset{S}{\otimes} U \ni \sum x_i \otimes u_i \mapsto \sum \sigma(x_i)(u_i) \in Y.$$

For this ρ we then have

$$\rho^*(x)(u) = \rho(x \otimes u) = \sigma(x)(u),$$

i.e. $\Phi(\rho) = \sigma$, thus Φ is also an epimorphism and consequently an isomorphism.

(2) By running down the left edge we obtain for $\rho \in \mathrm{Hom}_T(X \underset{S}{\otimes} U, Y)$:

$$\rho \mapsto \mathrm{Hom}(\xi \otimes \mu, \eta)(\rho) = \eta\rho(\xi \otimes \mu) \mapsto (\eta\rho(\xi \otimes \mu))^*$$

and similarly from the right edge

$$\rho \mapsto \rho^* \mapsto \mathrm{Hom}(\xi, \mathrm{Hom}(\mu, \eta))(\rho^*) = \mathrm{Hom}(\mu, \eta)\rho^*\xi.$$

If we apply the mappings first on the right to $x' \in X'$ and subsequently to $u' \in U'$ then we obtain

$$(\eta\rho(\xi \otimes \mu))^*(x')(u') = (\eta\rho(\xi \otimes \mu))(x' \otimes u') = \eta\rho(\xi x' \otimes \mu u')$$

$$\begin{aligned}(\mathrm{Hom}(\mu, \eta)\rho^*\xi)(x')(u') &= (\mathrm{Hom}(\mu, \eta)\rho^*)(\xi x')(u') = \eta(\rho^*(\xi x')(\mu u')) \\ &= \eta\rho(\xi x' \ \mu u').\end{aligned}$$

Consequently the diagram is commutative and the theorem is proved. □

10.3.4 COROLLARY. *For every* $U \in {}_S M_T$

$$-\underset{S}{\otimes} U : M_S \to M_T \qquad \text{and} \qquad \mathrm{Hom}_T(U, -) : M_T \to M_S$$

form a pair of adjoint functors.

(For the definition of adjoint functors, see Chapter 1.)

Proof. This follows from 10.3.3 for $\mu = 1_U$. □

10.3.5 *Remark.* 10.3.3 and 10.3.4 hold analogously also for ${}_S X$, ${}_R U_S$, ${}_R Y$ (permutation of the sides). In particular in the place of (1) in 10.3.3 the isomorphism

$$\mathrm{Hom}_R(U \underset{S}{\otimes} X, Y) \cong \mathrm{Hom}_S(X, \mathrm{Hom}_R(U, Y)),$$

appears and the adjoint functors in 10.3.4 are now

$$U \underset{S}{\otimes} - : {}_S M \to {}_R M,$$

$$\mathrm{Hom}_R(U, -) : {}_R M \to {}_S M.$$

Since important applications of the adjointness of ⊗ and Hom are treated later, we can here forego examples. In the next section the first application already follows.

10.4 FLAT MODULES AND REGULAR RINGS

Let $A \hookrightarrow R$ be a two-sided ideal of the ring R and let $\iota : A \to R$ be the inclusion mapping. We consider then

$$\iota \otimes 1 : A \underset{R}{\otimes} R/A \to R \underset{R}{\otimes} R/A \qquad (\text{where } 1 = 1_{R/A}).$$

PROPOSITION

(1) $$\iota \otimes 1 = 0.$$

(2) $$A \underset{R}{\otimes} R/A \cong A/A^2; \quad \textit{thus} \quad A \underset{R}{\otimes} R/A \neq 0 \quad \textit{for} \quad A \neq A^2.$$

Proof. (1) For $a \in A$, $\bar{r} \in R/A$ we have

$$(\iota \otimes 1)(a \otimes \bar{r}) = a \otimes \bar{r} = 1 \cdot a \otimes \bar{r} = 1 \otimes \overline{ar} = 1 \otimes \bar{0} = 0,$$

thus $\iota \otimes 1 = 0$.

(2) Now let $\hat{a} := a + A^2 \in A/A^2$ for $a \in A$. Since the mapping

$$A \times R/A \ni (a, \bar{r}) \mapsto \widehat{ar} \in A/A^2$$

is R-tensorial and surjective, there is an epimorphism

$$\lambda : A \underset{R}{\otimes} R/A \to A/A^2$$

with $\lambda(a \otimes r) = \hat{a}r$. Let

$$\sum_{i=1}^{n} a_i \otimes \bar{r}_i = \sum_{i=1}^{n} a_i r_i \otimes \bar{1} = \left(\sum_{i=1}^{n} a_i r_i\right) \otimes \bar{1} \in \operatorname{Ker}(\lambda),$$

then it follows that

$$\sum_{i=1}^{n} a_i r_i \in A^2,$$

thus

$$\sum_{i=1}^{n} a_i r_i = \sum_{j=1}^{k} a'_j a''_j \text{ with } a'_j, a''_j \in A.$$

Consequently we have

$$\left(\sum_{i=1}^{n} a_i r_i\right) \otimes \bar{1} = \left(\sum_{j=1}^{k} a'_j a''_j\right) \otimes \bar{1} = \sum_{j=1}^{k} (a'_j a''_j \otimes \bar{1})$$

$$= \sum_{j=1}^{k} a'_j \otimes \bar{a}''_j = \sum_{j=1}^{k} a_j \otimes \bar{0} = 0,$$

i.e. λ is also a monomorphism, thus in fact an isomorphism. In the case that $A^2 \neq A$ (e.g. $A = n\mathbb{Z} \hookrightarrow \mathbb{Z}$ with $n > 1$) $\iota : A \to R$ is thus a monomorphism but $\iota \otimes 1$ is not a monomorphism. □

On the other hand there are modules ${}_RM$ so that for every monomorphism $\alpha : A_R \to B_R$

$$\alpha \otimes 1_M : A \underset{R}{\otimes} M \to B \underset{R}{\otimes} M$$

is also a monomorphism. As we show in the following this property is satisfied, for example, by all projective modules. Such modules are of interest in many respects. They are now to be investigated.

10.4.1 *Definition.* ${}_RM$ is called a *flat module* if for every monomorphism

$$\alpha : A_R \to B_R$$

$\alpha \otimes 1_M$ is also a monomorphism.

10.4.2 COROLLARY. *Every isomorphic image of a flat module is flat.*

Proof. Let ${}_RM$ be flat and let $\varphi : {}_RM \to {}_RN$ be an isomorphism. Then we have the commutative diagram:

$$\begin{array}{ccc} A \underset{R}{\otimes} M & \xrightarrow{\alpha\otimes 1_M} & B \underset{R}{\otimes} M \\ \downarrow{\scriptstyle 1_A\otimes\varphi} & & \downarrow{\scriptstyle 1_B\otimes\varphi} \\ A \underset{R}{\otimes} N & \xrightarrow{\alpha\otimes 1_N} & B \underset{R}{\otimes} N. \end{array}$$

Since $1_A \otimes \varphi$ and $1_B \otimes \varphi$ are isomorphisms, $\alpha \otimes 1_N$ is a monomorphism if and only if $\alpha \otimes 1_M$ is a monomorphism. □

10.4.3 THEOREM. *Let*

$${}_RM = \coprod_{i\in I} M_i \qquad \left(\text{or } {}_RM = \bigoplus_{i\in I} M_i\right),$$

then we have: M is then flat if and only if all M_i, $i \in I$ are flat.

Proof. By 10.4.2 it suffices to consider the case $M = \coprod_{i\in I} M_i$ in which the elements are denoted as in Chapter 4 by (m_i) (with only finitely many $m_i \neq 0$). Then the diagram

$$\begin{array}{ccc} A \underset{R}{\otimes} \left(\coprod_{i\in I} M_i\right) & \xrightarrow{\alpha\otimes 1_M} & B \underset{R}{\otimes} \left(\coprod_{i\in I} M_i\right) \\ \downarrow & & \downarrow \\ \coprod_{i\in I} (A \underset{R}{\otimes} M_i) & \xrightarrow{\coprod(\alpha\otimes 1_{M_i})} & \coprod_{i\in I} (B \underset{R}{\otimes} M_i) \end{array}$$

is commutative; letting the vertical mappings be the isomorphisms defined in 10.2.4 (e.g. we have for the left isomorphism $a \otimes (m_i) \mapsto (a \otimes m_i)$). It follows that $\alpha \otimes 1_M$ is a monomorphism if and only if $(\alpha \otimes 1_{M_i})$ is a monomorphism and this is the case if and only if $\alpha \otimes 1_{M_i}$ is a monomorphism for every $i \in I$. Hence the assertion follows. □

10.4.4 THEOREM. *Every projective module is flat.*

Proof. Since, as we know, every projective module is isomorphic to a direct summand of a free module, it suffices by 10.4.2 and 10.4.3 to prove the assertion for ${}_RR$. In the commutative diagram

$$\begin{array}{ccc} A \underset{R}{\otimes} R & \xrightarrow{\alpha \otimes 1_R} & B \underset{R}{\otimes} R \\ \downarrow \cong & & \downarrow \cong \\ A & \xrightarrow{\alpha} & B \end{array}$$

$\alpha \otimes 1_R$ is then a monomorphism if and only if α is a monomorphism. □

If we consider this result, the question immediately and naturally arises whether the converse holds and as to what assumptions are necessary. In 1960 H. Bass characterized those rings R for which every flat R-module is projective. They are characterized by the following equivalent conditions:

(1) $R/\mathrm{Rad}(R)$ is semisimple and $\mathrm{Rad}(R)$ is right transfinitely nilpotent, i.e. to every sequence $a_1, a_2, a_3, \ldots$ of elements from $\mathrm{Rad}(R)$ there is an n with $a_1 a_2 \ldots a_n = 0$.

(2) R satisfies the minimal condition for principal right ideals (= cyclic right ideals).

(3) Every left R-module ${}_RM$ has a projective cover, i.e. there exists an epimorphism ${}_RP \to {}_RM$ with projective P and small kernel. A ring with these (and further equivalent) properties is called *left perfect.* By (1) resp. (2) every left resp. right artinian ring is left perfect. We shall later discuss thoroughly the theory of perfect rings.

A second related question concerns the rings R for which every R-module is flat. The main aim of the following considerations is to characterize these rings.

10.4.5 LEMMA. *Let B_R and ${}_RM$ be given.*

(a) *If $0 = \sum b_i \otimes m_i \in B \underset{R}{\otimes} M$, holds, then there are finitely generated submodules $B_0 \hookrightarrow B$, $M_0 \hookrightarrow M$ with $b_i \in B_0$, $m_i \in M_0$ and*

$$0 = \sum b_i \otimes m_i \in B_0 \underset{R}{\otimes} M_0.$$

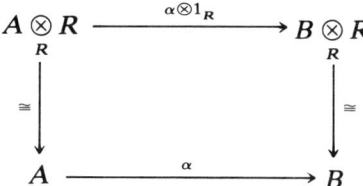

(b) *Let* $B_1 \hookrightarrow B$, $M_1 \hookrightarrow M$ *and let* $0 = \sum b_i \otimes m_i \in B_1 \underset{R}{\otimes} M_1$ *hold, then it follows that*

$$0 = \sum b_i \otimes m_i \in B \underset{R}{\otimes} M.$$

Proof. (a) As generating elements of B_0 resp. of M_0 we take firstly the b_i resp. m_i occurring in $\sum b_i \otimes m_i$, so that we have $\sum b_i \otimes m_i \in B_0 \underset{R}{\otimes} M_0$. In order to conclude that $\sum b_i \otimes m_i = 0 \in B_0 \underset{R}{\otimes} M_0$ further elements are needed. In the sense of 10.1.1 $\sum b_i \otimes m_i = 0 \in B \underset{R}{\otimes} M$, indicates that $\sum (b_i, m_i) \in K$ where $K = K(B, M)$ depends on B and M. With regard to the representation of $\sum (b_i, m_i)$ as an element in K there occur only finitely many first components from B resp. second components from M. These are subsumed as generating elements for B_0 resp. M_0 so that we then have $\sum (b_i, m_i) \in K(B_0, M_0)$, thus $0 = \sum b_i \otimes m_i \in B_0 \underset{R}{\otimes} M_0$.

(b) Let $\iota_{B_1}: B_1 \to B$ and $\iota_{M_1}: M_1 \to M$ be the inclusion mappings. Then we have

$$0 = (\iota_{B_1} \otimes \iota_{M_1})(0) = (\iota_{B_1} \otimes \iota_{M_1})(\textstyle\sum b_i \otimes m_i) = \sum b_i \otimes m_i \in B \underset{R}{\otimes} M. \qquad \square$$

10.4.6 COROLLARY. *If* ${}_RM$ *is a module such that every finitely generated submodule of* M *is contained in a flat submodule then* M *is flat.*

Proof. Let $\alpha: A_R \to B_R$ be a monomorphism and let $\sum a_i \otimes m_i \in \operatorname{Ker}(\alpha \otimes 1_M)$. Then by 10.4.5(a) there is a finitely generated submodule $M_0 \hookrightarrow M$ so that $\sum a_i \otimes m_i \in A \underset{R}{\otimes} M_0$ and $\sum a_i \otimes m_i \in \operatorname{Ker}(\alpha \otimes 1_{M_0})$.

Let $M_0 \hookrightarrow M_1 \hookrightarrow M$ and let M_1 be flat, then by 10.4.5(b) it follows that $\sum a_i \otimes m_i \in \operatorname{Ker}(\alpha \otimes 1_{M_1})$. Since M_1 is flat we must have $\sum a_i \otimes m_i = 0 \in A \underset{R}{\otimes} M_1$ and by 10.4.5(b) it follows that $\sum a_i \otimes m_i = 0 \in A \underset{R}{\otimes} M$, which was to be shown. $\square$

10.4.7 COROLLARY. *If for a homomorphism* $\alpha: A_R \to B_R$ *and a module* ${}_RM$

$$\alpha \otimes 1_M: A \underset{R}{\otimes} M \to B \underset{R}{\otimes} M$$

is not a monomorphism, then there is a finitely generated submodule $A_0 \hookrightarrow A$ *such that* $(\alpha \mid A_0) \otimes 1_M$ *is not a monomorphism.*

Proof. By assumption there is an element $0 \neq \sum a_i \otimes m_i \in \operatorname{Ker}(\alpha \otimes 1_M)$. Let A_0 be the submodule of A generated by the a_i appearing in $\sum a_i \otimes m_i$, then by 10.4.5(b) we have

$$0 \neq \sum a_i \otimes m_i \in A_0 \underset{R}{\otimes} M$$

and as before

$$((\alpha \mid A_0) \otimes 1_M)(\sum a_i \otimes m_i) = \sum \alpha(a_i) \otimes m_i = 0 \in B \underset{R}{\otimes} M. \qquad \square$$

In order to verify whether a module ${}_RM$ is flat, by virtue of this corollary, we can confine ourselves to monomorphisms $\alpha : A \to B$ in which A is finitely generated. The question arises as to whether we can still further restrict the class of necessary "test monomorphisms" $\alpha : A \to B$. We are led back in this situation to injective modules and to the application of Baer's Criterion.

The reduction to injective modules is facilitated by the help of an injective cogenerator of $\mathcal{M}_{\mathbb{Z}}$. Let D be an injective cogenerator, say $D = \mathbb{Q}/\mathbb{Z}$ (see 5.8.6) then for $X \in \mathcal{M}_{\mathbb{Z}}$ define

$$X^\circ := \operatorname{Hom}_{\mathbb{Z}}(X, D),$$

so that X° is again a $\mathbb{Z}$-module. For $X = {}_RM$ by setting (see 3.6)

$$(\varphi r)(m) = \varphi(rm), \qquad \varphi \in M^\circ,\ r \in R,\ m \in M$$

M° becomes a right R-module and can then be considered as a $\mathbb{Z}$-R-bimodule. For arbitrary $\mu : {}_RM \to {}_RN$ let

$$\mu^\circ := \operatorname{Hom}(\mu, 1_D) : N^\circ \to M^\circ,$$

then $^\circ$ is a contravariant functor of ${}_R\mathcal{M}$ into $\mathcal{M}_R$.

10.4.8 THEOREM. *The following are equivalent for* ${}_RM$:

(1) ${}_RM$ *is flat.*

(2) *For every finitely generated right ideal* $A \hookrightarrow R_R$ *with*

$$\iota_A : A_R \to R_R$$

as the inclusion mapping $\iota_A \otimes 1_M$ *is a monomorphism.*

(3) $M_R^\circ = \operatorname{Hom}_{\mathbb{Z}}(M, D)$ *is injective.*

Proof. The following are equivalent for a homomorphism $\alpha : A \to B$:

(a) $\alpha \otimes 1_M$ is a monomorphism.

(b) $\operatorname{Hom}(\alpha \otimes 1_M, 1_D) : (B \underset{R}{\otimes} M)^\circ \to (A \underset{R}{\otimes} M)^\circ$ is an epimorphism.

(c) $\operatorname{Hom}(\alpha, \operatorname{Hom}(1_M, 1_D)) = \operatorname{Hom}(\alpha, 1_{M^\circ}) : \operatorname{Hom}_R(B, M^\circ) \to \operatorname{Hom}_R(A, M^\circ)$ is an epimorphism.

But (a)⇔(b) holds by 5.8.4, and (b)⇔(c) by 10.3.3. If we demand the validity of (a), (b), (c) for every monomorphism α, then (a) implies that ${}_RM$ is flat and (c) implies that M_R° is injective; i.e. we have therefore proved (1)⇔(3). According to Baer's Criterion 5.7.1 M_T° is then injective if and only if (c) holds for all inclusions $\iota_A : A_R \to B_R$. If therefore we again return to (a), then it follows that M_R° is injective if and only if for every $A_R \hookrightarrow R_R$ $\iota_A \otimes 1_M$ is a monomorphism. Finally, by virtue of 10.4.7 we can restrict ourselves to finitely generated right ideals $A_R \hookrightarrow R_R$ and so we have (3)⇔(2). □

We now answer the question of those rings for which every module is flat. In this context we recall that the rings for which every module is projective resp. injective, are semisimple rings. Since, as was established before, every projective module is flat the semisimple rings are in any event subsumed by those rings which are characterized in the following theorem.

10.4.9 THEOREM. *The following conditions are equivalent for a ring R:*
(1) *Every module ${}_RM$ is flat.*
(2) *For every element $r \in R$ there exists an element $r' \in R$ with $rr'r = r$.*
(3) *Every cyclic right ideal of R is a direct summand of R_R.*
(4) *Every finitely generated right ideal of R is a direct summand of R_R.*

It is clear that condition (2) is symmetric with regard to sides so that the corresponding left-sided conditions are equivalent to those above.

10.4.10 *Definition.* A ring R, which satisfies the conditions of 10.4.9, is called a *regular ring.*

Proof of 10.4.9. "(1)⇒(2)": For $r \in R$ we consider the inclusion $\iota : rR \to R$. Then by assumption

$$\iota \otimes 1_{R/Rr}: rR \otimes_R (R/Rr) \to R \otimes_R (R/Rr)$$

is a monomorphism. Since

$$(\iota \otimes 1_{R/Rr})(r \otimes \bar{1}) = r \otimes \bar{1} = 1 \otimes r\bar{1} = 1 \otimes \bar{r} = 1 \otimes \bar{0} = 0$$

we must have $0 = r \otimes \bar{1} \in rR \otimes_R (R/Rr)$. As before we denote $\bar{y} := y + Rr \in R/Rr$ and let $\widehat{rz} := rz + rRr \in rR/rRr$. Then evidently

$$rR \times R/Rr \ni (rx, \bar{y}) \mapsto \widehat{rxy} \in rR/rRr$$

is an R-tensorial mapping and consequently

$$\tau: rR \underset{R}{\otimes} (R/Rr) \ni \sum rx_i \otimes \bar{y}_i \mapsto \sum \widehat{rx_i y_i} \in rR/rRr$$

is a homomorphism (of additive groups, and indeed even an isomorphism). As $0 = r \otimes \bar{1} \in rR \underset{R}{\otimes} (R/Rr)$

$$\tau(r \otimes \bar{1}) = \hat{r} = 0 \in rR/rRr,$$

thus $r \in rRr$, i.e. there is an $r' \in R$ with $rr'r = r$.

"(2)⇒(3)": From $rr'r = r$ it follows that $(rr')(rr') = (rr'r)r' = rr'$, thus $e := rr'$ is an idempotent so that

$$R_R = eR \oplus (1-e)R$$

follows. Further we have $eR = rr'R \hookrightarrow rR$ and on the other hand as $er = rr'r = r$ we have $rR \hookrightarrow eR$, thus altogether $rR = eR$.

"(3)⇒(4)": By induction on the number of generators we show that every finitely generated right ideal is generated by an idempotent. The beginning of the induction is provided by (3) for if $R_R = rR \oplus A$ with $1 = e_1 + e_2$, $e_1 \in rR$, $e_2 \in A$, then e_1, e_2 are orthogonal idempotents with $rR = e_1R$, $A = e_2R$ (see 7.2.3). Let now

$$B := r_1R + \ldots + r_nR \hookrightarrow R_R$$

be given. By the induction hypothesis there is an idempotent $e \in R$ with $eR = r_1R + \ldots + r_{n-1}R$. Then as $r_n = er_n + (1-e)r_n$ we have

$$r_nR \hookrightarrow er_nR + (1-e)r_nR$$

and consequently

$$B = eR + r_nR = eR + (1-e)r_nR.$$

As shown at the beginning of the induction, there is an idempotent $f \in R$ with

$$fR = (1-e)r_nR,$$

so that $eR + r_nR = eR + fR$ holds. As $f \in (1-e)r_nR$ we have $ef = 0$. We claim that $g := e + f(1-e)$ is an idempotent with

$$gR = eR + fR = r_1R + \ldots + r_nR.$$

First of all we have $gR \hookrightarrow eR + fR$. Further we have $geR = eR \hookrightarrow gR$ as well as

$$gfR = (ef + f^2 + fef)R = f^2R = fR \hookrightarrow gR,$$

(as $ef = 0$ and $f^2 = f$), thus $gR = eR + fR$. Finally

$$g^2 = (e + f(1-e))(e + f(1-e)) = e + f(1-e)f(1-e)$$
$$= e + f^2 - f^2e = e + f(1-e) = g,$$

i.e. g is an idempotent. It follows that

$$R_R = gR \oplus (1-g)R,$$

by which (4) is proved.

"(4)⇒(1)": By 10.4.8 it suffices to verify whether for every inclusion mapping

$$\iota_A : A_R \to R_R$$

of a finitely generated right ideal $A \hookrightarrow R_R$ and for an arbitrary module ${}_RM$ the mapping $\iota_A \otimes 1_M$ is a monomorphism. Since A is a direct summand in R_R there is an idempotent g with $A = gR$. Let

$$\sum a_i \otimes m_i = \sum ga_i \otimes m_i = \sum g^2a_i \otimes m_i$$
$$= \sum g \otimes ga_im_i = g \otimes (\sum ga_im_i) \in \operatorname{Ker}(\iota_A \otimes 1_M),$$

thus

$$g \otimes \sum a_im_i = 1 \otimes \sum ga_im_i = 0 \in R \underset{R}{\otimes} M.$$

Then it follows (by 10.2.5) that $\sum ga_im_i = 0$, thus also

$$\sum a_i \otimes m_i = g \otimes (\sum ga_im_i) = g \otimes 0 = 0.$$

Consequently $\iota_A \otimes 1_m$ is a monomorphism, hence (1) is proved. □

As mentioned before every semisimple ring is regular. However, there are also regular rings which are not semisimple. In order to construct such an example let K be a regular ring (e.g. a field) and let

$$R := \prod_{i=1}^{\infty} K_i \quad \text{with} \quad K_i = K \quad \text{for } i = 1, 2, 3, \ldots.$$

By means of componentwise defined addition and similarly defined multiplication

$$(k_i) \cdot (k_i') = (k_ik_i')$$

R becomes a ring. This ring is regular. Namely let $k_ik_i'k_i = k_i$ then it follows that

$$(k_i)(k_i')(k_i) = (k_i).$$

If K is a field then we can choose

$$k_i' = \begin{cases} k_i^{-1} & \text{for } k_i \neq 0 \\ 0 & \text{for } k_i = 0. \end{cases}$$

As we easily verify $A := \coprod_{i=1}^{\infty} K_i$ is a proper two-sided ideal in $R = \prod_{i=1}^{\infty} K_i$ which is large both in R_R and in ${}_RR$. Consequently A cannot be a direct summand in R_R (or in ${}_RR$). Hence R is not semisimple and neither $(R/A)_R$ nor ${}_R(R/A)$ are projective (for then $R \to R/A$ would split). Since every R-module is flat we have in $(R/A)_R$ a flat but not projective module.

In conclusion we direct attention to the concept of a pure homomorphism which "dualizes" the concept of a flat module.

Definition 10.4.11. A monomorphism is called *pure* if $\alpha \otimes 1_M$ is a monomorphism for every R-module ${}_RM$. If $A_R \hookrightarrow B_R$ and the inclusion mapping $\iota : A \to B$ is pure then A is called a *pure submodule* of B.

10.5 FLAT FACTOR MODULES OF FLAT MODULES

We investigate here the question of those conditions under which a factor module of a flat module is again flat. This question is particularly of interest in connection with perfect rings, which are treated in the next section.

10.5.1 LEMMA. *Let ${}_RM$ be flat, let $U \hookrightarrow {}_RM$, $A \hookrightarrow R_R$ and let $\iota : A \to R$ denote the inclusion mapping. Then the following are equivalent:*

(1) $\iota \otimes 1_{M/U} : A \otimes_R (M/U) \to R \otimes_R (M/U)$ *is a monomorphism.*

(2) $U \cap AM = AU$.

Proof. "(1)⇒(2)": Let $u = \sum a_i m_i \in U \cap AM$, then for $t = \sum a_i \otimes \bar{m}_i \in A \otimes_R (M/U)$ it follows that:

$$(\iota \otimes 1_{M/U})(t) = \sum a_i \otimes \bar{m}_i = 1 \otimes \sum a_i \bar{m}_i = 1 \otimes \bar{u} = 0 \in R \otimes_R (M/U),$$

thus by assumption $t = 0$. The relation

$$A \times (M/U) \ni (a, \bar{m}) \mapsto \widehat{am} := am + AU \in AM/AU$$

is evidently an R-tensorial mapping, by which a homomorphism

$$\lambda : A \otimes_R (M/U) \to AM/AU$$

is induced. From $t=0$ it follows that

$$0=\varphi(0)=\varphi(t)=\sum \widehat{a_i m_i}=\hat{u},$$

thus $u \in AU$.

"(2)⇒(1)": Let $t=\sum a_i \otimes \bar{m}_i \in A \underset{R}{\otimes} (M/U)$ with

$$(\iota \otimes 1_{M/U})(t)=\sum a_i \otimes \bar{m}_i = 1 \otimes \sum \overline{a_i m_i}=0,$$

thus $\sum a_i m_i \in U$. By assumption there is an equation

$$\sum a_i m_i = \sum a_j' u_j \in AU \quad \text{with} \quad u_j \in U.$$

Obviously it then follows that

$$\sum a_i \otimes m_i - \sum a_j' \otimes u_j \in \operatorname{Ker}(\iota \otimes 1_M).$$

Since by assumption M is flat, we thus have $\operatorname{Ker}(\iota \otimes 1_M)=0$, this implies that $\sum a_i \otimes m_i = \sum a_j' \otimes u_j$ and consequently for $\gamma: M \to M/U$:

$$\begin{aligned} t &= (1_A \otimes \gamma)(\sum a_i \otimes m_i) = \sum a_i \otimes \bar{m}_i \\ &= (1_A \otimes \gamma)(\sum a_j' \otimes u_j) = \sum a_j' \otimes \bar{u}_j = 0 \in A \underset{R}{\otimes} (M/U). \end{aligned}$$

Thus in fact $\iota \otimes 1_{M/U}$ is a monomorphism. □

We remark that for (1)⇒(2) we have not used the assumption that ${}_RM$ is flat but only for (2)⇒(1).

10.5.2 THEOREM. *Let ${}_RM$ be flat and let $U \hookrightarrow {}_RM$. Then the following are equivalent*:

(1) *M/U is flat.*

(2) *$U \cap AM = AU$ for every finitely generated right ideal $A \hookrightarrow R_R$.*

Proof. This follows from 10.5.1 and 10.4.8. □

As is easily seen the proof of 10.5.1 (1)⇒(2) is a generalization of 10.4.9 (1)⇒(2). Conversely we can deduce 10.4.9 (1)⇒(2) from 10.5.1 resp. 10.5.2. Namely in 10.5.2 let $M={}_RR$, $U=Rr$, $A=rR$, then we have

$$Rr \cap rR \cdot R = Rr \cap rR = rR \cdot Rr = rRr;$$

as $r \in Rr \cap rR$ it follows that there is an r' with $rr'r=r$.

Theorem 10.5.2 has an interesting application for flat factor modules of projective modules.

10.5.3 THEOREM. *Let $_RP$ be projective, $U \hookrightarrow \mathrm{Rad}(P)$ and let P/U be flat, then $U = 0$.*

Proof. (1) We establish the proof firstly for a free module $_RF$ in place of $_RP$. Let $\{x_i \mid i \in I\}$ be a basis of F and let $u \in U$ with a representation in terms of the basis as

$$u = \sum a_i x_i, \qquad a_i \in R.$$

By $A = \sum a_i R$ denote the right ideal generated by the coefficients a_i of u, which by definition is finitely generated. By 10.5.2 we have

$$U \cap AF = AU,$$

thus $u = \sum b_j u_j$ with $b_j \in A$, $u_j \in U$. By assumption we have $U \hookrightarrow \mathrm{Rad}(F) = \mathrm{Rad}(R)F$ (latter equation holds by 9.2.1). Thus (since $\mathrm{Rad}(R)$ is a two-sided ideal) in the representation of

$$u_j = \sum c_{jk} x_k$$

in terms of the basis all $c_{jk} \in \mathrm{Rad}(R)$. It follows that

$$u = \sum_i a_i x_i = \sum_j \sum_k b_j c_{jk} x_k$$

and on comparing coefficients we deduce that $a_i = \sum_j b_j c_{ji} \in A\ \mathrm{Rad}(R)$.

Since this holds for all generators a_i of A it follows that $A \hookrightarrow A\ \mathrm{Rad}(R)$, thus

$$A = A\ \mathrm{Rad}(R).$$

Then by 9.2.1 we must have $A = 0$, thus we also have $u = 0$. Since $u \in U$ was arbitrary, it follows that $U = 0$. Hence the proof is established for a free module.

(2) Now let P be a direct summand of a free module F, thus

$$F = P \oplus P_1,$$

and let $U \hookrightarrow \mathrm{Rad}(P)$ and also let P/U be flat. Let $\nu : F \to F/U$; then it follows that

$$F/U = \nu(F) = \nu(P) + \nu(P_1).$$

As $U \hookrightarrow P$ we have further

$$\nu(P) + \nu(P_1) = \nu(P) \oplus \nu(P_1),$$

and also

$$\nu(P) = P + U/U = P/U, \qquad \nu(P_1) = P_1 + U/U \cong P_1/P_1 \cap U \cong P_1.$$

Consequently we have

$$F/U \cong P/U \oplus P_1.$$

Since P/U, by assumption, and P_1 (by 10.4.4), as a projective module, are flat, by 10.4.2 and 10.4.3 F/U is also flat. Since $U \hookrightarrow \text{Rad}(P) \hookrightarrow \text{Rad}(F)$ it follows, as shown above, that $U = 0$. For an arbitrary projective module the assertion holds by 10.4.3. □

Since the 0-module is flat, as a direct corollary we obtain a result already proved in 9.6.3.

10.5.4 COROLLARY. *Let ${}_RP$ be projective and let* $\text{Rad}(P) = P$, *then it follows that* $P = 0$.

EXERCISES

(1)

Let a commutative ring S as well as S-modules A and U be given. Show:

$A \underset{S}{\otimes} U \cong U \underset{S}{\otimes} A.$

(2)

Let an arbitrary ring S and let S-modules $B_S \hookrightarrow A_S$, ${}_SV \hookrightarrow {}_SU$ be given. Let $L(B, V)$ denote the subgroup of $A \underset{S}{\otimes} U$ which is generated by the elements of the form $a \otimes v$, $b \otimes v$ with $a \in A$, $b \in B$, $v \in V$, $u \in U$. Show:

$$(A/B) \underset{S}{\otimes} (U/V) \cong (A \underset{S}{\otimes} U)/L(B, V).$$

(3)

(a) Let B be a right and V be a left ideal of a ring S and let $B + V$ denote the additive subgroup of S generated by B and V. Show:

$$(S/B) \underset{S}{\otimes} (S/V) \cong S/(B + V).$$

(b) Give an example of a ring S and ideal $B_S \neq S$, ${}_SV \neq S$ with $(S/B) \underset{S}{\otimes} (S/V) = 0$.

(4)

(a) For the ideals B_S, ${}_SV \hookrightarrow S$ show:

$$B \underset{S}{\otimes} (S/V) \cong B/BV$$

where BV denotes the additive subgroup of S which is generated by the elements of the form bv with $b \in B$, $v \in V$.

(b) Give an example of the case $B_S \neq 0$, ${}_SV \neq S$ and $B \otimes_S (S/V) = 0$.

(5)

(a) Let ι_B and ι_V be the inclusion mappings of the ideals B_S and ${}_SV$ in S. Show:

$$\operatorname{Im}(\iota_B \otimes \iota_V) \cong BV.$$

(b) Give an example of the case $B \otimes_S V \neq 0$, but $\operatorname{Im}(\iota_B \otimes \iota_V) = 0$.

(6)

Let $\mathbb{Q}$ be the additive group of the rational numbers. Show:

$$\mathbb{Q} \otimes_{\mathbb{Z}} \mathbb{Q} \cong \mathbb{Q}.$$

(7)

For an abelian group A show: $A \otimes_{\mathbb{Z}} A = 0 \Leftrightarrow A$ is divisible and every element of A has finite order (see Chapter 4, Exercise 10 and 11.)

(8)

Let $S := K[x, y]$ be the polynomial ring in the indeterminates x and y with coefficients in a field K. Let $B := xS + yS$ denote the ideal of S generated by x and y. Show: The element $x \otimes y - y \otimes x \in B \otimes_S B$ is not equal to 0.

(9)

For a set H and a module M_S let

$$M^H := \prod_{h \in H} M_h \quad \text{with} \quad M_h = M \quad \text{for all } h \in H.$$

As in Chapter 4 we denote the elements of M^H by (m_h). Show for M_S:

(a) For every set H there is exactly one homomorphism.

$$\varphi_H : M \otimes_S S^H \to M^H \quad \text{with} \quad \varphi_H(m \otimes (s_k)) = (ms_k).$$

(b) If the set H is finite, then φ_H is an isomorphism.

(c) $\operatorname{Im}(\varphi_H) = \bigcup B^H$ where B runs over all finitely generated submodules of M_S.

(d) M_S is then finitely generated if and only if for every set H φ_H is an epimorphism.

(10)

Construct sets I and J and also right resp. left S-modules A_i resp. U_j so that there holds:

$$\left(\prod_{i\in I} A_i\right) \underset{S}{\otimes} \left(\prod_{j\in J} U_j\right) \not\cong \prod_{i\in I, j\in J} (A_i \underset{S}{\otimes} U_j).$$

(11)

Let a unitary ring homomorphism $\rho: R \to S$ be given. Then every right S-module M_S becomes by the definition $mr := m\rho(r)$, $m \in M$, $r \in R$ a right R-module (see 3.2). The analogue holds on the left side. Let this be assumed in the following for right resp. left S-modules.

Show for ${}_SU$:

(a) The mapping $\lambda: U \ni u \mapsto 1 \otimes u \in S \underset{R}{\otimes} U$ is a monomorphism of the left R-modules ${}_RU$ and ${}_R(S \underset{R}{\otimes} U)$.

(b) The mapping

$$\mu: S \underset{R}{\otimes} U \ni \sum s_i \otimes u_i \mapsto \sum s_i u_i \in U$$

is an S-epimorphism and the kernel of μ is generated by the elements $s \otimes u - 1 \otimes su$.

(c) ${}_R(S \underset{R}{\otimes} U) = \operatorname{Im}(\lambda) \oplus \operatorname{Ker}(\mu)$.

(d) Further let ${}_RC$ be given and let

$$\kappa: C \ni c \mapsto 1 \otimes c \in S \underset{R}{\otimes} C.$$

Then

$$\operatorname{Hom}_S({}_S(S \underset{R}{\otimes} C), {}_SU) \ni \varphi \mapsto \varphi\kappa \in \operatorname{Hom}_R({}_RC, {}_RU)$$

is an isomorphism.

(e) Let $\rho: R \to S$ and ${}_RC$ be given. Further let an ${}_SX$ be given so that an R-homomorphism $\kappa': {}_RC \to {}_RX$ exists such that for every ${}_SU$ the mapping

$$\operatorname{Hom}_S({}_SX, {}_SU) \ni \varphi \mapsto \varphi\kappa' \in \operatorname{Hom}_R({}_RC, {}_RU)$$

is an isomorphism. Show that $S \underset{R}{\otimes} U$ and X are then S-isomorphic.

(f) Give an example of a $\rho: R \to S$ and a module ${}_RC$ so that $\kappa: C \ni c \mapsto 1 \otimes c \in S \otimes_R C$ is not a monomorphism.

(12)

Let R, S be rings and let ${}_RM_S$ be an R-S bimodule. Define the functors

$$F: \mathcal{M}_R \ni A \mapsto A \otimes_R M \in \mathcal{M}_S,$$

$$G: \mathcal{M}_S \ni X \mapsto \mathrm{Hom}_S(M, X) \in \mathcal{M}_R$$

and show:

(a) F is left adjoint to G.

(b) The following are equivalent:

(1) F preserves monomorphisms.

(2) G preserves injective objects (i.e. injective $X_S \Rightarrow$ injective $\mathrm{Hom}_S(M, X)_R$).

(3) ${}_RM$ is flat.

(c) The following are equivalent:

(1) G preserves epimorphisms.

(2) F preserves projective objects (i.e. projective $A_R \Rightarrow$ projective $(A \otimes_R M)_S$).

(3) M_S is projective.

For a unitary ring homomorphism $\rho: S \to R$ we have:

(d) Q_S is injective $\Rightarrow \mathrm{Hom}_S(P, Q)$ is injective as a right R-module.

(e) P_S is projective $\Rightarrow P \otimes_S R$ is projective as a right R-module.

(13)

(a) Let ${}_RM$ be free with basis $\{e_i \mid i \in I\}$. Show that for $U \hookrightarrow M$ the following are equivalent:

(1) M/U is flat.

(2) $u \in U \Rightarrow u \in A_u U$ where A_u is the right ideal generated by the coefficients of u with respect to the given basis.

(3) $u \in U \Rightarrow$ there is $\varphi: M \to U$ with $\varphi(u) = u$.

(4) $u_1, \ldots, u_n \in U \Rightarrow$ there is $\varphi: M \to U$ with $\varphi(u_i) = u_i$ for $i = 1, \ldots, n$.

(b) Show that the equivalence of (1), (3), (4), holds also for projective ${}_RM$.

(14)

Let R be commutative and let ${}_RM$ be semisimple. Show:

(a) If ${}_RM$ is injective then it is flat.

(b) If ${}_RM$ is flat and if it has only finitely many homogeneous components then it is injective.

(c) Now give an example in which ${}_RM$ is semisimple and flat but is not injective.

(15)

(a) Show: An abelian group is flat if and only if it is torsion-free.
(b) Construct an abelian group which is flat but not projective.

(16)

Let a module ${}_RM$ be called regular if every cyclic submodule of ${}_RM$ is a direct summand. Show:

(a) In a regular module every finitely generated submodule is a direct summand.

(b) If $M_i \mid i \in I)$ is a family of regular projective R-modules, then $M = \coprod_{i \in I} M_i$ is also regular (and projective).

(Hint: Show the assertion first for $|I| = 2$.)

(c) Question: Does the statement in (b) hold without the additional assumption "projective"?

(d) If R is left noetherian or if $R/\text{Rad}(R)$ is semisimple then every regular left R-module is already semisimple.

(17)

Let R be a ring, M an R-module and $S = \text{End}(M)$. Show:

(a) S is regular $\Leftrightarrow$ for every $\alpha \in S$ $\text{Im}(\alpha)$ and $\text{Ker}(\alpha)$ are direct summands in M.

(b) R is regular $\Rightarrow$ every projective R-module is regular.

(c) R is regular and M projective and finitely generated $\Rightarrow S = \text{End}(M)$ is regular.

(d) R is regular $\Rightarrow M_n(R)$ (= ring of $n \times n$ square matrices over R) is regular.

(18)

Show that for a commutative ring R the following are equivalent:

(1) R is regular.
(2) Every (cyclic) ideal $I \hookrightarrow R$ is idempotent (i.e. $I^2 = I$).
(3) Every irreducible ideal is a prime ideal.
(4) Every irreducible ideal is maximal.
(5) Every (cyclic) R-module M has zero radical (i.e. $\text{Rad}(M) = 0$).
(6) Every simple R-module is injective.

(19)

Let G be a finite group and let T be a ring. Show: The group ring GT is regular if and only if T is regular and $\text{Ord}(G)$ is a unit in T.

Chapter 11

Semiperfect Modules and Perfect Rings

In the historical development of the structure theory of "non-commutative" rings and modules the finite-dimensional algebras were first investigated. For this the essential resource of the theory of vector spaces was available. Then later it was shown—above all beginning with E. Noether—that frequently in the investigation of the structure only chain conditions are required and that the investigation can be pursued not only for rings and their ideals but also for modules. Thus, in particular, there is obtained a structure theory for artinian rings and for modules over such rings.

The most recent development goes further in this regard. New concepts, in particular categorical and homological concepts such as projectivity, injectivity, flatness, homological dimension, etc., give rise to the possibility of extending the structure theory in different directions. For example we have already become acquainted with the decomposition theorems of injective modules over noetherian and artinian rings. Now we shall require the existence of projective covers for certain modules and under this assumption develop in a simple manner a structure theory for a class of modules and rings which embraces properly the artinian case.

In this introduction we cannot present all of the results to follow in this chapter, but nevertheless we should like to present here a particularly significant result since it gives a good impression of the considerations to follow.

THEOREM (H. BASS, 1960). *The following conditions are equivalent for a ring R:*

(1) *Every module M_R has a projective cover (i.e. there exists an epimorphism $\xi: P \to M$ with projective domain P and small kernel in P).*

(2) *Every flat right R-module is projective.*

(3) *R satisfies the descending chain condition for cyclic left ideals.*

(4) *Every left R-module* $\neq 0$ *has a socle* $\neq 0$ *and* ${}_RR$ *satisfies the minimal condition for direct summands.*

(5) $R/\mathrm{Rad}(R)$ *is semisimple and* $\mathrm{Rad}(R)$ *is left t-nilpotent; i.e. to every sequence* $a_1, a_2, a_3, \ldots$ *of elements* $a_i \in \mathrm{Rad}(R)$ *there is a* $k \in \mathbb{N}$ *with* $a_k a_{k-1} \ldots a_1 = 0$.

A ring with these equivalent properties is called *right perfect.* As (5) shows every right or left artinian ring is right perfect. The conditions (1) and (2) are of particular interest for us, for they enable us to answer two of the questions that we earlier pursued. For these reasons the theorem is also noteworthy because the "outer" properties as in (1) and (2) turn out to be equivalent to the "inner" properties as in (3) and (5).

11.1 SEMIPERFECT MODULES, BASIC CONCEPTS

We had earlier established that every module does indeed possess an injective hull but not however a projective cover. In the case $R = \mathbb{Z}$ for example, only the projective = free $\mathbb{Z}$-modules have projective covers (which are then isomorphic to the free modules). Here the existence of "sufficiently many" projective covers will be assumed.

We begin with a theorem which under the assumption of the existence of the projective cover represents the counterpart dual to 5.6.4. Evidently this theorem could already have been proved in Chapter 5; nevertheless we should like to have collected here as far as possible all considerations involving the existence of projective covers.

11.1.1 THEOREM. *Let the module* N_R *have a projective cover. If*

$$\sigma : P \to N$$

is an epimorphism with projective domain P, then there is a direct decomposition $P = P_1 \oplus P_2$ *where* $P_2 \hookrightarrow \mathrm{Ker}(\sigma)$ *and*

$$\sigma_1 := \sigma | P_1 : P_1 \to N$$

is a projective cover.

Proof. Let $\tau : P_0 \to N$ be a projective cover of N, then there exists a commutative diagram

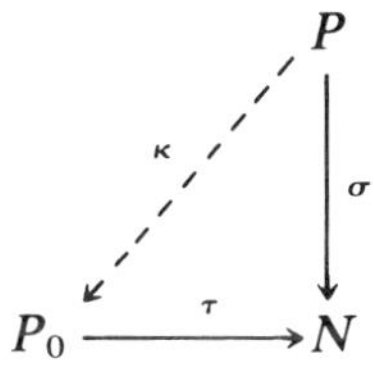

Since σ is an epimorphism, by 3.4.10 we have $P_0 = \mathrm{Im}(\kappa) + \mathrm{Ker}(\tau)$. Since $\mathrm{Ker}(\tau) \overset{\circ}{\hookrightarrow} P_0$ we have in fact $P_0 = \mathrm{Im}(\kappa)$, i.e. κ is an epimorphism. Moreover since P_0 is projective it follows by 5.3.1 that κ splits:

$$P = P_1 \oplus \mathrm{Ker}(\kappa).$$

Then

$$\kappa_1 := \kappa | P_1 : P_1 \to P_0$$

is an isomorphism. Since

$$\mathrm{Ker}(\tau\kappa_1) = \overset{-1}{\kappa_1}(\mathrm{Ker}(\tau)) \overset{\circ}{\hookrightarrow} P_1$$

(by 5.1.3)

$$\tau\kappa_1 = \sigma_1 : P_1 \to N$$

is also a projective cover of N. As $\mathrm{Ker}(\kappa) \hookrightarrow \mathrm{Ker}(\sigma)$ and $P = P_1 \oplus \mathrm{Ker}(\kappa)$ we have finally with $P_2 := \mathrm{Ker}(\kappa)$ the assertion for P_2. □

11.1.2 COROLLARY. *Let $U \hookrightarrow P$, let P be projective and let P/U have a projective cover. Then there is a decomposition $P = P_1 \oplus P_2$ with*

$$P_2 \hookrightarrow U \wedge P_1 \cap U \overset{\circ}{\hookrightarrow} P_1.$$

Proof. This follows from 11.1.1 for $\sigma = \nu : P \to P/U$. □

We notice also that from $P_2 = 0$ it follows that $P_1 = P$ and $P \cap U = U \overset{\circ}{\hookrightarrow} P$, i.e. if U contains no direct summand $\neq 0$ of P, then U is small in P.

If the existence of a projective cover is demanded for every epimorphic image of a fixed module M_R, then this already has such interesting consequences for the structure of M that we wish first to examine this situation.

11.1.3 *Definition.* Let R be an arbitrary ring and let M_R be a right R-module.

(a) M is called *semiperfect* : $\Leftrightarrow$ every epimorphic image of M has a projective cover.

(b) M is called *complemented* : $\Leftrightarrow$ every submodule of M has an addition complement (=adco, see 5.2.1) in M.

11.1.4 COROLLARY

(1) *Every epimorphic image of a semiperfect module is semiperfect.*

(2) *Every projective cover of a simple module is semiperfect.*

(3) *Every epimorphic image of a complemented module is complemented.*

Proof. (1) Clear by definition.

(2) Let $\xi: P \to E$ be the projective cover of a simple module E. Then $\operatorname{Ker}(\xi)$ is a small and maximal submodule of P. For arbitrary $U \underset{\neq}{\hookrightarrow} P$ we then have $U + \operatorname{Ker}(\xi) \underset{\neq}{\hookrightarrow} P$ and consequently $U \hookrightarrow \operatorname{Ker}(\xi)$. Thus we also have $U \overset{\circ}{\hookrightarrow} P$ and consequently $P \to P/U$ is a projective cover of P/U. Thus P is the projective cover of every epimorphic image $\neq 0$ of P, i.e. P is semiperfect.

(3) Let C be complemented, let $\gamma: C \to M$ be an epimorphism and let $B \hookrightarrow M$. We assert that $\gamma(\gamma^{-1}(B)^{\cdot})$ is a complement of B in M. Put $A := \gamma^{-1}(B)$. From $C = A + A^{\cdot}$ it follows that

$$M = \gamma(A) + \gamma(A^{\cdot}) = B + \gamma(A^{\cdot}).$$

Since $A^{\cdot}$ is an adco of A, we have $A \cap A^{\cdot} \overset{\circ}{\hookrightarrow} A^{\cdot}$. By 5.1.3(c) this implies $\gamma(A \cap A^{\cdot}) \overset{\circ}{\hookrightarrow} \gamma(A^{\cdot})$. Since also

$$\gamma(A \cap A^{\cdot}) = \gamma(\gamma^{-1}(B) \cap A^{\cdot}) = B \cap \gamma(A^{\cdot}),$$

the assertion is proved. □

Later we shall show that a finitely generated projective module P is already semiperfect if every simple image of P has a projective cover.

The next theorem shows that the investigation of semiperfect modules can be reduced essentially to the projective semiperfect modules.

11.1.5 THEOREM. *Let $\xi: P \to M$ be a projective cover of M, the following are equivalent*:

(1) *M is semiperfect.*

(2) *P is semiperfect.*

(3) *P is complemented.*

Proof. We show (2)⇒(1)⇒(3)⇒(2).

"(2)⇒(1)": Clear from the definition of semiperfect.

"(1)⇒(3)": Let $A \hookrightarrow P$, then consider the epimorphism

$$\sigma = \nu\xi: P \xrightarrow{\xi} M \xrightarrow{\nu} M/\xi(A).$$

By 11.1.1 there is a direct summand $P_1 \hookrightarrow P$ such that

$$\sigma_1 := \sigma | P_1 : P_1 \to M/\xi(A)$$

is a projective cover.

We assert that P_1 is an adco of A in P. From $\sigma(P_1)=M/\xi(A)$ it follows that $P=P_1+\mathrm{Ker}(\sigma)$. As $\mathrm{Ker}(\sigma)=\mathrm{Ker}(\nu\xi)=\xi^{-1}\,\mathrm{Ker}(\nu)=\xi^{-1}(\xi(A))=A+\mathrm{Ker}(\xi)$ it follows that $P=P_1+A+\mathrm{Ker}(\xi)$, since $\mathrm{Ker}(\xi)\overset{\circ}{\hookrightarrow}P$ we have $P=P_1+A$. For $U\hookrightarrow P_1$ let now $P=U+A$, then it follows that $\sigma(P)=\sigma(P_1)=\sigma_1(P_1)=\sigma_1(U)$ (since $\sigma(A)=0$), thus

$$P_1=\sigma_1^{-1}(\sigma_1(P_1))=\sigma_1^{-1}(\sigma_1(U))=U+\mathrm{Ker}(\sigma_1).$$

Since $\mathrm{Ker}(\sigma_1)\overset{\circ}{\hookrightarrow}P_1$ it follows that $P_1=U$, and hence P_1 is in fact an adco of A in P.

"(3)$\Rightarrow$(2)": Let $\sigma: P\to N$ be an epimorphism and let $U:=\mathrm{Ker}(\sigma)$, then let $U^{\cdot}$ be an adco of U in P. By 5.2.4 we have $U^{\cdot}\cap U=U^{\cdot}\cap\mathrm{Ker}(\sigma)\overset{\circ}{\hookrightarrow}U^{\cdot}$. We show that $U^{\cdot}$ is a direct summand of P, and thus is projective. Then it follows that

$$\sigma|U^{\cdot}: U^{\cdot}\to N$$

is a projective cover of N.

Let $U^{\cdot\cdot}$ be an adco of $U^{\cdot}$, then we assert: $P=U^{\cdot}\oplus U^{\cdot\cdot}$. For the proof let

$$\nu: P=U^{\cdot}+U^{\cdot\cdot}\to P/U^{\cdot}\cap U^{\cdot\cdot}$$

be the natural epimorphism from which, with the notation $\bar{P}:=\nu(P)$, $\bar{U}^{\cdot}:=\nu(U^{\cdot})$, $\bar{U}^{\cdot\cdot}:=\nu(U^{\cdot\cdot})$, we have evidently $\bar{P}=\bar{U}^{\cdot}\oplus\bar{U}^{\cdot\cdot}$. Further let $\pi:\bar{P}\to\bar{U}^{\cdot}$ be the projection onto $\bar{U}^{\cdot}$ corresponding to $\bar{P}=\bar{U}^{\cdot}\oplus\bar{U}^{\cdot\cdot}$. Then a commutative diagram exists

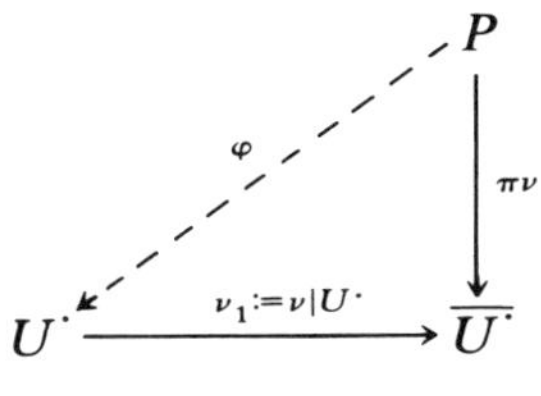

As $\pi\nu=\nu_1\varphi$ we have $\pi\nu(U^{\cdot})=\bar{U}^{\cdot}=\nu_1\varphi(U^{\cdot})$, thus $U^{\cdot}=\varphi(U^{\cdot})+\mathrm{Ker}(\nu_1)$. Since $\mathrm{Ker}(\nu_1)=U^{\cdot}\cap U^{\cdot\cdot}\overset{\circ}{\hookrightarrow}U^{\cdot}$ (see 5.2.4) it follows that $U^{\cdot}=\varphi(U^{\cdot})$, thus $P=U^{\cdot}+\mathrm{Ker}(\varphi)$. As $\mathrm{Ker}(\varphi)\hookrightarrow\mathrm{Ker}(\pi\nu)=U^{\cdot\cdot}$ and from the minimality of $U^{\cdot\cdot}$ it follows that $\mathrm{Ker}(\varphi)=U^{\cdot\cdot}$. On the other hand we have

$$U^{\cdot\cdot}=\mathrm{Ker}(\pi\nu)=\mathrm{Ker}(\nu_1\varphi)=\varphi^{-1}(\mathrm{Ker}(\nu_1))=\varphi^{-1}(U^{\cdot}\cap U^{\cdot\cdot}),$$

and since φ is an epimorphism, it follows that

$$0=\varphi(U^{\cdot\cdot})=\varphi\varphi^{-1}(U^{\cdot}\cap U^{\cdot\cdot})=U^{\cdot}\cap U^{\cdot\cdot},$$

which was to be shown. □

11.1.6 COROLLARY. *Every projective artinian module is semiperfect.*

Proof. Every artinian module is complemented. □

11.1.7 THEOREM. *If M_R is semiperfect, we have*

(a) *M is complemented.*

(b) *$M/\mathrm{Rad}(M)$ is semisimple.*

(c) *$\mathrm{Rad}(M)$ is small in M.*

Proof. (a) This follows from 11.1.4 and 11.1.5.

(b) Since $M/\mathrm{Rad}(M)$, as an epimorphic image of M, is again semiperfect, $M/\mathrm{Rad}(M)$ is complemented. Let $\Lambda \hookrightarrow M/\mathrm{Rad}(M)$, then for an adco $\Lambda^{\cdot}$ of Λ in $M/\mathrm{Rad}(M)$ we have:

$$M/\mathrm{Rad}(M) = \Lambda + \Lambda^{\cdot} \qquad \text{and} \qquad \Lambda \cap \Lambda^{\cdot} \overset{\circ}{\hookrightarrow} M/\mathrm{Rad}(M),$$

thus $\Lambda \cap \Lambda^{\cdot} \hookrightarrow \mathrm{Rad}(M/\mathrm{Rad}(M)) = 0$. Consequently we have $M/\mathrm{Rad}(M) = \Lambda \oplus \Lambda^{\cdot}$, i.e., every submodule is a direct summand and consequently $M/\mathrm{Rad}(M)$ is semisimple.

(c) Let $\xi : P \to M$ be a projective cover of M. Since $\mathrm{Ker}(\xi) \overset{\circ}{\hookrightarrow} P$, thus $\mathrm{Ker}(\xi) \hookrightarrow \mathrm{Rad}(P)$, it follows by 9.1.5 that $\xi(\mathrm{Rad}(P)) = \mathrm{Rad}(M)$, so that by 5.1.3 we have only to show that $\mathrm{Rad}(P) \overset{\circ}{\hookrightarrow} P$. Let $\nu : P \to P/\mathrm{Rad}(P)$, then by 11.1.2 there is a decomposition $P = P_1 \oplus P_2$, with $P_1 \cap \mathrm{Rad}(P) \overset{\circ}{\hookrightarrow} P_1$ and $P_2 \hookrightarrow \mathrm{Rad}(P)$. By 9.6.4 it follows that $P_2 = 0$, thus $P = P_1$ and

$$\mathrm{Rad}(P) = P \cap \mathrm{Rad}(P) \overset{\circ}{\hookrightarrow} P.$$

□

11.2 LIFTING OF DIRECT DECOMPOSITIONS

11.2.1 *Definition*

(a) Let $\alpha : A \to M$ be a homomorphism. We say that the *decomposition*

$$M = \bigoplus_{i \in I} M_i$$

can be *lifted with respect to* α, if a decomposition

$$A = \bigoplus_{i \in I} A_i$$

exists so that for all $i \in I$ we have: $\alpha(A_i) = M_i$.

(b) Let $B \hookrightarrow A$. We say that the *decomposition*

$$A/B = \bigoplus_{i \in I} M_i$$

can be *lifted to* A, if it can be lifted with respect to $\nu : A \to A/B$.

11.2.2 Theorem. *Let $\xi: P \to M$ be a projective cover and let*

$$M = \bigoplus_{i \in I} M_i.$$

For every $i \in I$ let there be given an epimorphism $\alpha_i: A_i \to M_i$ with projective A_i and $\operatorname{Ker}(\alpha_i) \hookrightarrow \operatorname{Rad}(A_i)$. Then the decomposition $M = \bigoplus_{i \in I} M_i$ can be lifted with respect to ξ.

Proof. Consider the commutative diagram

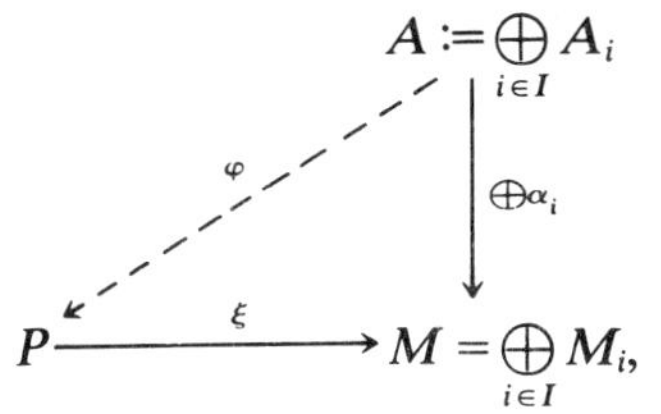

where φ exists since ξ is an epimorphism and A is projective. Since $\bigoplus \alpha_i$ is an epimorphism we have by 3.4.10

$$P = \operatorname{Im}(\varphi) + \operatorname{Ker}(\xi).$$

As $\operatorname{Ker}(\xi) \overset{\circ}{\hookrightarrow} P$ it follows that $P = \operatorname{Im}(\varphi)$, i.e. φ is an epimorphism. Since P is projective, φ splits:

$$A = P_0 \oplus \operatorname{Ker}(\varphi).$$

Since the diagram is commutative, it follows that $\operatorname{Ker}(\varphi) \hookrightarrow \operatorname{Ker}(\bigoplus \alpha_i) = \bigoplus \operatorname{Ker}(\alpha_i) \hookrightarrow \bigoplus \operatorname{Rad}(A_i) = \operatorname{Rad}(A)$, in which the last equation holds by 9.1.5. By 9.6.4 it then follows that $\operatorname{Ker}(\varphi) = 0$, thus φ is an isomorphism. Therefore we have

$$P = \bigoplus_{i \in I} \varphi(A_i)$$

with $\xi\varphi(A_i) = \alpha_i(A_i) = M_i$, $i \in I$. Hence we have lifted the decomposition $M = \bigoplus M_i$ with respect to ξ. □

From this there follows directly

11.2.3 Corollary. *Let $\xi: P \to M$ be a projective cover of the semiperfect module M. Then every direct decomposition of M can be lifted with respect to ξ.*

Proof. This follows from 11.2.2 since every direct summand of M possesses a projective cover. □

11.2.4 COROLLARY. *Let P be semiperfect and projective. Then every direct decomposition of the semisimple module $P/\mathrm{Rad}(P)$ can be lifted to P.*

Proof. This follows from 11.2.2 since by 11.1.7 $\mathrm{Rad}(P) \overset{\circ}{\hookrightarrow} P$ and every direct summand of $P/\mathrm{Rad}(P)$ possesses a projective cover. □

As a special case it follows that with respect to a right artinian ring R every direct decomposition of $R/\mathrm{Rad}(R)$ (as right R-module) can be lifted to R_R. If this lemma is not available then the lifting is done in the literature usually by calculations with idempotents.

11.3 MAIN THEOREM ON PROJECTIVE SEMIPERFECT MODULES

The following characterizations of a projective, semiperfect module are of great interest both with regard to the structure of such a module and also for determining whether a given module is semiperfect.

11.3.1 THEOREM. *The following are equivalent for a projective module R_R:*

(a) *P is semiperfect.*

(b) *P is complemented.*

(c) *There holds*

(1) *$P/\mathrm{Rad}(P)$ is semisimple;*

(2) *every direct summand of $(P/\mathrm{Rad}(P))_R$ is the image of a direct summand of P_R with regard to $P \to P/\mathrm{Rad}(P)$;*

(3) $\mathrm{Rad}(P) \overset{\circ}{\hookrightarrow} P$.

We have in (c) made condition (2) to be as weak as possible, since by 11.2.4 for the assertion (a)⇒(c) we have in any case a stronger statement. Since for applications the direction (c)⇒(a) is of interest it is desirable to formulate (c) as weakly as possible.

Proof. "(a)⇔(b)": This was shown in 11.1.5.

"(a)⇒(c)": This holds by 11.1.7 and 11.2.4.

It remains to prove (c)⇒(b): Let

$$\nu : P \to P/\mathrm{Rad}(P) =: \bar{P}$$

denote the natural homomorphism. Let now $A \hookrightarrow P$, then there is, since $\bar{P}$ is semisimple, a direct decomposition

$$\bar{P}_R = \nu(A) \oplus \Gamma.$$

By (2) there is a direct summand $P_2 \hookrightarrow P$ with $\nu(P_2) = \Gamma$. We claim that P_2

is a complement of A in P. From $\bar{P}=\nu(A)\oplus\nu(P_2)$ it follows that

$$P=A+P_2+\text{Rad}(P), \qquad A\cap P_2\hookrightarrow \text{Rad}(P),$$

thus since $\text{Rad}(P)\overset{\circ}{\hookrightarrow} P$

$$P=A+P_2, \qquad A\cap P_2\overset{\circ}{\hookrightarrow} P.$$

Since P_2 is a direct summand in P, it follows from $A\cap P_2\overset{\circ}{\hookrightarrow} P$ by 5.1.3(c) (with the help of the projection of P on P_2) that indeed $A\cap P_2\overset{\circ}{\hookrightarrow} P_2$. If we suppose that for $B\hookrightarrow P$ we have

$$A+B=P, \quad B\hookrightarrow P_2,$$

then by the modular law it follows that

$$A\cap P_2+B=P_2,$$

thus $B=P_2$ as $A\cap P_2\overset{\circ}{\hookrightarrow} P_2$. □

11.3.2 COROLLARY. *Let R be an arbitrary ring. Then we have*

(I) R_R *is semiperfect* $\Leftrightarrow$
  (1) $\bar{R}:=R/\text{Rad}(R)$ *is semisimple and*
  (2) *to every idempotent* $\varepsilon\in\bar{R}$ *there is an idempotent* $e\in R$ *with* $\varepsilon=\bar{e}$.
(II) R_R *is semiperfect* $\Leftrightarrow {}_RR$ *is semiperfect.*

Proof. (I) By 9.2.1 we have $\text{Rad}(R)\overset{\circ}{\hookrightarrow} R_R$, thus (3) in 11.3.1(c) is satisfied for an arbitrary ring and hence the condition is here superfluous. Further since the condition (1) here coincides with that in (c), we must only check whether the conditions (2) in 11.3.1 and in 11.3.2 follow mutually from one another.

"$\Rightarrow$": Let $\varepsilon\in\bar{R}$ be an idempotent. Corresponding to the decomposition $\bar{R}_R=\varepsilon\bar{R}\oplus(\bar{1}-\varepsilon)\bar{R}$ there is by 11.2.4 a decomposition

$$R_R=eR\oplus(1-e)R,$$

with an idempotent $e\in R$ and

$$\bar{e}\bar{R}=\varepsilon\bar{R}, \qquad (\bar{1}-\bar{e})\bar{R}=(\bar{1}-\varepsilon)\bar{R}.$$

Then it follows that $\varepsilon\bar{e}=\bar{e}$, $(\bar{1}-\varepsilon)(\bar{1}-\bar{e})=\bar{1}-\bar{e}$, thus $\varepsilon=\bar{e}$.

"$\Leftarrow$": Every direct summand of $\bar{R}_R$ is of the form $\varepsilon\bar{R}$ for an idempotent $\varepsilon\in\bar{R}$. Let now e be an idempotent of R with $\bar{e}=\varepsilon$, then eR is a direct summand of R_R with $\overline{eR}=\bar{e}\bar{R}=\varepsilon\bar{R}$.

(II) Conditions (1) and (2) in (I) are independent of the side. □

A ring R is called *semiperfect*, if it satisfies the (equivalent) conditions of 11.3.2. In particular this concept is by (II) independent of the side.

As already established, a projective artinian module is semiperfect. In particular a right artinian ring R_R is thus semiperfect and indeed so also on the left, independently of whether R is also left artinian. However there are also semiperfect rings which are not artinian. Let R be a local ring (7.1.2), then $R/\text{Rad}(R)$ is a skew field, thus in particular semisimple and $R/\text{Rad}(R)$ has only 1 as an idempotent $\neq 0$. By 11.3.2 R is consequently semiperfect.

For example the ring $R := K[[x]]$ of all power series $\sum_{i=0}^{\infty} k_i x^i$ in an indeterminate x and with coefficients from a field K is a local ring. In this case

$$\text{Rad}(R) = \left\{ \sum_{i=1}^{\infty} k_i x^i \,\middle|\, k_i \in K \right\} = xR$$

and this radical has no "nil-properties" of any kind. We emphasize this here, because this is a semiperfect ring which is not perfect (see 11.6).

11.3.3 THEOREM. *Let $(P_i | i \in I)$ be a family of semiperfect, projective R-modules. Then we have:*

$$P := \bigoplus_{i \in I} P_i$$

is semiperfect if and only if $\text{Rad}(P) \overset{\circ}{\hookrightarrow} P$.

Proof. By 11.3.1 the condition $\text{Rad}(P) \overset{\circ}{\hookrightarrow} P$ is necessary. In order to prove this it is sufficient we show that in 11.3.1(c) the conditions (1), (2), (3) are fulfilled. By assumption we have (3).

(1) By 9.1.5(d) we have

$$P/\text{Rad}(P) \cong \bigoplus_{i \in I} P_i/\text{Rad}(P_i).$$

Since by 11.3.1 $P_i/\text{Rad}(P_i)$ is semisimple for every $i \in I$, $P/\text{Rad}(P)$ is also semisimple.

(2) First of all we establish that every simple submodule E of $P/\text{Rad}(P)$ possesses a projective cover. As $P/\text{Rad}(P) \cong \bigoplus P_i/\text{Rad}(P_i)$ E is isomorphic to a simple submodule E' of $\bigoplus P_i/\text{Rad}(P_i)$. If we decompose every $P_i/\text{Rad}(P_i)$ into a direct sum of simple submodules and we apply 8.1.2(b) then it follows that E' is isomorphic to a simple submodule of one of the $P_i/\text{Rad}(P_i)$. Since this, as a direct summand of the semiperfect module $P_i/\text{Rad}(P_i)$, has a projective cover, the module E isomorphic to it has a projective cover.

Let now $P/\mathrm{Rad}(P) = \Lambda_1 \oplus \Lambda_2$. Since $P/\mathrm{Rad}(P)$ is semisimple, every Λ_k is a direct sum of simple submodules.

$$\Lambda_k = \bigoplus_{j\in J_k} E_j^k, \qquad k = 1, 2.$$

Let $\xi_j^k : A_j^k \to E_j^k$ be a projective cover, then

$$\alpha_k := \bigoplus_{j\in J_k} \xi_j^k : A_k := \bigoplus_{j\in J_k} A_j^k \to \Lambda_k = \bigoplus_{j\in J_k} E_j^k$$

is an epimorphism with projective domain A_k and we have as $\mathrm{Ker}(\xi_j^k) \overset{\circ}{\hookrightarrow} A_j^k$, hence $\mathrm{Ker}(\xi_j^k) \hookrightarrow \mathrm{Rad}(A_j^k)$ and thus

$$\mathrm{Ker}(\alpha_k) = \bigoplus_{j\in J_k} \mathrm{Ker}(\xi_j^k) \hookrightarrow \mathrm{Rad}(A_k), \qquad k = 1, 2.$$

If in 11.2.2 we put $\xi = \nu : P \to P/\mathrm{Rad}(P)$, then the assumptions of 11.2.2 are satisfied and it follows that the decomposition $P/\mathrm{Rad}(P) = \Lambda_1 \oplus \Lambda_2$ can be lifted to P.

11.3.4 COROLLARY
(a) *Every direct sum of finitely many semiperfect R-modules is semiperfect.*
(b) *If R_R is semiperfect then every finitely generated R-module is semiperfect.*

Proof. (a) Let $M_1, \ldots, M_n$ be semiperfect and let

$$\xi_i : P_i \to M_i, \qquad i = 1, \ldots, n$$

be a projective cover. By 11.1.5 P_i is semiperfect and by 11.3.3 so also is $P := \bigoplus_{i=1}^{n} P_i$ for $\mathrm{Rad}(P) = \bigoplus_{i=1}^{n} \mathrm{Rad}(P_i)$ is itself, as a finite sum of small submodules $\mathrm{Rad}(P_i)$, small in P. Since P is semiperfect, $M_1 \oplus \ldots \oplus M_n$ is also semiperfect as an epimorphic image of P.

(b) By (a) every finitely generated free module is semiperfect and then also every epimorphic image of it. □

We give now another interesting characterization of the semiperfect modules which will be useful later.

11.3.5 THEOREM. *The following are equivalent for a projective module:*
(1) *P is semiperfect.*
(2) *P satisfies the conditions:*
 (a) *Every proper submodule of P is contained in a maximal submodule of P; and*
 (b) *every simple factor-module of P has a projective cover.*

Proof. "(1)⇒(2)": By definition of "semiperfect" (b) is satisfied. For the proof of (a) let $U \underset{\neq}{\hookrightarrow} P$; since P/U is semiperfect, P/U has by 11.1.7 a small radical, which consequently is a proper submodule of P/U. Since the radical is the intersection of all maximal submodules, there exists at least one maximal submodule of P/U of the form X/U with $U \hookrightarrow X \hookrightarrow P$. Since X/U is maximal in P/U and we have $P/X \cong (P/U)/(X/U)$, then X is maximal in P.

"(2)⇒(1)": We establish this proof in three steps.

Step 1. We are to show that $\mathrm{Rad}(P) \overset{\circ}{\hookrightarrow} P$. Suppose that $U + \mathrm{Rad}(P) = P$ with $U \underset{\neq}{\hookrightarrow} P$, then by (a) there exists a maximal submodule $X \hookrightarrow P$ with $U \hookrightarrow X$. From this it follows that $U + \mathrm{Rad}(P) \hookrightarrow X \neq P$, contradiction!

Step 2. We are now to show that $\bar{P} := P/\mathrm{Rad}(P)$ is semisimple. Let $\nu: P \to \bar{P}$ be the natural epimorphism. Suppose that $\mathrm{Soc}(\bar{P}) \neq \bar{P}$, then it follows that $\nu^{-1}(\mathrm{Soc}(\bar{P})) \neq P$ and by (a) a maximal submodule $X \hookrightarrow P$ exists with $\nu^{-1}(\mathrm{Soc}(\bar{P})) \hookrightarrow X$. Since P/X by (b) has a projective cover, we deduce from 11.1.2 that

$$P = P_1 \oplus P_2 = P_1 + X$$

with $P_2 \hookrightarrow X$ and $P_1 \cap X \overset{\circ}{\hookrightarrow} P_1$, thus $P_1 \cap X \hookrightarrow \mathrm{Rad}(P)$. Therefore it follows that

$$(*) \qquad \bar{P} = \nu(P_1) \oplus \nu(X).$$

Since X is maximal in P (thus $\mathrm{Rad}(P) \hookrightarrow X$), $P/X \cong (P/\mathrm{Rad}(P))/(X/\mathrm{Rad}(P)) = \bar{P}/\nu(X) \cong \nu(P_1)$ is simple, thus $\nu(P_1) \hookrightarrow \mathrm{Soc}(\bar{P}) \hookrightarrow \nu(X)$; contradiction to $(*)$!

Step 3. Now let

$$\bar{P} = \bigoplus_{i \in I} E_i \qquad \text{with simple } E_i.$$

There follows for every $j \in I$

$$E_j \cong \bar{P}/\bigoplus_{\substack{i \in I \\ i \neq j}} E_i \cong P/\nu^{-1}\Big(\bigoplus_{\substack{i \in I \\ i \neq j}} E_i\Big).$$

By (b) projective covers

$$\alpha_j : A_j \to E_j, \qquad j \in I.$$

exist.

Since every simple module, which has a projective cover, is obviously semiperfect, all E_j and by 11.1.5 also all A_j are semiperfect.

For $\nu: P \to P/\mathrm{Rad}(P)$ (in place of ξ) and for $P/\mathrm{Rad}(P) = \bigoplus E_i$ (in place of M) the assumptions of 11.2.2 are satisfied. As in the proof of 11.2.2 it follows that

$$\varphi: A := \bigoplus A_i \to P$$

is an isomorphism. Thus $P = \bigoplus \varphi(A_i)$ is a direct sum of the semiperfect modules $\varphi(A_i)$. As $\mathrm{Rad}(P) \overset{\circ}{\hookrightarrow} P$ it follows from 11.3.3 that P is semiperfect. □

11.4 DIRECTLY INDECOMPOSABLE SEMIPERFECT MODULES

It was established in 11.2.4 for a projective semiperfect module P_R that every decomposition of the semisimple module $\bar{P} := P/\mathrm{Rad}(P)$ into a direct sum

$$\bar{P} = \bigoplus_{i \in I} E_i$$

of simple modules E_i, $i \in I$ can be lifted to P. Let

$$\nu: P \to \bar{P} = P/\mathrm{Rad}(P),$$

then a decomposition

$$P = \bigoplus_{i \in I} P_i \quad \text{with} \quad \nu(P_i) = E_i,\ i \in I.$$

exists. As $\mathrm{Rad}(P) = \bigoplus \mathrm{Rad}(P_i)$ (see 9.1.5) we have $\mathrm{Rad}(P_i) = \mathrm{Rad}(P) \cap P_i$. Therefore it follows that

$$E_i = \nu(P_i) = P_i + \mathrm{Rad}(P)/\mathrm{Rad}(P) \cong P_i/P_i \cap \mathrm{Rad}(P) = P_i/\mathrm{Rad}(P_i).$$

Since E_i is simple $\mathrm{Rad}(P_i)$ is a maximal submodule of P_i.

We wish now to investigate projective modules in which the radical is a maximal submodule. In this regard a module $M_R \neq 0$ is called indecomposable, if it is not the sum of two proper submodules. If $M_R = 0$ or if M_R is the sum of two proper submodules, then M_R is called decomposable (for "directly indecomposable" see 6.6.1).

11.4.1 THEOREM. *Let $P_R \neq 0$ be projective. Then the following are equivalent*:

(a) *P is indecomposable,*
(b) *P is semiperfect and directly indecomposable,*
(c) $\mathrm{Rad}(P)$ *is a maximal and a small submodule of P,*
(d) $\mathrm{Rad}(P)$ *is the largest proper submodule of P,*
(e) $\mathrm{End}(P_R)$ *is local.*

Proof. "(a)⇒(b)": By 11.1.5 it has only to be shown that P is complemented. But by (a) every submodule of P different from P has P itself as adco.

"(b)⇒(c)": By 11.1.7 we have $\mathrm{Rad}(P) \overset{\circ}{\hookrightarrow} P$. Since by 11.2.4 $P/\mathrm{Rad}(P)$ is directly indecomposable, $P/\mathrm{Rad}(P)$ is not only semisimple but also simple, thus $\mathrm{Rad}(P)$ is maximal in P.

"(c)⇒(d)": Let $U \hookrightarrow P$, $U \not\hookrightarrow \mathrm{Rad}(P)$. Since $\mathrm{Rad}(P)$ is a maximal submodule, it follows that $U + \mathrm{Rad}(P) = P$. As $\mathrm{Rad}(P) \overset{\circ}{\hookrightarrow} P$ it follows that $U = P$. Thus (d) also holds.

"(d)⇒(e)": If $\varphi : P \to P$ is an epimorphism then it must split. By (d) it follows that φ is an automorphism. If $\varphi_1, \varphi_2 \in \mathrm{End}(P_R)$ are not invertible then they cannot in consequence be epimorphisms. Then we have

$$\mathrm{Im}(\varphi_1 + \varphi_2) \hookrightarrow \mathrm{Im}(\varphi_1) + \mathrm{Im}(\varphi_2) \hookrightarrow \mathrm{Rad}(P),$$

thus $\varphi_1 + \varphi_2$ is also not invertible, i.e., $\mathrm{End}(P_R)$ is local.

"(e)⇒(a)": From $P = A + B$ we obtain a commutative diagram

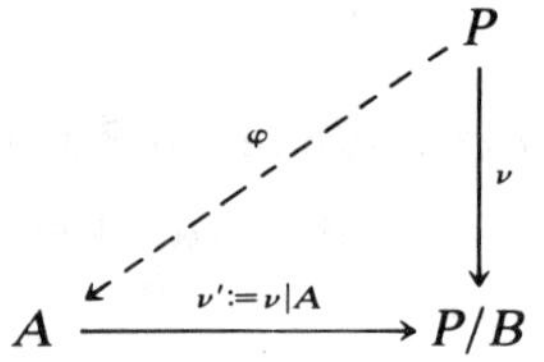

and for $\gamma := \iota_A \varphi$, where $\iota_A : A \to P$ is the inclusion mapping we then have

$$\mathrm{Im}(\gamma) \hookrightarrow A,\ \mathrm{Im}(1_P - \gamma) \hookrightarrow B \qquad (\text{since } x + B = \varphi(x) + B \text{ for all } x \in P).$$

As $1_P = \gamma + (1_P - \gamma)$, and since $\mathrm{End}(P_R)$ is local, γ or $1 - \gamma$ must be an automorphism, thus we have $A = P$ or $B = P$. □

11.4.2 COROLLARY. If P_R *is a projective, semiperfect module then a decomposition*

$$P = \bigoplus_{i \in I} P_i,$$

exists, in which the P_i *satisfy the properties of* 11.4.1. *The decomposition is unique in the sense of the Krull–Remak–Schmidt Theorem* (7.3.1).

For later use we wish to write down explicitly the result in the case of a ring, bearing in mind 7.2.3.

11.4.3 COROLLARY. *Let R be a semiperfect ring. Then there exists a decomposition, unique in the sense of* 7.3.1,

$$R_R = e_1 R \oplus \ldots \oplus e_n R$$

with the following properties:

(1) $e_i, \ldots, e_n$ *are orthogonal idempotents* $\neq 0$ *with*

$$1 = \sum_{i=1}^{n} e_i.$$

(2) $\mathrm{Rad}(e_iR)$ *is the largest proper right ideal in* e_iR *and* $\mathrm{Rad}(e_iR) = e_i\mathrm{Rad}(R)$.

(3) e_iR *is indecomposable*.

(4) $\mathrm{End}(e_iR)$ *is local and* $\mathrm{End}(e_iR) \cong e_iRe_i$.

Proof. By 11.4.2 and 7.2.3 all is immediately clear except for the two following aspects, which hold for arbitrary idempotents $e \in R$: "$\mathrm{Rad}(eR) = e\mathrm{Rad}(R)$". By 9.1.5 we have $\mathrm{Rad}(eR) \hookrightarrow \mathrm{Rad}(R)$. Since $x = ex$ for every element $x \in eR$ it follows that $\mathrm{Rad}(eR) \hookrightarrow e\mathrm{Rad}(R)$. On the other hand by 9.2.1 we have $e\mathrm{Rad}(R) \hookrightarrow \mathrm{Rad}(eR)$.

"$\mathrm{End}(eR) \cong eRe$": Multiplication of eR by an element $eae \in eRe$ evidently involves the endomorphism

$$(eae)' : eR \ni er \mapsto eaer \in eR$$

of eR. We obtain therefore a ring homomorphism

$$\psi : eRe \ni eae \mapsto (eae)' \in \mathrm{End}(eR).$$

ψ is a "monomorphism": From $eaer = eber$ for all $er \in eR$ there follows for $r = 1$: $eae = ebe$.

ψ is an "epimorphism": Let $\alpha \in \mathrm{End}(eR)$; since eR is a direct summand in R_R, α can be extended to an epimorphism of R_R, i.e. to a left multiplication by an element $a \in R$:

$$\alpha(er) = a(er) = eaer,$$

the latter equality since $aer \in eR$. Thus we have $\alpha = (eae)'$. □

Example. For $R := \mathbb{Z}/n\mathbb{Z}$, $n > 1$, we wish to set out explicitly the decomposition existing by 11.4.3. At the end of section 9.1 the radical and socle of R were determined. We utilize here the previously employed notation. Let

$$n_i := \frac{n}{p_i^{m_i}}, \qquad i = 1, \ldots, k,$$

then obviously we have $GCD(n_1, \ldots, n_k) = 1$. Consequently there are $a_1, \ldots, a_k \in \mathbb{Z}$ with $a_1n_1 + \ldots + a_kn_k = 1$. Therefore it follows that $(a_i, p_i) = 1$. Let now

$$e_i := a_in_i + n\mathbb{Z} \in \mathbb{Z}/n\mathbb{Z}.$$

Then, it is asserted, $R_R = e_1R \oplus \ldots \oplus e_kR$ is the decomposition appearing in 11.4.3. First of all it is clear that we have

$$e_1 + \ldots + e_k = 1 \in R.$$

Further since $n | n_i n_j$ for $i \neq j$ we have

$$e_i e_j = 0 \quad \text{for} \quad i \neq j.$$

Then it follows from $e_1 + \ldots + e_j = 1$ on multiplication by e_i that

$$e_i^2 = e_i.$$

By 7.2.3 we then have

$$R_R = e_1R \oplus \ldots \oplus e_kR.$$

It still remains to be shown that the e_iR have local endomorphism rings.

The ring epimorphism

$$\mathbb{Z} \ni z \mapsto e_i \bar{z} e_i = e_i \bar{z} \in e_i R e_i$$

has, by definition of the e_i (note that $(a_i, p_i) = 1$), the kernel $p_i^{m_i}\mathbb{Z}$, thus we have

$$e_i R e_i \cong \mathbb{Z}/p_i^{m_i}\mathbb{Z}.$$

As indicated in 9.1, we have further

$$\operatorname{Rad}(\mathbb{Z}/p_i^{m_i}\mathbb{Z}) = p_i\mathbb{Z}/p_i^{m_i}\mathbb{Z},$$

and since

$$(\mathbb{Z}/p_i^{m_i}\mathbb{Z})/(p_i\mathbb{Z}/p_i^{m_i}\mathbb{Z}) \cong \mathbb{Z}/p_i\mathbb{Z}.$$

this is a maximal ideal in $\mathbb{Z}/p_i^{m_i}\mathbb{Z}$. Then $\operatorname{Rad}(e_iRe_i)$ is also a maximal ideal and therefore the largest proper ideal of e_iRe_i. By 11.4.1 it follows that e_iRe_i is a local ring.

11.5 PROPERTIES OF NIL IDEALS AND OF t-NILPOTENT IDEALS

For the investigation of perfect rings properties of nil ideals and of t-nilpotent ideals are needed, which here are collected together.

There is first the question of "lifting" orthogonal idempotents modulo a nil ideal. To this effect we recall that an element $e \in R$ is called an idempotent if $e^2 = e$ holds. An ideal A of R is called a nil ideal if every $a \in A$ is nilpotent, i.e. there exists an $n \in \mathbb{N}$ (depending on a) with $a^n = 0$. In 9.3.8 it was shown that every nil ideal is contained in $\operatorname{Rad}(R)$.

As groundwork for the lifting of idempotents we prove the following simple lemma.

11.5.1 LEMMA. *Let b be an arbitrary element of a ring R and let R_0 be the subring of R generated by $1 \in R$ and b.*

(a) *For arbitrary $m, n \in \mathbb{N}$ we have*

$$R = b^n R + (1-b)^m R + (b-b^2)R, \qquad b^n R \cap (1-b)^m R = b^n (1-b)^m R.$$

(b) *If $b - b^2$ is nilpotent then there is an idempotent $e \in R_0$ such that we have*

$$e = br_0, \qquad e - b = (b-b^2)s_0 \quad \textit{with} \quad r_0, s_0 \in R_0.$$

Proof. (a) If $\mathbb{Z}[x]$ is the polynomial ring in the indeterminate x with coefficients in $\mathbb{Z}$. We have

$$1 - x^n - (1-x)^m \in (x - x^2)\mathbb{Z}[x],$$

for $x(1-x) = x - x^2$ divides $1 - x^n - (1-x)^m$, since $x = 0$ and $x = 1$ are zeroes of $1 - x^n - (1-x)^m$. Consequently there is a $z_0 \in \mathbb{Z}[x]$ such that we have

$$1 = x^n + (1-x)^m + (x - x^2)z_0.$$

From the ring epimorphism $\mathbb{Z}[x] \mapsto R_0$ with $x \mapsto b$ it follows that

$$1 = b^n + (1-b)^m + (b-b^2)r_0, \qquad r_0 \in R_0,$$

thus we have

$$R_0 = b^n R_0 + (1-b)^m R_0 + (b-b^2)R_0$$

and then also $R = b^n R + (1-b)^m R + (b-b^2)R$.

For the proof of the second equation in (a) it is immediately clear that $b^n(1-b)^m R \hookrightarrow b^n R \cap (1-b)^m R$. Conversely let $d = b^n r = (1-b)^m s \in b^n R \cap (1-b)^m R$ $(r, s \in R)$. Then from

$$d = (1-b)^m s = \left(1 - \binom{m}{1} b + \binom{m}{2} b^2 - + \dots\right) s$$

$$= s - b\left(\binom{m}{1} - \binom{m}{2} b + - \dots\right) s$$

there follows an equation of the form

$$s = d + br_0 s = b^n r + br_0 s \quad \text{with} \quad r_0 \in R_0.$$

If we substitute, in this equation on the right for s, the same equation again, then it follows that

$$s = b^n r + br_0(b^n r + br_0 s) = b^n r_1 + b^2 r_0^2 s \quad \text{with} \quad r_2 \in R,$$

where we use the fact that R_0 is commutative. If we continue inductively in this manner, then after finitely many steps we obtain an equation of the form $s = b^n t$ with $t \in R$. Hence it follows that

$$d = (1-b)^m s = (1-b)^m b^n t,$$

thus $d \in b^n(1-b)^m R$, which was to be shown.

(b) Let $(b-b^2)^n = 0$, then $(b-b^2)R_0$ is a nilpotent ideal in R_0 (since R_0 is commutative), thus we have $(b-b^2)R_0 \hookrightarrow \mathrm{Rad}(R_0)$ and consequently $(b-b^2)R_0$ is small in R_0.

From

$$R_0 = b^n R_0 + (1-b)^n R_0 + (b-b^2)R_0$$

it then follows that

$$R_0 = b^n R_0 + (1-b)^n R_0,$$

thus as

$$b^n R_0 \cap (1-b)^n R_0 = b^n(1-b)^n R_0 = (b-b^2)^n R_0 = 0$$

we have in fact $R_0 = b^n R_0 \oplus (1-b)^n R_0$.

Then by 7.2.3 an idempotent $e \in R_0$ exists with

$$eR_0 = b^n R_0,\ (1-e)R_0 = (1-b)^n R_0,$$

thus we have $e = br_0$, $r_0 \in R_0$. Further it follows from (a) that

$$e - b = (1-b) - (1-e) \in bR_0 \cap (1-b)R_0 = (b-b^2)R_0,$$

thus $e - b = (b-b^2)s_0$, $s_0 \in R_0$. □

11.5.2 *Definition.* (a) Let $A \hookrightarrow {}_R R_R$ and let $\nu: R \to R/A$ be the natural ring epimorphism. We say that an idempotent $\varepsilon \in R/A$ can be *lifted to R* if an idempotent $e \in R$ exists with $\nu(e) = \varepsilon$.

(b) We say that a set $\{\varepsilon_i | i \in I\}$ of orthogonal idempotents $\varepsilon_i \in R/A$ can be *lifted to R* if a set $\{e_i | i \in I\}$ of orthogonal idempotents $e_i \in R$ exists with $\nu(e_i) = \varepsilon_i$ for all $i \in I$.

11.5.3 THEOREM. *Let $A \hookrightarrow {}_R R_R$ be a nil ideal. Then every finite or countably infinite set of orthogonal idempotents $\varepsilon_i \in R/A$ can be lifted to R.*

Proof by induction. Beginning of the induction: An idempotent $\varepsilon \in R/A$ can be lifted to R. Again let $\nu: R \to R/A$ and let $b \in R$ with $\nu(b) = \varepsilon$, then it follows that

$$\nu(b-b^2) = \nu(b) - \nu(b)^2 = \varepsilon - \varepsilon = 0,$$

thus $b-b^2 \in \mathrm{Ker}(\nu)=A$. By 11.5.1(b) there is an idempotent $e \in R$ with $e-b=(b-b^2)s_0 \in A$. It then follows that

$$0=\nu(e-b)=\nu(e)-\nu(b)=\nu(e)-\varepsilon,$$

thus $\nu(e)=\varepsilon$.

For the induction step now let

$$\varepsilon_1, \varepsilon_2, \varepsilon_3, \ldots$$

be finitely, or countably infinitely, many orthogonal idempotents from R/A. Let $e_1, \ldots, e_n$ be already determined as required. Then let $c \in R$ with $\nu(c)=\varepsilon_{n+1}$ and let

$$b := \left(1-\sum_{i=1}^{n} e_i\right)c\left(1-\sum_{i=1}^{n} e_i\right).$$

From the orthogonality of the $e_1, \ldots, e_n$ we have therefore

$$e_ib = be_i = 0, \qquad i=1, \ldots, n,$$

and also

$$\nu(b)=\left(1-\sum_{i=1}^{n} \varepsilon_i\right)\varepsilon_{n+1}\left(1-\sum_{i=1}^{n} \varepsilon_i\right)=\varepsilon_{n+1}.$$

By the initial induction step and 11.5.1(b) there is an idempotent e_{n+1} with

$$\nu(e_{n+1})=\nu(b)=\varepsilon_{n+1}, \qquad e_{n+1}=br_0=r_0b.$$

As $e_ib = be_i = 0$ it follows that

$$e_ie_{n+1}=e_{n+1}e_i=0, \qquad i=1, \ldots, n,$$

by which the proof is completed. □

We come now to the investigation of t-nilpotent ideals and repeat first the definition previously given at the beginning of this chapter.

11.5.4 *Definition.* A set A of elements of a ring R is called *left, resp. right, t-nilpotent,* if for every family

$$(a_1, a_2, a_3, \ldots), \qquad a_i \in A$$

a $k \in \mathbb{N}$ exists with

$$a_ka_{k-1}\ldots a_1=0, \qquad a_1a_2\ldots a_k=0.$$

It is clear then that every left or right t-nilpotent ideal is a nil ideal. On the other hand not every t-nilpotent ideal is indeed nilpotent. The t-nilpotent ideals come between the nilpotent ideals and the nil ideals.

11.5.5 THEOREM. *The following are equivalent for a right ideal* $A \hookrightarrow R_R$:

(a) *A is left t-nilpotent.*

(b) *For every module* M_R *with* $MA = M$ *we have* $M = 0$.

(c) *For every module* M_R *we have* $MA \overset{\circ}{\hookrightarrow} M$.

(d) $R^{(\mathbb{N})}A \overset{\circ}{\hookrightarrow} R^{(\mathbb{N})}$ *as right modules.*

Proof. "(a)⇒(b)": Suppose we have $MA = M$ and $M \neq 0$. Then an $m_1 a_1 \neq 0$ exists with $m_1 \in M$, $a_1 \in A$. Let $m_1 = \sum m_i' a_i'$. Then $m_1 a_1 = \sum m_i' a_i' a_1$ and hence there exists $m_2 a_2 a_1 \neq 0$, $m_2 \in M$, $a_2 \in A$.

Let $m_2 = \sum m_i'' a_i''$. Then $m_2 a_2 a_1 = \sum m_i'' a_i'' a_2 a_1$, so there exists $m_3 a_3 a_2 a_1 \neq 0$. Inductively therefore we obtain a sequence $(a_1, a_2, a_3, \ldots)$, $a_i \in A$ with $a_n a_{n-1} \ldots a_1 \neq 0$ for every $n \in \mathbb{N}$. Contradiction to the t-nilpotence!

"(b)⇒(c)": Assume $MA + U = M$. Then $(M/U)A = M/U$, so $M/U = 0$ by assumption, whence $U = M$ which was to be shown.

"(c)⇒(d)": (d) is a special case of (c).

"(d)⇒(a)": Let $F := R^{(\mathbb{N})}$ as a right module with basis $x_1, x_2, x_3, \ldots$. Along with the sequence $(a_1, a_2, a_3, \ldots)$ with $a_i \in A$ we consider the submodule

$$U = \sum_{i=1}^{\infty} u_i R$$

of F with $u_1 := x_i - x_{i+1} a_i$, $i \in \mathbb{N}$. Obviously we then have $FA + U = F$, thus by assumption $U = F$.

In particular we then have $x_1 \in U$, thus there is a representation

$$x_1 = \sum_{i=1}^{k} u_i r_i = x_1 r_1 + x_2(r_2 - a_1 r_1) + x_3(r_3 - a_2 r_2) + \ldots + \\ + x_k(r_k - a_{k-1} r_{k-1}) - x_{k+1} a_k r_k.$$

Hence by equating coefficients we have

$$r_1 = 1, \qquad r_2 = a_1, \qquad r_3 = a_2 a_1, \ldots, \qquad r_k = a_{k-1} a_{k-2} \ldots a_1,$$

and also $a_k r_k = a_k a_{k-1} \ldots a_1 = 0$. □

11.5.6 COROLLARY. *The following are equivalent for a ring R:*

(1) *Rad(R) is left t-nilpotent.*

(2) *Every projective right R-module has a small radical.*

(3) *As a right R-module* $R^{(\mathbb{N})}$ *has a small radical.*

Proof. "(1)⇒(2)": This is a special case of (a)⇒(c) in 11.5.5, if we observe that by 9.2.1 $\mathrm{Rad}(P) = P\,\mathrm{Rad}(R)$ for a projective module P.

"(2)$\Rightarrow$(3)": Clear.
"(3)$\Rightarrow$(1)": (d)$\Rightarrow$(a) in 11.5.5 on observing 9.2.1(g). □

A further interesting characterization of t-nilpotent ideals arises with the help of the annihilator conditions.

11.5.7 THEOREM. *The following are equivalent for a right ideal* $A \hookrightarrow R_R$:

(a) *A is left t-nilpotent.*
(b) *For every module $_RM$ with $\boldsymbol{r}_M(A)=0$ we have $M=0$.*
(c) For every module $_R$M we have $\mathbf{r}_M(A) \overset{*}{\hookrightarrow} {_R}M$.

Proof. "(a)$\Rightarrow$(b)": Assume we have $\boldsymbol{r}_M(A)=0$ and $M\neq 0$. Then to every $0\neq m\in M$ there is an $a\in A$ with $am\neq 0$. For a fixed $0\neq m_0\in M$ we obtain inductively therefore a sequence $(a_1, a_2, a_3, \ldots)$, $a_i\in A$ with

$$a_na_{n-1}\ldots a_1m_0\neq 0 \quad \text{for every} \quad n\in\mathbb{N},$$

thus also $a_na_{n-1}\ldots a_1\neq 0$ for every $n\in\mathbb{N}$. This contradicts the assumption.

"(b)$\Rightarrow$(c)": Assume that for $X\hookrightarrow M$ we have

$$\boldsymbol{r}_M(A)\cap X=0,$$

then it follows that $\boldsymbol{r}_X(A)=0$, thus $X=0$.

"(c)$\Rightarrow$(a)": We show that 11.5.5(b) is satisfied. For $M_R\neq 0$ we shall show that $MA\neq M$. Let $U:=\boldsymbol{r}_R(M)$, then U is a proper two-sided ideal in R. Further let

$$H:=\{x\,|\,x\in R\wedge Ax\subset U\},$$

then it follows that $U\subset H$ and

$$H/U=\boldsymbol{r}_{R/U}(A).$$

By (c) we have $H/U\overset{*}{\hookrightarrow} R/U$, thus $U\subset H$ and consequently $MH\neq 0$, but $MAH\subset MU=0$. We deduce therefore that $MA\neq M$. □

11.6 PERFECT RINGS

As announced in the preamble to this chapter we now come to the investigation of perfect rings and first of all repeat the definition.

11.6.1 *Definition.* A ring is called *right perfect* $(=R_R$ perfect$)$:$\Leftrightarrow$. Every right R-module has a projective cover.

11.6.2 COROLLARY. *The following are equivalent for a ring R:*

(a) *R is right perfect.*

(b) *$R^{(\mathbb{N})}$ is semiperfect as a right R-module.*

(c) *R is semiperfect and every free right R-module has a small radical.*

Proof. "(a)⇒(b)": Clear by definition.

"(b)⇒(c)": As a direct summand of $R_R^{(\mathbb{N})}$, R is semiperfect. By 11.1.7 $R_R^{(\mathbb{N})}$ has a small radical so that by 11.5.6 every projective right R-module has a small radical.

"(c)⇒(a)": If R is semiperfect and if every free right R-module has a small radical then by 11.3.3 every free right R-module is semiperfect. Since every right R-module is the image of a free right R-module, every right R-module is semiperfect, i.e. R is right perfect. □

In order to have an example of a perfect ring we take note that a right artinian ring is perfect on both sides.

Thus let R_R be artinian. Referring to 11.3.2 it was there established that a right artinian ring is semiperfect. Since by 9.3.10 we have for every right R-module and left R-module M

$$\mathrm{Rad}(M) \overset{\circ}{\hookrightarrow} M$$

the assertion follows from 11.6.2.

In this section we shall prove the theorem mentioned in the preamble which we repeat for the sake of completeness.

11.6.3 THEOREM. *The following conditions are equivalent for a ring R:*

(1) *R is right perfect.*

(2) *Every flat right R-module is projective.*

(3) *R satisfies the descending chain condition for cyclic left ideals.*

(4) *Every left R-module $\neq 0$ possesses a socle $\neq 0$ and R contains no infinite set of orthogonal idempotents.*

(5) *$R/\mathrm{Rad}(R)$ is semisimple and $\mathrm{Rad}(R)$ is left t-nilpotent.*

Proof. We shall show successively (1)⇒(2)⇒(3)⇒(4)⇒(5)⇒(1).

"(1)⇒(2)": By assumption every right R-module, thus in particular every flat right R-module has a projective cover. By 10.5.3 (with $M = P/U$) every flat right R-module is then projective.

"(2)⇒(3)": Let

$$Ra_1 \hookleftarrow Ra_2 \hookleftarrow Ra_3 \hookleftarrow \dots$$

be a chain of left ideals of R. As $a_{i+1} \in Ra_i$ there is a $b_{i+1} \in R$ with

$a_{i+1} = b_{i+1}a_i$. It follows inductively, if we put $b_1 = a_1$, that

$$a_n = b_n b_{n-1} \dots b_1, \qquad n \in \mathbb{N}.$$

The previous chain can consequently be presented in the form

$$Rb_1 \hookleftarrow Rb_2b_1 \hookleftarrow Rb_3b_2b_1 \hookleftarrow \dots$$

and it is uniquely determined by the sequence $b_1, b_2, b_3, \dots$

We show in three steps: There is a left ideal $A \hookrightarrow {}_RR$ and an $m \in \mathbb{N}$ with

$$Rb_nb_{n-1} \dots b_1 = Ra_n = A$$

for all $n \geq m$. Then evidently this is equivalent to having the original sequence stationary.

Step 1. Let $F := R^{(\mathbb{N})} \in M_R$ with the basis

$$x_i := (\underbrace{0 \dots 0\,1}_{i \text{ places}}\,0 \dots), \qquad i \in \mathbb{N}.$$

Further let

$$B := \sum_{i=1}^{\infty} (x_i - x_{i+1}b_i)R \hookrightarrow F_R,$$

then we have to show that F/B is flat. By 10.5.2 we show that for every finitely generated left ideal $L \hookrightarrow {}_RR$ we have

$$B \cap FL = BL.$$

We always have $BL \hookrightarrow B \cap FL$ and in order to prove the reverse inclusion let $d \in B \cap FL$, thus

$$d = \sum_{i=1}^{n} (x_i - x_{i+1}b_i)r_i = \sum_{i=1}^{h} f_il_i, \qquad f_i \in F,\ l_i \in L.$$

Since L is a left ideal, we obtain

$$d = \sum_{i=1}^{h} f_il_i = \sum_{j=1}^{k} x_jl'_j \quad \text{with} \quad l'_j \in L.$$

Equating coefficients yields

$$r_1 = l'_1, \qquad r_2 - b_1r_1 = l'_2, \qquad r_3 - b_2r_2 = l'_3, \dots,$$

from which it follows successively that all $r_i \in L$, thus we have $d \in BL$ which was to be shown.

Step 2. By virtue of assumption (2) it now follows that F/B is projective. Then the epimorphism

$$\nu: F \to F/B$$

splits and we deduce that $F = B \oplus U$. Now let

$$\pi: F \ni b + u \mapsto u \in F, \qquad b \in B, u \in U,$$

be the corresponding projection of F onto $U \hookrightarrow F$ (with codomain F!).

Then we have

$$\pi(x_k - x_{k+1}b_k) = 0, \qquad k \in \mathbb{N},$$

thus $\pi(x_k) = \pi(x_{k+1})b_k$. If we now put $z_k := \pi(x_k)$ then it follows that

$$z_k = z_{k+1}b_k, \qquad k \in \mathbb{N},$$

from which by successive substitution we obtain

$$z_k = z_{m+1}b_m b_{m-1} \ldots b_k, \qquad m \geq k.$$

As $\pi^2 = \pi$ we have finally $\pi(z_k) = z_k$, $k \in \mathbb{N}$.

Assertion. Let $\boldsymbol{r}_R(z_k)$ be the right annihilator of z_k in R, then we have

$$\boldsymbol{r}_R(z_k) = \{r \mid r \in R \wedge b_m b_{m-1} \ldots b_k r = 0 \quad \text{for an } m \geq k\}.$$

That the set appearing on the right is contained in $\boldsymbol{r}_R(z_k)$ follows immediately from $z_k = z_{m+1}b_m b_{m-1} \ldots b_k$. Now let $r \in \boldsymbol{r}_R(z_k)$, i.e. $z_k r = 0$. Then from

$$x_k = y_k + z_k, \qquad y_k \in B, z_k \in U$$

there follows an equation of the form

$$x_k r = y_k r = \sum_{j=1}^{m} (x_j - x_{j+1}b_j)r_j, \qquad r_j \in R, m \geq k.$$

Equating coefficients yields

$$\begin{aligned} r_1 = r_2 = \ldots = r_{k-1} &= 0, \\ r_k &= r, \\ r_{k+1} &= b_k r_k, \\ r_{k+2} &= b_{k+1} r_{k+1}, \\ &\vdots \\ r_m &= b_{m-1} r_{m-1}, \\ 0 &= b_m r_m. \end{aligned}$$

By substitution it follows that

$$0 = b_m r_m = b_m b_{m-1} r_{m-1} = \dots = b_m b_{m-1} \dots b_k r,$$

by which the assertion is proved.

Step 3. In the sense of the definition of $F = R^{(\mathbb{N})}$ now let

$$z_k = (s_i^k) = (s_1^k s_2^k \dots), \qquad k \in \mathbb{N}.$$

Let A be the left ideal of R generated by the coefficients s_i^1 of z_1:

$$A := \sum_i R s_i^1;$$

since almost all $s_i^1 = 0$, A is finitely generated.

Assertion. There exists m_0 with $Rb_n b_{n-1} \dots b_1 = A$ for all $n \geq m_0$. If this is shown then obviously we have done with the proof of (2)$\Rightarrow$(3).

In Step 2

$$z_1 = z_{m+1} b_m b_{m-1} \dots b_1, \qquad m \in \mathbb{N}$$

was established, consequently we have for all $i \in \mathbb{N}$

$$s_i^1 = s_i^{m+1} b_m b_{m-1} \dots b_1,$$

from which it follows that

$$A \hookrightarrow Rb_m b_{m-1} \dots b_1, \qquad m \in \mathbb{N}.$$

We have to show that for sufficiently large m the reverse inclusion holds. From $\pi = \pi^2$ it follows that $z_j = \pi(x_j) = \pi^2(x_j) = \pi(z_j)$, and thus we obtain

$$z_1 = (s_i^1) = \left(\sum_{j=1}^{h} s_i^j s_j^1 \right),$$

where h is so chosen that we have $s_j^1 = 0$ for $j \geq h$; in addition we note that from

$$z_1 = \sum_{j=1}^{h} x_j s_j^1$$

it follows that

$$z_1 = \pi(z_1) = \sum_j \pi(x_j) s_j^1 = \sum_j z_j s_j^1 = \sum_i \sum_j x_i s_i^j s_j^1.$$

Equating coefficients yields

$$s_i^1 = \sum_{j=1}^{h} s_i^j s_j^1, \qquad i \in \mathbb{N}.$$

From $z_j = z_{h+1}b_hb_{h-1}\ldots b_j$, $h \geqslant j$ (Step 2) we obtain

$$s_i^j = s_i^{h+1}b_hb_{h-1}\ldots b_j.$$

If we insert this into the preceding equation we deduce that

$$s_i^1 = \sum_{j=1}^{h} s_i^{h+1}\, b_hb_{h-1}\ldots b_js_j^1 = s_i^{h+1}\sum_{j=1}^{h} b_hb_{h-1}\ldots b_js_j^1.$$

On the other hand since we have $s_i^1 = s_i^{h+1}b_hb_{h-1}\ldots b_1$ it follows that

$$s_i^{h+1}\left(\sum_{j=1}^{h} b_hb_{h-1}\ldots b_js_j^1 - b_hb_{h-1}\ldots b_1\right) = 0, \qquad i \in \mathbb{N}$$

thus

$$z_{h+1}\left(\sum_{j=1}^{h} b_hb_{h-1}\ldots b_js_j^1 - b_hb_{h-1}\ldots b_1\right) = 0.$$

By Step 2 it follows that there is an $m_0 \geqslant h+1$ with

$$b_{m_0}b_{m_0-1}\ldots b_{h+1}\left(\sum_{j=1}^{h} b_h\ldots b_js_j^1 - b_hb_{h-1}\ldots b_1\right) = 0.$$

This implies that $b_{m_0}b_{m_0-1}\ldots b_1 \in A$ and consequently we also have

$$Rb_n\ldots b_1 \hookrightarrow A \quad \text{for} \quad n \geqslant m_0.$$

"(3)⇒(4)": Let $0 \neq m \in {}_RM$, then we have to show that Rm contains a simple submodule. Suppose this were not the case, then every submodule $\neq 0$ of Rm must contain a proper submodule $\neq 0$. Then there is therefore an infinite chain

$$Rm \supsetneqq Rr_1m \supsetneqq Rr_2r_1m \supsetneqq \ldots.$$

Consequently we have

$$R \supsetneqq Rr_1 \supsetneqq Rr_2r_1 \supsetneqq \ldots$$

in contradiction to the descending chain condition for cyclic left ideals.

Now we show that R cannot contain an infinite set of orthogonal idempotents. Namely if $e_1, e_2, e_3, \ldots$ are orthogonal idempotents $\neq 0$ in R then, as we shall immediately establish,

$$R \supsetneqq R(1-e_1) \supsetneqq R(1-e_1-e_2) \supsetneqq \ldots$$

is a proper descending chain of cyclic left ideals in contradiction to the assumption. Since

$$(1-e_1-e_2-\ldots-e_n)(1-e_1-\ldots-e_{n-1}) = 1-e_1-\ldots-e_n$$

we have

$$R(1-e_1\ldots-e_{n-1}) \hookleftarrow R(1-e_1-\ldots-e_n);$$

suppose

$$(1-e_1-\ldots-e_{n-1}) = r(1-e_1-\ldots-e_n),$$

then it would follow on pre-multiplication by e_n that $e_n = 0$ ↯.

"(4)⇒(5)": First of all in order to show that Rad(R) is left t-nilpotent, by 11.5.7, it must be shown that for every left R-module M

$$r_M(\mathrm{Rad}(R)) \overset{*}{\hookrightarrow} M.$$

But we always have

$$\mathrm{Soc}({}_RM) \hookrightarrow r_M(\mathrm{Rad}(R))$$

(for $\mathrm{Rad}(R)\mathrm{Soc}(M) \hookrightarrow \mathrm{Rad}(\mathrm{Soc}(M)) = 0$), and because $\mathrm{Soc}(M) \overset{*}{\hookrightarrow} M$ holds by assumption, the assertion follows.

Since Rad(R) is left t-nilpotent and so is certainly nil it follows from 11.5.3 that $R/\mathrm{Rad}(R)$ cannot contain an infinite set of orthogonal idempotents.

For the further considerations we remark first of all that the left ideals of $R/\mathrm{Rad}(R)$ coincide with the R-submodules of ${}_R(R/\mathrm{Rad}(R))$, so that every left ideal $\neq 0$ of $R/\mathrm{Rad}(R)$ contains a simple left ideal. For brevity we put $T := R/\mathrm{Rad}(R)$.

Assertion. Every simple left ideal $E \hookrightarrow {}_TT$ is a direct summand in ${}_TT$.

Proof. Since $\mathrm{Rad}(R/\mathrm{Rad}(R)) = \mathrm{Rad}(T) = 0$, E is not small in ${}_TT$, thus an $A \underset{\neq}{\hookrightarrow} T$ exists with $E + A = {}_TT$. Since E is simple, it follows that $E \cap A = 0$ (since otherwise $E \hookrightarrow A \Rightarrow A = T$), thus $E \oplus A = {}_TT$.

We now construct a sequence of orthogonal idempotents which in accordance with the assertion at the beginning must break off after finitely many steps. Let $E_1 \hookrightarrow {}_TT$, then there is an idempotent e_1 with $E_1 = Te_1$ and

$${}_TT = E_1 \oplus A_1 \quad (= Te_1 \oplus T(1-e_1)).$$

If $A_1 = 0$ then ${}_TT$ is simple and we are done. If $A_1 \neq 0$ then by assumption there is a simple left ideal $E_2 \hookrightarrow A_1$. Let ${}_TT = E_2 \oplus U_2$, then it follows by the modular law that $A_1 = E_2 \oplus (A_1 \cap U_2)$. If now we put $A_2 := A_1 \cap U_2$ it follows therefore that

$${}_TT = E_1 \oplus E_2 \oplus A_2.$$

If we continue inductively in this way, then we obtain a sequence of decompositions

$$_TT = E_1 \oplus \ldots \oplus E_n \oplus A_n, \qquad n = 1, 2, 3, \ldots$$

with $A_{n-1} = E_n \oplus A_n$, $n = 2, 3, \ldots$, which then only breaks off if $A_n = 0$ occurs. But then $_TT$ is semisimple and the proof is complete.

By 7.2.3, to the sequence of direct decompositions there corresponds a sequence of orthogonal idempotents

$$e_1, \ldots, e_n, a_n, \qquad n = 1, 2, 3, \ldots$$

with $a_{n-1} = e_n + a_n$, $n = 2, 3, \ldots$ (i.e. with respect to the splitting of a_{n-1} into the idempotents e_n and a_n the orthogonal idempotents $e_1, \ldots, e_{n-1}$ do not change!) Since as asserted the sequence $e_1, e_2, e_3, \ldots$ must break off, the case $a_n = 0$ must hold, thus $A_n = Ta_n = 0$ happens.

"(5)⇒(1)": By 11.3.2 and 11.5.3 R is semiperfect. By 11.5.6 for every free right R-module F_r we have

$$\mathrm{Rad}(F_R) \hookrightarrow\!\!\!\!\!\!\circ\;\; F_R.$$

Then it follows by 11.3.3 that every free and therefore every right R-module is semiperfect. But this implies that R_R is perfect. □

Hence the proof of Theorem 11.6.3 is complete. The rings characterized by this theorem are of interest in various respects. We shall return later many times to them. Here let it be emphasized once more that for every right R-module over a right perfect ring R all statements concerning semiperfect modules are at our disposal. In particular for every projective module over such a ring we have the decomposition property 11.4.2 (Krull–Remak–Schmidt).

11.6.4 Corollary. *For a right perfect ring R we have:*

(a) *Every noetherian left R-module is artinian.*

(b) *Every artinian right R-module is noetherian.*

(c) *If R_R is noetherian then R_R is artinian.*

Proof. (a) Let $_RM$ be noetherian, then every submodule and every factor module is again noetherian. Consequently the socle of any factor module of M is finitely generated. Since by 11.6.3(4) the socle of an arbitrary left R-module is large in the module, by 9.4.4 $_RM$ is artinian.

(b) Let M_R be artinian and let $U \hookrightarrow M_R$. Then U is artinian, thus $U/\mathrm{Rad}(U)$ is semisimple and artinian and consequently finitely generated. Since R is right perfect we have (by 11.1.7) that $\mathrm{Rad}(U) \hookrightarrow\!\!\!\!\!\!\circ\;\; U$, and by 9.4.1 it follows that U is finitely generated. But this means that M is noetherian.

(c) By 9.3.7 Rad(R) is nilpotent. Since moreover $R/\mathrm{Rad}(R)$ is semi-simple, 9.3.11 yields the assertion. □

11.7 A THEOREM OF BJÖRK

By 11.6.3 a ring is right perfect if and only if it satisfies the descending chain condition for cyclic left ideals. The question then arises as to whether it also satisfies the descending chain condition for finitely generated left ideals. That this is in fact the case, is the content of the following theorem (J.-E. Björk, [32]).

11.7.1 THEOREM. *Let R be an arbitrary ring. Every R-module which satisfies the descending chain condition for cyclic submodules also satisfies this condition for finitely generated submodules.*

Proof. We recall first of all that the descending chain condition for cyclic, resp. finitely generated, submodules is equivalent to the minimal condition for cyclic, resp. finitely generated, submodules. As an abbreviation we denote the descending chain condition for cyclic, resp. finitely generated, submodules by (C) resp. (F). The proof is set out for right R-modules and is subdivided into several steps.

Step 1. *Assertion*: In the set of the submodules of an arbitrary module, satisfying (F), there is a maximal element. The proof is obtained with the help of Zorn's Lemma. Let M be an arbitrary module and let $\mathcal{F}$ be the set of submodules of M that satisfy (F). Then we have $0 \in \mathcal{F}$ and $\mathcal{F}$ is ordered by $\subset$. Let $\mathcal{K} \neq \varnothing$ be a chain from $\mathcal{F}$. Then

$$V := \bigcup_{U \in \mathcal{K}} U$$

is an upper bound of $\mathcal{K}$ in $\mathcal{F}$. For if $v_1, \ldots, v_n \in V$, then a $U \in \mathcal{K}$ exists (since $\mathcal{K}$ is a chain) with $v_1, \ldots, v_n \in U$. Consequently every finitely generated submodule of V is already contained in a $U \in \mathcal{K}$. Hence every descending chain of finitely generated submodules of V is already contained in a $U \in \mathcal{K}$ and consequently is stationary. Thus in fact we have $V \in \mathcal{F}$. By Zorn's Lemma a submodule $A \hookrightarrow M$, maximal in $\mathcal{F}$ then exists. If $A = M$ then we are done. Therefore in the following let A be properly contained in M.

Step 2. Now let M_R be a module which satisfies (C). Since $A \subsetneq M$ the set of cyclic submodules mR, $m \in M$ with $mR \not\subset A$ is not empty. By assumption there is in this set a minimal element y_0R.

Assertion: $U_0 := A + y_0R$ satisfies (F) in contradiction to the maximality of A. In order to see this let

$$U_1 \hookleftarrow U_2 \hookleftarrow U_3 \hookleftarrow \dots$$

be a descending chain of finitely generated submodules of U_0. If we have $U_i \hookrightarrow A$ for an i then by assumption on A this chain is stationary. Hence let $U_i \not\subset A$ for all $i = 1, 2, \dots$. In every U_i there is then a cyclic submodule uR with $uR \not\subset A$. Then by assumption with respect to $ur \not\subset A$ there exists a minimal cyclic submodule $y_iR \subset U_i$. In this sense for every $i = 1, 2, \dots$ let a fixed y_i be chosen.

Step 3. *Assertion*: If $U_i = A_i + y_iR$ holds with $A_i \hookrightarrow A$, then it follows that $U_i = A_i + y_{i+1}R$ for $i = 0, 1, 2, \dots$. As $y_{i+1} \in U_{i+1} \hookrightarrow U_i$ it follows that $y_{i+1} = a + y_ir$ with $a \in A$, $r \in R$. As $y_{i+1} \notin A$ we also have $y_ir \notin A$ and consequently $y_irR \not\subset A$. Since y_iR is minimal in U_i with respect to $uR \not\subset A$, it follows that $y_irR = y_iR$, thus there is an $r' \in R$ with $y_irr' = y_i$. Then it follows that $y_{i+1}r' = ar' + y_irr' = ar' + y_i$, thus we have $y_i = ar' - y_{i+1}r'$ with $ar' \in A_i$. Hence altogether as $y_{i+1} \in U_i$ we deduce that $U_i = A_i + y_iR = A_i + y_{i+1}R$.

Step 4. By induction we show:

$$U_i = A_i + y_iR, \qquad i = 1, 2, 3, \dots$$

with

$$A \hookleftarrow A_i \hookleftarrow A_{i+1}$$

and with A_i being finitely generated.

Proof. By assumption every U_{i+1}, $i = 0, 1, 2, \dots$ is finitely generated. Let $v_1, \dots, v_n$ be a set of generators of U_{i+1}. Let now $U_i = A_i + y_iR$ with $A_i \hookrightarrow A$, then it follows by the third step that $U_{i+1} \hookrightarrow U_i = A_i + y_{i+1}R$, thus we have $v_j = a_j + y_{i+1}r_j$ with $a_j \in A_i$, $r_j \in R$. It follows that $a_j = v_j - y_{i+1}r_j$ and so $a_1, \dots, a_n, y_{i+1}$ is a set of generators of U_{i+1}. With $A_{i+1} := a_1R + \dots + a_nR$ the assertion then holds. As a beginning for the induction $U_0 = A + y_0R$ is available.

Step 5. Since the condition (F) is satisfied for A, the sequence

$$A_1 \hookleftarrow A_2 \hookleftarrow A_3 \hookleftarrow \dots$$

is stationary. Thus there is an n with

$$A_n = A_{n+i}, \qquad i = 1, 2, 3, \dots.$$

Then it follows that

$$U_n = A_n + y_{n+1}R = A_{n+1} + y_{n+1}R = U_{n+1}$$

and by induction we deduce that $U_n = U_{n+i}$, $i = 1, 2, 3, \ldots$. Thus the sequence $U_1 \hookleftarrow U_2 \hookleftarrow U_3 \hookleftarrow \ldots$ is also stationary. Hence the assertion given in the second step is proved and the proof of Björk's theorem is complete. □

11.7.2 COROLLARY. *The following are equivalent for a ring R:*

(1) *R is right perfect,*

(2) *every left R-module satisfies the descending chain condition for finitely generated submodules.*

Proof. "(1)⇒(2)": Every descending chain of cyclic submodules of a module $_RM$ can be written in the form

$$Rm \hookleftarrow Rr_1m \hookleftarrow Rr_2r_1m \hookleftarrow Rr_3r_2r_1m \hookleftarrow \ldots .$$

Since by 11.6.3 $_RR$ satisfies (C), the chain

$$R \hookleftarrow Rr_1 \hookleftarrow Rr_2r_1 \hookleftarrow Rr_3r_2r_1 \hookleftarrow \ldots$$

is stationary and consequently the preceding chain is also stationary. Thus (C) holds for $_RM$ and by 11.7.2 then (F) also holds.

"(2)⇒(1)": By assumption $_RR$ satisfies the condition (F), thus also (C) and then (1) follows by 11.6.3. □

11.7.3 COROLLARY. *If R is right perfect and B is a two-sided ideal of A then R/B is also right perfect.*

Proof. By 11.7.2 (C) is satisfied for $_R(R/B)$. Since B is a two-sided ideal, R/B is also an (R/B)-left module and the submodules of $_R(R/B)$ and $_{R/B}(R/B)$ coincide. Thus (C) is also satisfied for $_{R/B}(R/B)$. By 11.6.3 it follows that R/B is right perfect. □

Obviously in this proof 11.7.1 is not used, but only 11.7.2 for the condition (C). Interesting corollaries of 11.7.1 in which not only (C) but also (F) must be used are still outstanding.

EXERCISES

(1)

For an integral domain R with quotient field K show:

(a) If R is not a field then K as an R-module does not have a projective cover.

(b) If R is not local and if M_R is indecomposable then M_R does not have a projective cover.

(c) If R is not local and M_R is semiperfect then $M = 0$.

(2)

(a) If A and $A \oplus B$ have projective covers then so also has B.

(b) Let R be an integral domain with exactly n maximal ideals ($n \geq 2$). Show:

(1) The R-module $M := R/\mathrm{Rad}(R)$ is semisimple and has 2^n submodules.

(2) Only two submodules of M have projective covers.

(3)

Let R be a local principal ideal domain, but not a field. Show for M_R:

(a) Then M has a projective cover if and only if it is the direct sum of a projective and a finitely generated R-module.

(b) Then M is semiperfect if and only if it is finitely generated. (Hint: The quotient field is countably generated as an R-module.)

(4)

(a) Give an example of a complemented module M with a non-complemented submodule U.

(b) If $M = A + B$ and if A and B are complemented then so also is M.

(c) If M is finitely generated and if every maximal submodule in M has an adco, then M is itself complemented.

(5)

(1) Show the following are equivalent for a module M_R with $\mathrm{Rad}(M) \neq M$:

(a) M is indecomposable.

(b) For every $X \mathrel{\underset{\neq}{\hookrightarrow}} M$ M/X is directly indecomposable.

(c) $\mathrm{Rad}(M)$ is a maximal and a small submodule of M.

(d) $\mathrm{Rad}(M)$ is the largest proper submodule of M.

(e) For all $m \in M$ either $mR \mathrel{\overset{\circ}{\hookrightarrow}} M$ holds or $mR = M$.

(2) Let $M \neq 0$ be semiperfect and let $\xi: P \to M$ be a projective cover. Show that M then satisfies the equivalent properties of (1) if and only if P satisfies the equivalent properties in 11.4.1.

(6)

(1) Show that a projective module P_R is semiperfect if and only if it satisfies the two following conditions:

(i) Every submodule which is not small contains a direct summand different from zero.

(ii) Every submodule contains a maximal direct summand.

(2) Show that a finitely generated module M with property (ii) already satisfies the maximal condition for direct summands.

(7)

Show that the following are equivalent for a projective module P_R:

(a) $S = \mathrm{End}(P_R)$ is semiperfect.

(b) P is semiperfect and satisfies the maximal condition for direct summands.

(c) P is semiperfect and finitely generated.

(8)

(1) Show that the following are equivalent for a ring R:

(a) Every finitely generated right ideal has an adco in R_R.

(b) Every cyclic right ideal has an adco in R_R.

(c) $\bar{R} = R/\mathrm{Rad}(R)$ is a regular ring and to every idempotent $\varepsilon \in \bar{R}$ there is an idempotent $e \in R$ with $\bar{e} = \varepsilon$.

(2) If ${}_RR$ is injective then R satisfies the equivalent conditions in (1).

(3) R is semiperfect if and only if the equivalent conditions in (1) are satisfied and R contains no infinite set of orthogonal idempotents.

(9)

Show: If A is a two-sided ideal of a ring R which is left or right t-nilpotent and which is finitely generated as a left or right ideal then A is nilpotent.

(10)

For a ring R define the left R-module $K = (R_R)^\circ = \mathrm{Hom}_{\mathbb{Z}}(R, \mathbb{Q}/\mathbb{Z})$. Show:

(a) ${}_RK$ is an injective cogenerator.

(b) A right ideal $A \hookrightarrow R_R$ is left t-nilpotent if and only if we have: $l_{K^{\mathbb{N}}}(A) \overset{*}{\hookrightarrow} K^{\mathbb{N}}$.

(c) If $K^{\mathbb{N}}$ has a large socle then the radical of R is left t-nilpotent. Give an example in which the converse does not hold.

(11)

Show for a ring R: (a) A projective module P_R has a small radical if and only if $P.\mathrm{Rad}(P)$ as a right module has a projective cover.

(b) The radical of R is left t-nilpotent if and only if the right R-module $(R/\mathrm{Rad}(R))^{(\mathbb{N})}$ has a projective cover.

(c) R is right perfect if and only if every semisimple right R-module has a projective cover.

(12)

It is to be shown: If $R^{(\mathbb{N})}$ is a direct summand in $R^{\mathbb{N}}$ (as a right R-module) then R is right perfect. For this purpose let $Ra_1 \hookleftarrow Ra_2 \hookleftarrow \dots$ be a descending sequence of cyclic left ideals, let $b_1 := a_1$, $b_{i+1}a_i = a_{i+1}$, and define

$$z_k := (0, \dots, 0, b_k, b_{k+1}b_k, b_{k+2}b_{k+1}b_k, \dots) \in R^{\mathbb{N}} \quad \text{with } b_k \text{ in the } k\text{th place.}$$

Show:

(1) $z_1 = (a_1, a_2, \dots, a_k, 0, \dots) + z_{k+1}a_k$ for all $k \in \mathbb{N}$.

(2) If we decompose $z_k = u_k + v_k \in R^{(\mathbb{N})} \oplus V = R^{\mathbb{N}}$, then we have $v_1 = v_{k+1}a_k$ for all $k \in \mathbb{N}$.

(3) If the co-ordinates of u_1 from the place m are equal to zero then we have $Ra_m = Ra_{m+1} = \dots$

(13)

Let K be a field, let V by a vector space over K with countably infinite basis $x_1, x_2, \dots$ and let $S = \mathrm{End}_K(V)$. Further let

$$V_0 := 0, \qquad V_n := \sum_{i=1}^{n} x_i K, \qquad n \geq 1,$$

then define

$$N := \{f \in S \mid \dim(\mathrm{Im}\, f) < \infty \wedge f(V_{n+1}) \subset V_n \text{ for all } n \geq 0\}.$$

Show:

(1) N is a subgroup of S with $N^2 \subset N$.

(2) N is left t-nilpotent but not right t-nilpotent.

(3) $A \subset S$ is a subring admitting multiplication by a scalar then $R := A + N$ is a subring of S with $\mathrm{Rad}(R) = N$ and $R/\mathrm{Rad}(R) \cong K$ (as rings).

(4) R is a local ring which is right perfect but not left perfect.

(5) $\mathrm{Soc}(R_R) = 0$.

(Hint: First show $l_S(N) = 0$.)

(14)

Show: A ring R is semisimple if and only if the endomorphism ring of $F_R = R^{(\mathbb{N})}$ is regular.

(Hint: First show: $\mathrm{End}(F_R)$ is regular $\Rightarrow$ R is right perfect because in the proof of Theorem 11.6.3 B is the image of an appropriate endomorphism of F.)

Chapter 12

Rings with Perfect Duality

12.1 INTRODUCTION TO AND FORMULATION OF THE MAIN THEOREM

Let a ring R be called a ring with *perfect duality* if the right and left R-modules have the same duality properties as vector spaces over a field, thus the best possible duality properties. In this respect we have to put the finitely generated or finitely cogenerated R-modules in the place of the finite-dimensional vector spaces.

The question arises as to the characterization of rings with perfect duality. This question originated from J. Dieudonné, who asked it for artinian rings (1958). Here it is considered for arbitrary rings. In order to be able to formulate the answer (12.1.1) we must first of all develop some concepts.

Let R be an arbitrary ring. By the dual module to an arbitrary module M_R we understand

$$M^* := \operatorname{Hom}_R(M_R, R_R),$$

in which in consequence of the definition

$$(r\varphi(m) := r\varphi(m), \qquad r \in R,\ \varphi \in M^*,\ m \in M,$$

M^* is a left R-module (see 3.8.2).

If $M = {}_RM$ is a left R-module, then M^* is a right R-module in consequence of the definition

$$(\varphi r)(m) = \varphi(m)r \quad \text{resp.} \quad (m)(\varphi r) = ((m)\varphi)r,$$

according as the homomorphisms are written on the left or right of the argument.

For every $A \hookrightarrow M_R$ the orthogonal complement A° of A in M^* is defined by

$$A^\circ := \{\varphi \mid \varphi \in M^* \wedge \varphi(A) = 0\}$$

Then as we see easily, we have $A^\circ \hookrightarrow {}_RM^*$.

For $X \hookrightarrow {}_RM^*$ on the other hand let

$$X^\perp := \{m \mid m \in M \wedge \forall \xi \in X[\xi(m) = 0]\}.$$

Then it follows that $X^\perp \hookrightarrow M_R$.

For every module M_R there exists the homomorphism

$$\Phi_M : M_R \to M_R^{**},$$

defined by

$$\Phi_M(m)(\varphi) := \varphi(m), \qquad m \in M, \varphi \in M^*.$$

In the most favourable case Φ_M is an isomorphism and M is then called reflexive (3.8.3). It is well known that every finite-dimensional vector space is reflexive.

The most important results of this chapter will now be presented. They serve as guiding principles for the following considerations in which we prove these results step by step. In the formulation of these results we understand by an R-module, either a right or a left R-module.

12.1.1 MAIN THEOREM. *The following are equivalent for a ring R:*

(1) *Every finitely generated R-module is reflexive.*
(2) *Every cyclic R-module is reflexive.*
(3) *Every finitely cogenerated R-module is reflexive.*
(4) *For every R-module M and every submodule A of M we have $A = A^{\circ\perp}$.*
(5) *R_R and ${}_RR$ are cogenerators.*
(6) *R_R is a cogenerator and ${}_RR$ is injective.*
(7) *${}_RR$ is a cogenerator and R_R is injective.*
(8) *R_R and ${}_RR$ are injective and to every simple R-module there is an isomorphic ideal in R.*†

12.1.2 *Definition.* A ring, which satisfies the conditions of 12.1.1, is called a *ring with perfect duality.*

12.1.3 COROLLARY. *If R is a ring with perfect duality then R is semiperfect and both R_R and ${}_RR$ are finitely cogenerated.*

† The latter property is also named in the literature after the author.

In the Main Theorem conditions (1) to (4) have to do with concepts of duality whereas conditions (5) to (8) connect cogenerator and injectivity properties.

The corollary asserts that for rings with perfect duality all results over semiperfect rings (from Chapter 11) and over finitely cogenerated modules are at our disposal.

In the following considerations we shall not only provide the lemmas for the proof of the preceding results but we shall also prove results which are of independent interest and such as are needed in the next chapter.

12.2 DUALITY PROPERTIES

Let R be an arbitrary ring and let $f: A_R \to M_R$ be an arbitrary R-homomorphism. Then let

$$f^*: {}_RM^* \to {}_RA^*$$

be defined by

$$f^*(\varphi) := \varphi f, \qquad \varphi \in M^*.$$

As

$$f^*(r_1\varphi_1 + r_2\varphi_2) = (r_1\varphi_1 + r_2\varphi_2)f = r_1(\varphi_1 f) + r_2(\varphi_2 f) = r_1 f^*(\varphi_1) + r_2 f^*(\varphi_2)$$

f^*, as constructed, is an R-homomorphism. We call f^* *the dual homomorphism to f*. We now bring together some simple properties of dual homomorphisms.

12.2.1 PROPOSITION. *Let $f: A_R \to M_R$, $f_0: M_R \to A_R$ and $g: M_R \to W_R$ be homomorphisms. Then we have*

(a) $(gf)^* = f^* g^*$, $f_0 f = 1_A \Rightarrow f^* f_0^* = 1_A^* = 1_{A^*}$.

(b) *f is an epimorphism $\Rightarrow f^*$ is a monomorphism.*

(c) *If R_R is injective, then we have: f is a monomorphism $\Rightarrow f^*$ is an epimorphism.*

(d) *If R_R is a cogenerator, then we have: f^* is a monomorphism $\Rightarrow f$ is an epimorphism, f^* is an epimorphism $\Rightarrow f$ is a monomorphism.*

(e) *If* $0 \to A \xrightarrow{f} M \xrightarrow{g} W \to 0$ *is a split exact sequence* (3.9.1) *then*

$$0 \to W^* \xrightarrow{g^*} M^* \xrightarrow{f^*} A^* \to 0$$

is also a split exact sequence.

(f) *If R_R is injective then along with the exact sequence*

$$0 \to A \xrightarrow{f} M \xrightarrow{g} W \to 0$$

the sequence

$$0 \to W^* \xrightarrow{g^*} M^* \xrightarrow{f^*} A^* \to 0$$

is also exact.

Proof. (a)

$$(gf)^*(\omega) = \omega(gf) = (\omega g)f = f^*(\omega g) = f^*(g^*(\omega)) = (f^*g^*)(\omega) \Rightarrow (gf)^* = f^*g^*.$$

$$1_A^*(\alpha) = \alpha 1_A = \alpha = 1_{A^*}(\alpha) \Rightarrow 1_A^* = 1_{A^*} \Rightarrow (f_0 f)^* = f^* f_0^* = 1_A^* = 1_{A^*}.$$

(b) From $f^*(\varphi) = \varphi f = 0$ it follows that $\varphi(f(m)) = 0$ for all $m \in M$. If f is an epimorphism then it follows that $\varphi(M) = 0$, thus $\varphi = 0$.

(c) Let $\alpha \in A^*$ be given. Since R_R is injective a $\varphi \in M^*$ exists with $\alpha = \varphi f = f^*(\varphi)$.

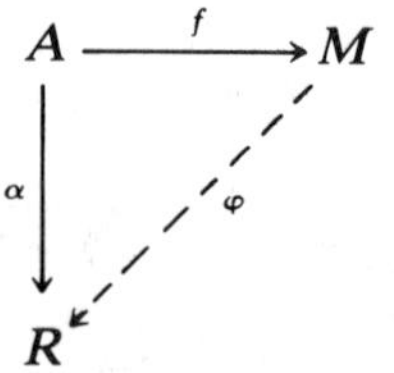

(d) Let f^* be a monomorphism, i.e. from $f^*(\varphi) = \varphi f = 0$ it follows that $\varphi = 0$. Suppose f were not an epimorphism, then, since R_R is a cogenerator, a $\tau \in (M/\mathrm{Im}(f))^*$, $\tau \neq 0$, would exist. Let now $\nu: M \to M/\mathrm{Im}(f)$ then letting $\varphi := \tau\nu \in M^*$, $\varphi \neq 0$ and $\varphi(\mathrm{Im}(f)) = 0$, then $\varphi f = 0$, thus $f^*(\varphi) = 0$, thus $\varphi = 0$ ↯.

Let f^* be an epimorphism. Suppose f were not a monomorphism. Since R_R is a cogenerator, there is an $\alpha \in A^*$ with $\alpha(\mathrm{Ker}(f)) \neq 0$. Let $\varphi \in M^*$ with $\alpha = f^*(\varphi) = \varphi f$, then it follows that $0 \neq \alpha(\mathrm{Ker}(f)) = \varphi(f(\mathrm{Ker}(f))) = 0$ ↯.

(e) Since the sequence splits, there is (by 3.9.3) a homomorphism $f_0: M \to A$, $g_0: W \to M$ with $f_0 f = 1_A$, $gg_0 = 1_W$. Therefore it follows from (a) that

$$1_{A^*} = f^* f_0^*, \qquad 1_{W^*} = g_0^* g^*.$$

Thus f^* is an epimorphism, g^* is a monomorphism and in the sequence

$$0 \to W^* \xrightarrow{g^*} M^* \xrightarrow{f^*} A^* \to 0$$

both $\mathrm{Im}(g^*)$ and the $\mathrm{Ker}(f^*)$ are direct summands in M^*. It remains to show the exactness at the position M^* where only the exactness of the original sequence but not however its splitting is used.

From $gf = 0$ it follows that

$$(f^*g^*)(\omega) = \omega g f = 0 \quad \text{for all} \quad \omega \in W^*,$$

thus we have $\mathrm{Im}(g^*) \subset \mathrm{Ker}(f^*)$. Now let $f^*(\varphi) = \varphi f = 0$, i.e. $\varphi(f(m)) = 0$ for every $m \in M$. Consequently we have

$$\mathrm{Ker}(g) = \mathrm{Im}(f) \hookrightarrow \mathrm{Ker}(\varphi).$$

In the diagram

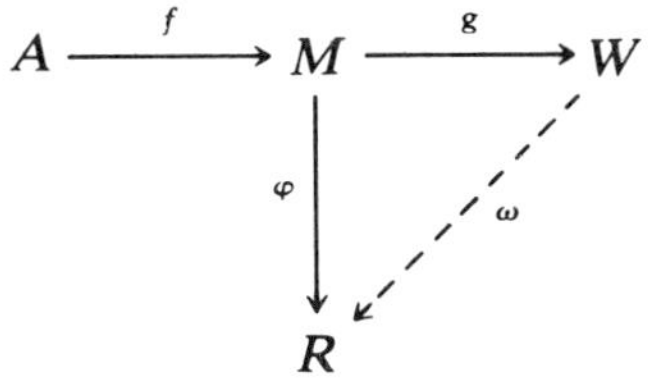

an $\omega \in W^*$ exists by 3.4.7 with $\varphi = \omega g = g^*(\omega)$, thus we also have

$$\mathrm{Ker}(f^*) \hookrightarrow \mathrm{Im}(g^*).$$

(f) By (b) g^* is a monomorphism and by (c) f^* is an epimorphism. As in the preceding proof it follows that $\mathrm{Im}(g^*) = \mathrm{Ker}(f^*)$. □

Obviously the corresponding statements hold on changing the sides. These considerations are now to be applied to

$$\Phi_M : M_R \to M_R^{**}.$$

There then holds

$$\Phi_M^* : {}_RM^{***} \to {}_RM^* \quad \text{with} \quad \Phi_M^*(\tau) = \tau\Phi_M,\ \tau \in M^{***}.$$

Further we have to consider

$$\Phi_{M^*} : {}_RM^* \to {}_RM^{***}.$$

12.2.2 LEMMA. *Let R be an arbitrary ring and let M_R be an arbitrary right R-module. Then we have:*

(a) $$\Phi_M^* \Phi_{M^*} = 1_{M^*}$$

and consequently

Φ_{M^*} *is a monomorphism,*

Φ_M^* *is an epimorphism,*

$${}_RM^{***} = \mathrm{Im}(\Phi_{M^*}) \oplus \mathrm{Ker}(\Phi_M^*).$$

(b) *If Φ_M is an epimorphism then Φ_{M^*} is an isomorphism, i.e. M^* is reflexive.*

(c) *For an arbitrary homomorphism* $f: A_R \to M_R$

$$\begin{array}{ccc} A & \xrightarrow{\quad f \quad} & M \\ \Big\downarrow{\scriptstyle \Phi_A} & & \Big\downarrow{\scriptstyle \Phi_M} \\ A^{**} & \xrightarrow{\quad f^{**} \quad} & M^{**} \end{array}$$

is commutative.

(d) R_R *is a cogenerator if and only if for every module* M_R Φ_M *is a monomorphism.*

(e) *Let* R_R *and* ${}_RR$ *be injective. Then for every exact sequence*

$$0 \to A \xrightarrow{f} M \xrightarrow{g} W \to 0$$

the sequence

$$0 \to A^{**} \xrightarrow{f^{**}} M^{**} \xrightarrow{g^{**}} W^{**} \to 0$$

is also exact and the diagram

$$\begin{array}{ccccccccc} 0 & \longrightarrow & A & \xrightarrow{f} & M & \xrightarrow{g} & W & \longrightarrow & 0 \\ & & \Big\downarrow{\scriptstyle \Phi_A} & & \Big\downarrow{\scriptstyle \Phi_M} & & \Big\downarrow{\scriptstyle \Phi_W} & & \\ 0 & \longrightarrow & A^{**} & \xrightarrow{f^{**}} & M^{**} & \xrightarrow{g^{**}} & W^{**} & \longrightarrow & 0 \end{array}$$

is commutative.

Proof. (a) Let $\varphi \in M^*$, then first of all we have

$$(\Phi_M^* \Phi_{M^*})(\varphi) = \Phi_M^*(\Phi_{M^*}(\varphi)) = \Phi_{M^*}(\varphi)\Phi_M.$$

For $m \in M$ it follows that

$$(\Phi_{M^*}(\varphi)\Phi_M)(m) = \Phi_{M^*}(\varphi)(\Phi_M(m)) = \Phi_M(m)(\varphi) = \varphi(m),$$

thus $(\Phi_M^* \Phi_{M^*})(\varphi) = \varphi$, and thus $\Phi_M^* \Phi_{M^*} = 1_{M^*}$.

(b) If Φ_M is an epimorphism then it follows by 12.2.1(b) that Φ_M^* is a monomorphism. By (a) Φ_M^* is then an isomorphism and Φ_{M^*} is the inverse isomorphism.

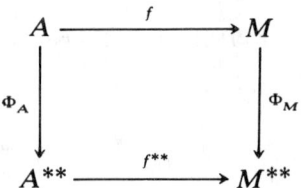

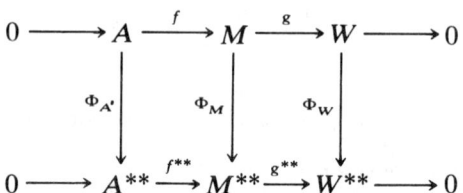

(c) For $a \in A$ and $\varphi \in M^*$ we have

$$\begin{aligned}((f^{**}\Phi_A)(a))(\varphi) &= (\Phi_A(a)f^*)(\varphi)\\ &= \Phi_A(a)(f^*(\varphi)) = \Phi_A(a)(\varphi f)\\ &= \varphi(f(a)) = \Phi_M(f(a))(\varphi)\\ &= ((\Phi_M f)(a))(\varphi),\end{aligned}$$

thus $f^{**}\Phi_A = \Phi_M f$.

(d) From $\Phi_M(m) = 0$ it follows that $\varphi(m) = 0$ for all $\varphi \in M^*$, thus $m \in \bigcap_{\varphi \in M^*} \operatorname{Ker}(\varphi)$. If R_R is a cogenerator then we have

$$\bigcap_{\varphi \in M^*} \operatorname{Ker}(\varphi) = 0$$

thus $m = 0$. The converse is clear.

(e) From 12.2.1(f) the exactness of

$$0 \to A^{**} \xrightarrow{f^{**}} M^{**} \xrightarrow{g^{**}} W^{**} \to 0$$

follows and from 12.2.2(c) the commutativity of the indicated diagrams. □

From our considerations so far the following theorem, which is of interest, immediately arises and later has other important applications.

12.2.3 THEOREM

(a) *Let R be an arbitrary ring. If*

$$0 \to A \xrightarrow{f} M \xrightarrow{g} W \to 0$$

is a split exact sequence of right (or left) R-modules then we have: M is reflexive if and only if A and W are reflexive.

(b) *Let R_R and ${}_RR$ be injective cogenerators. If*

$$0 \to A \xrightarrow{f} M \xrightarrow{g} W \to 0$$

is an exact sequence of right (or left) R-modules then we have: M is reflexive if and only if A and W are reflexive.

Proof. (a) By assumption and 3.9.3(b) there are homomorphisms f_0, g_0 with

$$f_0 f = 1_A, \qquad g g_0 = 1_W,$$

for which
$$0 \leftarrow A \xleftarrow{f_0} M \xleftarrow{g_0} W \leftarrow 0$$
is also exact and splits. The diagram

$$\begin{array}{ccccccccc}
0 & \rightleftarrows & A & \underset{f_0}{\overset{f}{\rightleftarrows}} & M & \underset{g_0}{\overset{g}{\rightleftarrows}} & W & \rightleftarrows & 0 \\
& & \downarrow \Phi_A & & \downarrow \Phi_M & & \downarrow \Phi_W & & \\
0 & \rightleftarrows & A^{**} & \underset{f_0^{**}}{\overset{f^{**}}{\rightleftarrows}} & M^{**} & \underset{g_0^{**}}{\overset{g^{**}}{\rightleftarrows}} & W^{**} & \rightleftarrows & 0
\end{array}$$

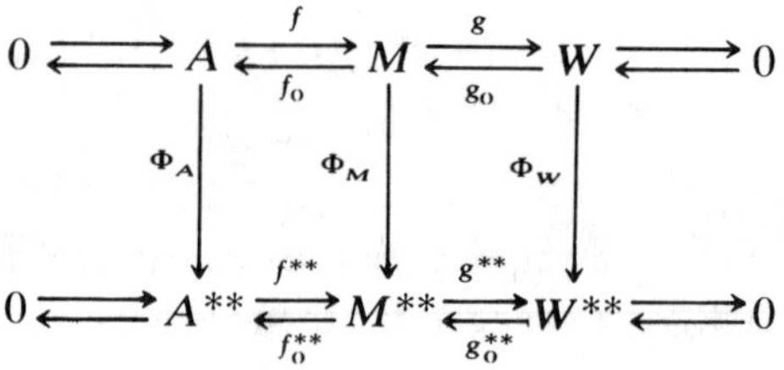

then has by 12.2.1(e) split exact rows and by 12.2.2(c) is commutative.

Let M be reflexive. Since now Φ_M is an isomorphism and f is a monomorphism, as $\Phi_M f = f^{**} \Phi_A$, Φ_A must also be a monomorphism. Analogously we see that Φ_W is also a monomorphism. The assertion that Φ_A and Φ_W are in fact isomorphisms now follows from 3.9.2.

Let now A and W be reflexive. In order to be able to apply 3.9.2 again, it must be shown that Φ_M is a monomorphism. Let $m \in \text{Ker}(\Phi_M)$, then since
$$g^{**}\Phi_M = \Phi_W g$$
and since Φ_W is an isomorphism it follows that
$$m \in \text{Ker}(g) = \text{Im}(f).$$
Thus there is an $a \in A$ with
$$f(a) = m.$$
Therefore it follows that
$$0 = \Phi_M(m) = \Phi_M(f(a)) = (\Phi_M f)(a) = (f^{**}\Phi_A)(a).$$
Since f^{**} and Φ_A are both monomorphisms it follows that $a = 0$, thus also $m = 0$. Since consequently Φ_M is a monomorphism, the assertion follows from 3.9.2.

(b) By assumption 12.2.2(e) holds. In the diagram in 12.2.2(e) Φ_A, Φ_M and Φ_W are monomorphisms by 12.2.2(d). The assertion follows then from 3.9.2. □

12.2.4 COROLLARY. *Let R be an arbitrary ring.*

(a) *If $A \cong M$ then it follows: M is reflexive if and only if A is reflexive.*

(b) *Let $M_R = \bigoplus_{i=1}^{n} M_i$, then we have: M is reflexive if and only if all M_i, $i = 1, \ldots, n$ are reflexive.*

(c) *Every finitely generated projective R-module is reflexive.*

Proof. (a) Let $f: A_R \to M_R$ be the given isomorphism. Then

$$0 \to A \xrightarrow{f} M \to 0 \to 0$$

is a split exact sequence and the assertion follows from 12.2.3(a).

(b) It suffices to prove the assertion for $n=2$ since it then follows for arbitrary n entirely by induction. For $n=2$ it follows from 12.2.3(a) on reflecting upon the split exact sequence

$$0 \to M_1 \xrightarrow{\iota} M_1 \oplus M_2 \xrightarrow{\pi} M_2 \to 0$$

in which τ is the inclusion of M_1 in $M = M_1 \oplus M_2$ and π is the projection of M onto M_2.

(c) Since R_R is reflexive, by (b) every finitely generated free R-module is reflexive. (We see this also directly as for vector spaces on using a dual basis.) Consequently every finitely generated projective module is reflexive as a direct summand of a finitely generated free module. □

12.2.5 COROLLARY. *For an arbitrary R-module we have*:

(a) *If M is reflexive, then all modules M^*, M^{**}, ... are also reflexive.*

(b) *If M^* is not reflexive then none of the modules M^{**}, M^{***}, ... is reflexive.*

Proof. (a) This follows from 12.2.2(b).

(b) By (a) it suffices to show: If M^{***} is reflexive then so also is M^*. By 12.2.2(a) M^* is isomorphic to the direct summand $\text{Im}(\Phi_{M^*})$ of M^{***}. If M^{***} is reflexive then it follows by 12.2.4 that M^* is reflexive. □

To conclude these duality considerations we prove a lemma which gives information on the reflexivity of cyclic modules.

12.2.6 LEMMA. *For $A \hookrightarrow R_R$ we have*:

(a) $h: {}_R(R/A)^* \ni \varphi \mapsto \varphi(\bar{1}) \in {}_R(\boldsymbol{l}_R(A))$ *with* $\bar{1} := 1 + A \in R/A$ *is an isomorphism.*

(b) $\Phi_{R/A}$ *is a monomorphism* $\Leftrightarrow \boldsymbol{r}_R \boldsymbol{l}_R(A) = A$.

(c) *Let* $\rho: R_R \to \boldsymbol{l}_R(A)^*$ *be defined by*

$$\rho(r)(x) := xr, \qquad r \in R, x \in \boldsymbol{l}_R(A),$$

then ρ is a homomorphism for which the diagram

$$\begin{array}{ccc} R & \xrightarrow{\nu} & R/A \\ \rho \big\downarrow & & \big\downarrow \Phi_{R/A} \\ l_R(A)^* & \xrightarrow{h^*} & (R/A)^{**} \end{array}$$

is commutative (in this, h is the isomorphism from (a)).

(d) $\Phi_{R/A}$ *is an epimorphism if and only if ρ is an epimorphism.*

Proof. (a) It is clear that h is an R-homomorphism. Let $h(\varphi)=\varphi(\bar{1})=0$, then it follows that

$$\varphi(\bar{r})=\varphi(\bar{1}r)=\varphi(\bar{1})r=0, \qquad r\in R,$$

thus $\varphi=0$, i.e. h is a monomorphism. Let $x\in \boldsymbol{l}_R(A)$ then $\varphi\in(R/A)^*$ with $h(\varphi)=\varphi(\bar{1})=x$ is defined by

$$\varphi: R/A\ni \bar{r}\mapsto xr\in R.$$

Thus h is also an epimorphism.

(b) For $\bar{r}\in R/A$ we have

$$\bar{r}\in \operatorname{Ker}(\Phi_{R/A})\Leftrightarrow \Phi_{R/A}(\bar{r})(\varphi)=\varphi(\bar{r})=\varphi(\bar{1})r=0 \quad \text{for all} \quad \varphi\in(R/A)^*$$

$\Leftrightarrow r\in \boldsymbol{r}_R\boldsymbol{l}_R(A)$ (by (a)). Therefore the assertion follows.

(c) For $r\in R$ and $\varphi\in(R/A)^*$ we have

$$(h^*\rho(r))(\varphi)=(\rho(r)h)(\varphi)=\rho(r)(\varphi(\bar{1}))=\varphi(\bar{1})r$$

and also

$$(\Phi_{R/A}\nu(r))(\varphi)=\Phi_{R/A}(\bar{r})(\varphi)=\varphi(\bar{r})=\varphi(\bar{1})r,$$

thus we conclude that $h^*\rho=\Phi_{R/A}\nu$.

(d) Since h^* is an isomorphism and ν is an epimorphism the assertion follows from $h^*\rho=\Phi_{R/A}\nu$. □

Having become conversant with duality concepts we apply ourselves first of all to other considerations which likewise are needed for the proof of the Main Theorem and which to some extent find application also in the next chapter.

12.3 CHANGE OF SIDE

We treat here of the question as to the manner in which given properties of M_R carry over to ${}_SM$ where $S := \operatorname{End}(M_R)$.

12.3.1 LEMMA. *Let R be an arbitrary ring, let $x, y \in M_R$, let $S := \mathrm{End}(M_R)$, let $yR \cong xR$ and let xR be contained in an injective submodule of M_R. Then Sx is isomorphic to a submodule of Sy (as a submodule of ${}_SM$). If Sy is simple then it follows that $Sy \cong Sx$.*

Proof. Let $\varphi : yR \to xR$ be the given isomorphism. Further let $xR \hookrightarrow Q_R \hookrightarrow M_R$ with injective Q_R. Then a commutative diagram exists:

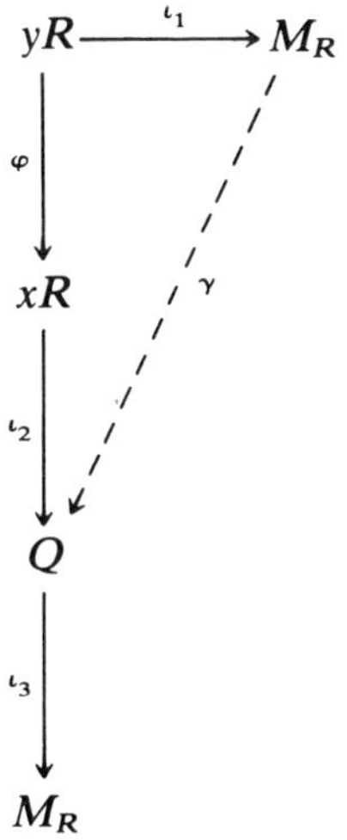

where ι_1, ι_2, ι_3 are the corresponding inclusion mappings. For $s_0 := \iota_3\gamma \in S$ we then have

$$\varphi(yr) = s_0yr, \qquad r \in R.$$

Let $r_0, r_1 \in R$ be determined by

$$\varphi(y) = s_0y = xr_0, \qquad \varphi(yr_1) = s_0yr_1 = x,$$

then it follows that

$$\hat{\varphi}: Sx \ni sx \mapsto sxr_0 = ss_0y \in Sy$$

is an S-homomorphism.

Suppose $sxr_0 = 0$ then it follows that

$$sxr_0r_1 = ss_0yr_1 = sx = 0,$$

thus $\hat{\varphi}$ is a monomorphism. Since from $xR \cong yR$ we have either $x = y = 0$ or $x \neq 0 \wedge y \neq 0$ it follows finally from the simplicity of Sy that $Sy \cong Sx$. □

12.3.2 LEMMA. *Let $S := \text{End}(M_R)$, let $x \in M$, let xR be simple and let xR be contained in an injective submodule Q of M_R. Then Sx is a simple submodule of ${}_SM$.*

Proof. We show that for arbitrary $s_0x \neq 0$, $s_0 \in S$ it follows that

$$Ss_0x = Sx.$$

Since xR is simple and $s_0x \neq 0$, s_0xR is also simple and

$$xR \ni xr \mapsto s_0xr \in s_0xR$$

is an isomorphism. Let

$$\tau : s_0xR \to xR$$

be the inverse isomorphism, then a φ exists so that the diagram

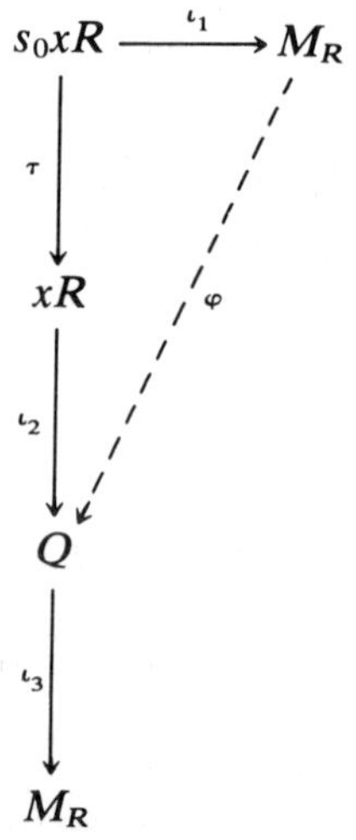

is commutative, where τ_1, τ_2, τ_3, are the corresponding inclusion mappings. If we put $t_0 := \iota_3\varphi$ then we have $t_0 \in S$ and $t_0s_0x = \tau s_0x = x$, thus $Ss_0x = Sx$, which was to be shown. □

12.4 ANNIHILATOR PROPERTIES

In this section we examine the annihilator properties of R. As abbreviation in place of $\boldsymbol{l}_R(A)$ resp. $\boldsymbol{r}_R(A)$ we write only $\boldsymbol{l}(A)$ resp. $\boldsymbol{r}(A)$. We have already become acquainted with such an annihilator property in 12.2.6 where it was a question of characterizing the reflexivity of cyclic modules.

12.4.1 LEMMA. *If C_R is a cogenerator then we have for every $A \hookrightarrow R_R$:*

$$\boldsymbol{rl}_C(A) = A.$$

Proof. From the definition of annihilator it follows that

$$A \hookrightarrow \boldsymbol{rl}_C(A).$$

Let $r \in R$, $r \notin A$, then by assumption there is a

$$\tau : (R/A)_R \to C_R \quad \text{with} \quad \tau(r+A) \neq 0.$$

Let now

$$\nu : R \to R/A,$$

then it follows that

$$0 = \tau\nu(A) = \tau\nu(1)A,$$

thus $\tau\nu(1) \in \boldsymbol{l}_C(A)$, and also $\tau\nu(1)r = \tau\nu(r) = \tau(r+A) \neq 0$, thus $r \notin \boldsymbol{rl}_C(A)$. Hence we have $\boldsymbol{rl}_C(A) \hookrightarrow A$; thus in conclusion the assertion. □

Mostly this lemma is applied in the case $C_R = R_R$ where $\boldsymbol{rl}(A) = A$ is then briefly written.

12.4.2 THEOREM

(a) *If R_R is injective then we have:*

(1) *for arbitrary $A \hookrightarrow R_R$, $B \hookrightarrow R_R$: $\boldsymbol{l}(A \cap B) = \boldsymbol{l}(A) + \boldsymbol{l}(B)$;*

(2) *for arbitrary finitely generated $C \hookrightarrow {}_RR$: $\boldsymbol{lr}(C) = C$.*

(b) *If conditions* (1) *and* (2) *in* (a) *are satisfied then every homomorphism of a finitely generated right ideal of R into R is obtained by left multiplication by an element from R.*

Proof. (a) Obviously we always have $\boldsymbol{l}(A) + \boldsymbol{l}(B) \hookrightarrow \boldsymbol{l}(A \cap B)$. Let now $x \in \boldsymbol{l}(A \cap B)$,

$$\varphi : A + B \ni a + b \mapsto xb \in R$$

is an R-homomorphism (for $a + b = a_1 + b_1 \Rightarrow a - a_1 = b_1 - b \in A \cap B \Rightarrow xb_1 = x(a - a_1) + xb = xb$).

Since R_R is injective, there is a $y \in R$ with $\varphi(a+b) = y(a+b) = xb$. In particular $0 = \varphi(a) = ya$ holds for all $a \in A$, thus $y \in \boldsymbol{l}(A)$. Further it follows that for all $b \in B$

$$\varphi(b) = yb = xb,$$

thus $z := x - y \in \boldsymbol{l}(B)$. Therefore it follows that $x = y + z \in \boldsymbol{l}(A) + \boldsymbol{l}(B)$, from which (1) is proved.

For the proof of (2) let $C = Rc_1 + \ldots + Rc_n \hookrightarrow {}_RR$. Trivially we have

$$\boldsymbol{r}\Big(\sum_{i=1}^{n} Rc_i\Big) = \bigcap_{i=1}^{n} \boldsymbol{r}(Rc_i)$$

and by successive applications of (1) it follows that

$$\boldsymbol{lr}\Big(\sum_{i=1}^{n} Rc_i\Big) = \boldsymbol{l}\Big(\bigcap_{i=1}^{n} \boldsymbol{r}(Rc_i)\Big) = \sum_{i=1}^{n} \boldsymbol{lr}(Rc_i).$$

In order to obtain (2) it must only be shown that

$$\boldsymbol{lr}(Rc) = Rc, \qquad c \in R.$$

Trivially we have $Rc \hookrightarrow \boldsymbol{lr}(Rc)$. Let now $b \in \boldsymbol{lr}(Rc)$, then it follows that $\boldsymbol{r}(c) \hookrightarrow \boldsymbol{r}(b)$ and hence

$$cR \ni cr \mapsto br \in R$$

is a homomorphism which, since R_R is injective, is obtained by left multiplication by an $a \in R$. In particular we then have $ac = b$, i.e. $b \in Rc$, thus $\boldsymbol{lr}(Rc) \hookrightarrow Rc$, which was still to be shown.

(b) The assertion is established by induction on the number n of the generating elements of a finitely generated right ideal.

$n = 1$: Let $\varphi : aR \to R_R$ be a homomorphism. Since from $ar = 0$ there follows also $\varphi(ar) = 0 = \varphi(a)r$, we have $\boldsymbol{r}(a) \hookrightarrow \boldsymbol{r}(\varphi(a))$. Hence we obtain

$$\boldsymbol{r}(Ra) \hookrightarrow \boldsymbol{r}(R\varphi(a))$$

and by (2) it follows that

$$R\varphi(a) = \boldsymbol{lr}(R\varphi(a)) \hookrightarrow \boldsymbol{lr}(Ra) = Ra.$$

Thus there is a $c \in R$ with $\varphi(a) = ca$ and consequently we have

$$\varphi(ar) = \varphi(a)r = car,$$

which was to be shown.

Inference from n to $n+1$: Let

$$\varphi : \sum_{i=1}^{n+1} a_iR \to R_R$$

be a homomorphism, then by induction assumption there are $c_1, c_2 \in R$ so that we have

$$\varphi\Big(\sum_{i=1}^{n} a_ir_i\Big) = c_1 \sum_{i=1}^{n} a_ir_i, \qquad \varphi(a_{n+1}r_{n+1}) = c_2a_{n+1}r_{n+1}.$$

By (1) it then follows that

$$c_1 - c_2 \in \boldsymbol{l}\left(\sum_{i=1}^{n} a_i R \cap a_{n+1} R\right) = \boldsymbol{l}\left(\sum_{i=1}^{n} a_i R\right) + \boldsymbol{l}(a_{n+1} R),$$

i.e. there are

$$s \in \boldsymbol{l}\left(\sum_{i=1}^{n} a_i R\right), \qquad t \in \boldsymbol{l}(a_{n+1} R)$$

with $c_1 - c_2 = s - t$. Let $c := c_1 - s = c_2 - t$ then it follows that

$$\varphi\left(\sum_{i=1}^{n+1} a_i r_i\right) = \varphi\left(\sum_{i=1}^{n} a_i r_i\right) + \varphi(a_{n+1} r_{n+1})$$

$$= (c_1 - s) \sum_{i=1}^{n} a_i r_i + (c_2 - t) a_{n+1} r_{n+1} = c \sum_{i=1}^{n+1} a_i r_i,$$

thus φ is obtained by left multiplication by c. Hence (b) is also proved. □

12.4.3 COROLLARY. *If R_R is noetherian and if the conditions* (1) *and* (2) *in* 12.4.2 *are satisfied then R_R is injective.*

Proof. Since R_R is noetherian every right ideal of R is finitely generated. Then the assertion follows from 12.4.2(b) and Baer's Criterion. □

12.5 INJECTIVITY AND THE COGENERATOR PROPERTY OF A RING

The cogenerator property of R_R is in general independent of the injectivity of R_R (see Exercises 13, 14). In order to obtain the equivalence of these properties additional conditions are required.

As a preparation for the corresponding theorem we need a lemma.

12.5.1 LEMMA. *Let R be an arbitrary ring.*

(a) *Let P_1, P_2 be projective right R-modules with small radicals. Then we have:*

$$P_1 \cong P_2 \Leftrightarrow P_1/\mathrm{Rad}(P_1) \cong P_2/\mathrm{Rad}(P_2).$$

(b) *Let Q_1, Q_2 be injective right R-modules with large socles. Then we have:*

$$Q_1 \cong Q_2 \Leftrightarrow \mathrm{Soc}(Q_1) \cong \mathrm{Soc}(Q_2).$$

Proof. (a) Let $\varphi: P_1 \to P_2$ be the given isomorphism. Since

$$\varphi(\text{Rad}(P_1)) = \text{Rad}(P_2)$$

φ induces an isomorphism

$$\hat{\varphi}: P_1/\text{Rad}(P_1) \ni p_1 + \text{Rad}(P_1) \mapsto \varphi(p_1) + \text{Rad}(P_2) \in P_2/\text{Rad}(P_2).$$

The converse follows from 5.6.3 for $P_1 \to P_1/\text{Rad}(P_1)$ and $P_2 \to P_2/\text{Rad}(P_2)$ are projective covers.

(b) Dual to (a). □

We come now to a theorem which is of independent interest. It can be considered as a one-sided weakening of the Main Theorem as mentioned at the beginning.

12.5.2 THEOREM. *The following are equivalent for a ring* R:

(1) R_R *is a cogenerator and there are only finitely many isomorphism classes of simple right* R*-modules.*

(2) R_R *is a cogenerator and every simple left* R*-module is isomorphic to a left ideal of* R.

(3) *Every module* M_R *with* $\mathbf{r}_R(M) = 0$ (i.e. M_R *faithful*) *is a generator.*

(4) *Every cogenerator of* M_R *is a generator.*

(5) R_R *is injective and finitely cogenerated.*

(6) R_R *is injective, semiperfect and has a large socle.*

Remark. A ring with the properties of the theorem is denoted in the literature as a *right PF-ring*. G. Azumaya posed the so far unanswered question as to whether a right *PF*-ring is also a left *PF*-ring.

Proof of 12.5.2. We show (2)⇒(3)⇒(4)⇒(5)⇒(6) together with (6)⇒(2) ∧ (1), (1)⇒(6).

"(2)⇒(3)": Since R_R is a generator it suffices by 3.3.2 to show that

$$T := \sum_{\varphi \in \text{Hom}_R(M,R)} \text{Im}(\varphi) = R.$$

Since $M^* = \text{Hom}_R(M, R)$ is a left R-module, T is also a left ideal. Now let $z \in \mathbf{r}_R(T)$, then it follows for every $m \in M$ and $\varphi \in M^*$ that

$$\varphi(mz) = \varphi(m)z = 0,$$

thus $Mz \subset \bigcap_{\varphi \in M^*} \text{Ker}(\varphi)$.

Since R_R is a cogenerator we have

$$\bigcap_{\varphi \in M^*} \operatorname{Ker}(\varphi) = 0,$$

thus $Mz = 0$. Since by assumption $r_R(M) = 0$, it follows that $z = 0$, thus $r_R T = 0$. Suppose $T \neq R$, then there is a maximal left ideal $A \hookrightarrow {}_R R$ with $T \hookrightarrow A \hookrightarrow R$. By assumption there is an $Rx \hookrightarrow {}_R R$ and an isomorphism

$$\sigma \colon R/A \cong Rx.$$

Then it follows for all $a \in A$ that

$$0 = \sigma(0) = \sigma(\bar{a}) = a\sigma(\bar{1}),$$

thus $0 \neq \sigma(\bar{1}) \in r_R(A) \hookrightarrow r_R(T) = 0$ ↯.

Consequently $T = R$ must hold, which was to be shown.

"(3)⇒(4)": From 12.4.1 it follows for $A = 0$ that every cogenerator is faithful. Then by (3) it is a generator.

"(4)⇒(5)": The cogenerator

$$C_0 = \coprod_{j \in J} Q_j$$

which is minimal in the sense of 5.8.6(b), is by assumption a generator.

By 5.8.2 R_R is then isomorphic to a direct summand of a direct sum of copies of C_0 and, by the definition of C_0, also of copies of Q_j, $j \in J$. Since $R = 1R$ is cyclic, R is isomorphic to a direct summand of a finite direct sum

$$Q := \coprod_{i=1}^{n} Q_i \quad \text{with} \quad Q_i \in \{Q_j \mid j \in J\}.$$

Since Q is injective, R_R is injective. By the definition of $Q_j = I(E_j)$ (see 5.8.6) and by 9.4.3 Q is finitely cogenerated and then so also is R_R as an isomorphic image of a direct summand of Q.

"(5)⇒(6)": It has only to be shown that R_R is semiperfect. From 9.4.3 we have

$$R_R = \bigoplus_{i=1}^{n} I(E_i) \quad \text{with simple } E_i.$$

As $E_i \overset{*}{\hookrightarrow} I(E_i)$, $I(E_i)$ is directly indecomposable. By 7.2.8 End$(I(E_i))$ is then local and by 11.4.1 $I(E_i)$ is semiperfect and then by 11.3.4 so also is R_R.

"(6)⇒(2) ∧ (1)": By assumption we have

$$\operatorname{Soc}(R_R) \overset{*}{\hookrightarrow} R_R.$$

Since by 12.3.2

$$\operatorname{Soc}(R_R) \hookrightarrow \operatorname{Soc}({}_R R)$$

and since by 9.2.1(a) and (b)

$$\mathrm{Soc}({}_R R) \hookrightarrow r_R(\mathrm{Rad}(R)),$$

it follows that

$$r_R(\mathrm{Rad}(R)) \overset{*}{\hookrightarrow} R_R.$$

Now let E be a simple left R-module and let $U \hookrightarrow {}_R R$ with $E \cong R/U$. Since R is semiperfect (by 11.3.2 on both sides!) there is by 11.1.2 a decomposition

$${}_R R = R_1 \oplus R_2$$

with $R_2 \hookrightarrow U$, $R_1 \cap U \overset{\circ}{\hookrightarrow} R_1$. Let now

$$R_1 = Re_1, \qquad R_2 = Re_2$$

with idempotents e_1, $e_2 = 1 - e_1$. As $R_2 \hookrightarrow U$ it follows from the modular law that we have

$$U = (R_1 \cap U) \oplus R_2,$$

in which $R_1 \cap U \hookrightarrow \mathrm{Rad}(R_1) \hookrightarrow \mathrm{Rad}(R)$ holds.

As $e_2^2 = e_2 \neq 1$ (since $R_2 \hookrightarrow U \neq R$) and $r = e_2 r + (1 - e_2) r$ for $r \in R$ it follows that

$$r_R(Re_2) = (1 - e_2)R \neq 0.$$

As $R_1 \cap U \hookrightarrow \mathrm{Rad}(R)$ we have

$$r_R(\mathrm{Rad}(R)) \hookrightarrow r_R(R_1 \cap U) \hookrightarrow R_R.$$

Since, as established at the beginning, $r_R(\mathrm{Rad}(R)) \overset{*}{\hookrightarrow} R_R$ it follows that $r_R(R_1 \cap U) \overset{*}{\hookrightarrow} R_R$ from which it follows that

$$0 \neq r_R(R_1 \cap U) \cap r_R(Re_2) = r_R((R_1 \cap U) + Re_2) = r_R(U).$$

Now let

$$0 \neq a \in r_R(U),$$

then it follows that $l_R(a) = U$, since U is maximal in ${}_R R$. Therefore it follows that

$$Ra \cong R/U \cong E,$$

i.e. for every simple left R-module R contains an isomorphic left ideal.

Let now

$$Ra_1, \ldots, Ra_n$$

with $Ra_i \hookrightarrow {}_R R$ be a set of representatives for the isomorphism classes of simple left R-modules. Since R is semiperfect, by 9.3.4, this is finite. Since R_R has a large socle every right ideal a_iR contains at least one simple right ideal which can be written in the form a_ib_iR, $i = 1, \ldots, n$. If we now put $c_i := a_ib_i$ then it follows, since Ra_i is simple, that $Ra_i \cong Rc_i$ and consequently

$$Rc_1, \ldots, Rc_n$$

is a set of representatives for the isomorphism classes of simple left R-modules. If we now suppose that $c_iR \cong c_jR$ then it follows by 12.3.1 that $Rc_i \cong Rc_j$ thus $i = j$. Consequently (by 9.3.4)

$$c_1R, \ldots, c_nR$$

is a set of representatives for the isomorphism classes of simple right R-modules. Since R_R is injective we deduce from 5.8.6 that R_R is a cogenerator. Therefore (1) and (2) are proved.

"(1)$\Rightarrow$(6)": Since R_R is a cogenerator there is (by 5.8.6) a set of representatives of the isomorphism classes of simple right R-modules of the form

$$a_1R, \ldots, a_nR$$

with $a_iR \hookrightarrow Q_i \hookrightarrow R_R$ where Q_i is an injective hull of a_iR. Since a_iR is simple and $a_iR \overset{*}{\hookrightarrow} Q_i$, Q_i is directly indecomposable. Since Q_i is a direct summand of R_R, Q_i is projective. By 7.2.8 and 11.4.1 $F_i := Q_i/\mathrm{Rad}(Q_i)$ is simple and $\mathrm{Rad}(Q_i) \overset{\circ}{\hookrightarrow} Q_i$. From 12.5.1 it then follows that $F_1, \ldots, F_n$ form again a set of representatives for the isomorphism classes of simple right R-modules. Since

$$\nu_i\colon Q_i \to Q_i/\mathrm{Rad}(Q_i) = F_i$$

is a projective cover of F_i, we deduce from 11.3.5 that R_R and then also ${}_RR$ (by 11.3.2) are semiperfect. Let

$$R_R = \bigoplus_{i=1}^{n} A_i$$

be a decomposition of R_R in the sense of 11.4.2, then for a suitable j $A_i/\mathrm{Rad}\, A_i \cong F_j$ thus by 12.5.1 $A_i \cong Q_j$. Consequently R_R as a finite direct sum of injective modules is itself injective. Since Q_j is the injective hull of the simple ideal a_jR, we have $a_jR \overset{*}{\hookrightarrow} Q_j$ and consequently $a_jR = \mathrm{Soc}(Q_j) \overset{*}{\hookrightarrow} Q_j$. As $A_i \cong Q_j$ we then also have $\mathrm{Soc}(A_i) \overset{*}{\hookrightarrow} A_i$ and $\mathrm{Soc}(A_i)$ is simple. By 5.1.8 and 9.1.5 it follows that

$$\mathrm{Soc}(R_R) = \bigoplus_{i=1}^{n} \mathrm{Soc}(A_i) \overset{*}{\hookrightarrow} R_R.$$

Therefore (1)$\Rightarrow$(6) is proved. □

12.5.3 COROLLARY. *If R_R is a noetherian cogenerator then R_R is injective and semiperfect and has a large socle.*

Proof. Since R_R is noetherian $\mathrm{Soc}(R_R)$ is finitely generated and has thus only finitely many homogeneous components. Since R_R is a cogenerator it follows that only finitely many isomorphism classes of simple right R-modules exist. The assertion then follows from 12.5.2. □

12.6 PROOF OF THE MAIN THEOREM

We now prove the Main Theorem 12.1.1 as given in the introduction by the following scheme.

(5)⇔(8),
(5) ∧ (8)⇒(6) ∧ (7),
(6)⇒(5), (7)⇒(5),
(5) ∧ (8)⇒(1) ∧ (3),
(1)⇒(2)⇒(8),
(3)⇒(5),
(5)⇔(4).

"(5)⇒(8)": By (5) 12.5.2(2) is satisfied on both sides. By 12.5.2 (8) then follows.

"(8)⇒(5)": Clear by 5.8.5(a).

"(5) ∧ (8)⇒(6) ∧ (7)": Clear.

"(6)⇒(5)": It must be shown that $_RR$ is a cogenerator. To this end we first show that R_R is complemented. Let $A \hookrightarrow R_R$ and let $B \hookrightarrow {}_RR$ be an intersection complement of $\boldsymbol{l}(A)$, i.e. $\boldsymbol{l}(A) \cap B = 0$ with B maximal in this equality. Since $_RR$ is injective it follows by 12.4.2 (on interchanging the sides) that

$$R = \boldsymbol{r}(0) = \boldsymbol{r}(\boldsymbol{l}(A) \cap B) = \boldsymbol{rl}(A) + \boldsymbol{r}(B).$$

Since R_R is a cogenerator it follows by 12.4.1 that

$$R = A + \boldsymbol{r}(B).$$

In this $\boldsymbol{r}(B)$ is minimal: Let $U \hookrightarrow \boldsymbol{r}(B)$, then $A + U = R \Rightarrow \boldsymbol{l}(A) \cap \boldsymbol{l}(U) = \boldsymbol{l}(A+U) = \boldsymbol{l}(R) = 0$; as $U \hookrightarrow \boldsymbol{r}(B)$ we have $B \hookrightarrow \boldsymbol{lr}(B) \hookrightarrow \boldsymbol{l}(U)$ and so $\boldsymbol{l}(U) = B$ from the maximality of B in $\boldsymbol{l}(A) \cap B = 0$; from $B = \boldsymbol{l}(U) \Rightarrow \boldsymbol{r}(B) = \boldsymbol{rl}(U) = U$ by 12.4.1. Since R by 11.1.5 is thus semiperfect, there are by 9.3.4 only finitely many isomorphism classes of simple right R-modules. Therefore 12.5.2(1) is satisfied. By 12.5.2(2) and since $_RR$ is injective, $_RR$ is a cogenerator.

"(7)⇒(5)": Analogously.

We have therefore proved (5)$\Leftrightarrow$(6)$\Leftrightarrow$(7)$\Leftrightarrow$(8); this is the part of the Main Theorem not referring to the Duality properties.

"(5) $\wedge$ (8)$\Leftrightarrow$(1) $\wedge$ (3)": Since every finitely generated module is an epimorphic image of a finitely generated free module, (1) follows from 12.2.4(c) and 12.2.3(b). Now let A_R be finitely cogenerated, then there is by 9.4.3 (since every Q_i from $I(A) = Q \oplus \ldots \oplus Q_n$ can be mapped monomorphically into R_R) a monomorphism of A into a finitely generated free R-module. 12.2.4(c) and 12.2.3(b) yields as before the assertion.

"(1)$\Rightarrow$(2)": Clear.

"(2)$\Rightarrow$(8)": By Baer's Criterion we have to show that ${}_RR$ is injective. Let $B \hookrightarrow {}_RR$; then we have by 12.2.6 (applied to ${}_R(R/B))\boldsymbol{lr}(B) = B$. If we now apply 12.2.6 to $A = \boldsymbol{r}(b) \hookrightarrow R_R$ then it follows that to every homomorphism τ of ${}_RB = \boldsymbol{lr}(B)$ into ${}_RR$ there is an $r_0 \in R$ with $\tau(x) = xr_0$. By Baer's Criterion ${}_RR$ is injective. Analogously we see that R_R is injective. Now let E_R be simple and let $A \hookrightarrow R_R$ with

$$R/A \cong E_R.$$

By 12.2.6 as $A = \boldsymbol{rl}(A)$ we have $\boldsymbol{l}(A) \neq 0$. Let $0 \neq x \in \boldsymbol{l}(A)$ then it follows that

$$xR \cong R/A \cong E,$$

which had still to be shown. Analogously for R_R simple.

"(3)$\Rightarrow$(5)": If E_R is simple and Q_R is an injective hull of E_R then Q_R is finitely cogenerated. Since Q_R is reflexive there is a $\varphi \in Q^*$ with $\varphi(E) \neq 0$. As $E^* \hookrightarrow Q$ it would follow in the case $\mathrm{Ker}(\varphi) \neq 0$ that on the other hand $E \hookrightarrow \mathrm{Ker}(\varphi)$. Thus we have $\mathrm{Ker}(\varphi) = 0$, i.e. φ is a monomorphism. Therefore it follows that R_R is a cogenerator. Analogously for ${}_RR$. (In this proof in place of (3) we have only used the fact that the injective hulls of the simple modules are torsionless.)

"(5)$\Rightarrow$(4)": By definition we have $A \hookrightarrow A^{\circ\perp}$. Since R_R is a cogenerator, for every $m \in M$, $m \notin A$ there is a $\varphi \in M^*$ with $\varphi(m) \neq 0$ and $\varphi(A) = 0$. Therefore it follows that $\varphi \in A^\circ$ and $m \notin A^{\circ\perp}$, thus $A^{\circ\perp} \hookrightarrow A$. Analogously for the left side.

"(4)$\Rightarrow$(5)": Let M_R be arbitrary. Then $0^\circ = M^*$ follows and

$$0 = 0^{\circ\perp} = \bigcap_{\varphi \in M^*} \mathrm{Ker}(\varphi),$$

thus R_R is a cogenerator. Analogously for ${}_RR$.

Hence the Main Theorem is completely proved. □

It remains finally to prove Corollary 12.1.3. This follows directly from 12.5.2 on using the fact that a ring with perfect duality satisfies the conditions of 12.5.2 on both sides (e.g. we see immediately that 12.1.1(5)$\Rightarrow$ 12.5.2(2)).

12.6.1 APPENDIX. *Let R be a ring with perfect duality. Then we have for every R-module M: For every $A \hookrightarrow M$ the homomorphism*

$$\psi: M^*/A^\circ \ni \varphi + A^\circ \mapsto \varphi|A \in A^*$$

is an isomorphism.

Proof. By definition ψ is a monomorphism. As we see easily ψ is an isomorphism for all M_R and all $A \hookrightarrow M_R$ if and only if R_R is injective. Namely if ψ is an epimorphism for all $A \hookrightarrow R_R$ then this means that Baer's Criterion is satisfied, i.e. that R_R is injective. Conversely if R_R is injective then every element of A^* can be continued to one such of M^*. □

In conclusion the following properties are to be pointed out. By 12.2.5 we have for every R-module M: If M is reflexive then so also is M^* and if M^{**} is reflexive then so also M^*. If R is a ring with perfect duality then in fact from the reflexivity of M^* the reflexivity of M follows. Namely let M^* be reflexive, then it follows by 12.2.2(a) that Φ_M^* is an isomorphism. By 12.2.1(d) Φ_M is then an isomorphism, thus M_R is reflexive.

In the next chapter we return to the duality properties. The quasi-Frobenius rings considered there are rings with perfect duality, which are artinian on both sides (it suffices to assume noetherian on one side). However there are rings with perfect duality that satisfy no chain conditions (see Exercise 11). In the case of artinian rings further additions can be made to the characterizations of perfect duality, as, for example, that the duals of all simple modules are again simple (see, for this, Exercise 12).

EXERCISES

(1)

Show:

(a) If M_R is reflexive, if $A \hookrightarrow M_R$, $A^{\circ\perp} = A$ and $\iota^*: M^* \to A^*$ is surjective then A is also reflexive.

(b) If M_R is reflexive, if $A \hookrightarrow M_R$, $A^{\circ\perp} = A$ and $\iota^*: M^{**} \to A^{\circ *}$ is surjective (corresponding to $\iota: A^\circ \to M^*$) then A/M is also reflexive.

(c) Construct a reflexive module M_R and a submodule $A \hookrightarrow M_R$ so that neither A nor M/A is reflexive.

(2)

Let $(M_i \mid i \in I)$ be a non-empty family of right R-modules. Show:

(a) There is a commutative diagram

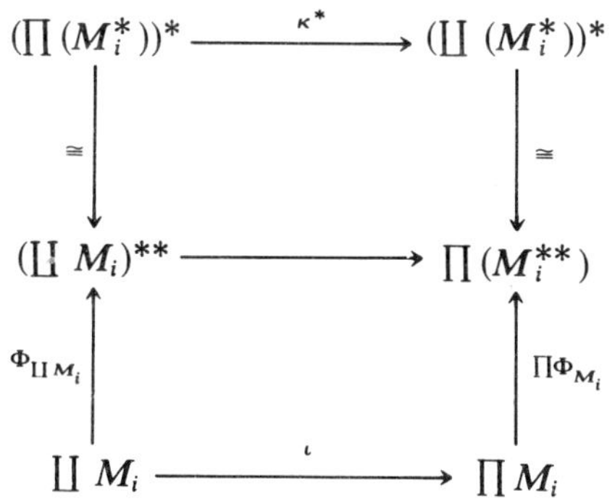

in which ι and $\kappa : \coprod(M_i^*) \to \prod(M_i^*)$ are inclusion mappings.

(b) If I is finite then we have $\coprod M_i$ is reflexive if and only if all M_i are reflexive.

(c) If ${}_RR$ is a cogenerator or is injective and if $\coprod M_i$ is reflexive then almost all (i.e. all up to finitely many) M_i are equal to zero.

(Hint: In the first case κ^* is a monomorphism, in the second an epimorphism).

(3)

For an arbitrary M let Y be a finitely generated submodule of M^* and let $\alpha \in Y^*$. Show: If R_R is a cogenerator then there is an $m \in M$ with $\alpha = \Phi_M(m) \mid Y$.

(4)

Let M_R be given. If $((m_i, U_i) \mid i \in I)$ is a non-empty family with $m_i \in M$ and $U_i \hookrightarrow M$, then $m \in M$ is called a *solution* (*of the family*) if $m - m_i \in U_i$ for all $i \in I$ holds. The module M_R is called *linearly compact* if every finitely soluble family $((m_i, U_i) \mid i \in I)$ (i.e. soluble for every finite subset $I_0 \subset I$) has a solution. Show:

(a) If R_R is a cogenerator then M_R is linearly compact if and only if M_R is reflexive and ${}_RR$ is injective with respect to M^* (the latter means: For every monomorphism $\alpha : {}_RY \to {}_RM^*$ and homomorphism $\beta : {}_RY \to {}_RR$ there is a $\gamma : {}_RM^* \to {}_RR$ with $\beta = \gamma\alpha$; see Chapter 5, Exercise 21).

(b) R is a ring with perfect duality if and only if R_R is a cogenerator and R_R is linearly compact.

(c) If R is a ring with perfect duality then an R-module is reflexive if and only if it is linearly compact.

(5)

Show:

(a) Every artinian module is linearly compact.

(b) Every linearly compact module is complemented (i.e. for every submodule an addition complement exists).

(Hint: The proof for the existence of an intersection complement may here be dualized).

(c) The converse holds neither in (a) nor in (b).

(d) If $M = {}_i\bigoplus_I M_i$ is linearly compact then almost all M_i are equal to zero.

(e) If M is linearly compact and if $A \hookrightarrow M$ then A and M/A are also linearly compact.

(f) If M is linearly compact and if Rad(M) is small in M (resp. Soc(M) is large in M) then M is finitely generated (resp. finitely cogenerated).

(g) If R is a non-local principal ideal domain then every linearly compact R-module is artinian.

(6)

A non-empty family $((m_i, U_i)|i \in I)$ with $m_i \in M$ and $U_i \hookrightarrow M$ is called *projective* if I is directed (i.e. is provided with an ordering $\leqslant$ so that for arbitrary $i, j \in I$ there is a $k \in I$ with $i \leqslant k$, $j \leqslant k$) and for $i \leqslant j$ both $U_j \hookrightarrow U_i$ and $m_j - m_i \in U_i$ hold. Show:

(a) M is linearly compact$\Leftrightarrow$every projective family from M has a solution in M.

(b) If $A \hookrightarrow M$ and if A and M/A are linearly compact then so also is M.

(7)

Show: If R is injective on both sides then for every finitely generated R-module M, M^* is reflexive.

(Hint: Use Exercise 1(a).)

(8)

If R is an integral domain with quotient field K then we define $\text{Rank}(M_R) := \dim_K\left(M \underset{R}{\otimes} K\right)$. Show:

(a) $\text{Rank}(M) = \text{Rank}(M/T(M))$, where $T(M)$ is the torsion submodule of M.

(Hint: K_R is flat.)

(b) $\text{Rank}(M) = 0 \Leftrightarrow M = T(M)$.

(c) $\text{Rank}(M) < \infty$ and $A \hookrightarrow M \Rightarrow \text{Rank}(A) < \infty \wedge \text{Rank}(M/A) < \infty \wedge \text{Rank}(M) = \text{Rank}(A) + \text{Rank}(M/A)$.

(d) $T(M) = 0 \Leftrightarrow$there is a free submodule A of M with $A \overset{*}{\hookrightarrow} M$. If $A \cong R^{(I)}$, then $\text{Rank}(M) = \text{Card}(I)$.

(e) If M is generated by n elements then we have $\mathrm{Rank}(M) \leqslant n$.

(f) $\mathrm{Rank}(M) < \infty \Rightarrow \mathrm{Rank}(M^*) \leqslant \mathrm{Rank}(M) \wedge$ reflexive M^*.

(9)

Let R_R be a cogenerator. Show:

(a) $\mathrm{Soc}({}_RR) \overset{*}{\hookrightarrow} {}_RR$.

(Hint: For $0 \neq x \in R$ choose a maximal right ideal which contains $\boldsymbol{r}_R(x)$.)

(b) $\mathrm{Soc}({}_RR) \hookrightarrow \mathrm{Soc}(R_R)$.

If in addition, $\mathrm{Soc}(R_R)$ has only finitely many homogeneous components then we have further

(c) $\boldsymbol{r}_R(\mathrm{Rad}(R)) = \mathrm{Soc}({}_RR) = \mathrm{Soc}(R_R) = \boldsymbol{l}_R(\mathrm{Rad}(R))$.

(d) $\boldsymbol{r}_R\boldsymbol{l}_R(\mathrm{Rad}(R)) = \mathrm{Rad}(R) = \boldsymbol{l}_R\boldsymbol{r}_R(\mathrm{Rad}(R))$.

(10)

Let T be a commutative ring and let M_T be a T-module. Then a commutative ring $R := \mathrm{Id}(M_T)$ is defined in the following manner:

(1) $R := M \times T$ as a set.

(2) Addition in R is componentwise: $(m, t) + (m', t') := (m + m', t + t')$.

(3) Multiplication in R: $(m, t)(m', t') := (mt' + m't, tt')$.

The unit element of this ring is then $(0, 1)$. Show:

(a) $x = (m, t)$ is invertible resp. nilpotent in $R \Leftrightarrow t$ is invertible resp. nilpotent in T.

(b) $\mathrm{Rad}(R) = M \times \mathrm{Rad}(T)$.

(c) $\mathrm{Soc}(R) = \mathrm{Soc}(M) \times (\mathrm{Soc}(T) \cap \boldsymbol{r}_T(M))$.

(d) R is perfect resp. semiperfect if and only if T is.

(e) R is noetherian resp. artinian if and only if T and M_T are.

(11)

Let T be a commutative ring and let M_T be a faithful T-module (i.e. $\boldsymbol{r}_T(M) = 0$). For the ring $R = \mathrm{Id}(M_T)$ defined in Exercise 10 show:

(a) R_R is injective $\Leftrightarrow M_T$ is injective and to every $\varphi \in \mathrm{End}(M_T)$ there is a $t \in T$ with $\varphi(m) = mt$ for all $m \in M$.

(b) R_R is a cogenerator $\Leftrightarrow \mathrm{R_R}$ is injective and M_T is a cogenerator.

(c) Let T be a complete discrete valuation ring with quotient field K and let $M_T := K/T$. Show: R is a ring with perfect duality but R is not noetherian.

(12)

(a) Let R be a commutative local ring with finitely generated socle and let E be a simple R-module. Show:

$$E^* \cong E^n, \qquad E^{**} \cong E^{n^2}, \qquad \ldots,$$

where $n := \mathrm{Le}(\mathrm{Soc}(R))$ (Definition 3.5.4).

(b) Show: If K is a field and if $M_K := K^n$ $(n \geq 1)$ then the ring $R = \mathrm{Id}(M_K)$, defined in Exercise 10, is commutative, local and artinian and $\mathrm{Le}(\mathrm{Soc}(R)) = n$.

(c) Show: If R is a commutative local ring then the following are equivalent for the simple R-module E:

(1) E is reflexive.
(2) E^* is simple.
(3) $\mathrm{Soc}(R)$ is simple.

(d) Let T be a non-complete discrete valuation ring with quotient field K and let $M_T := K/T$. Show: $R = \mathrm{Id}(M_T)$ (see Exercise 10) satisfies the conditions in (c), but R is not a ring with perfect duality.

(13)

Show:

(a) R is semisimple $\Leftrightarrow \mathrm{Rad}(R) = 0$ and for every simple right R-module there is an isomorphic right ideal in R.

(b) If R is an infinite product of fields then R_R is injective but not a cogenerator (see also Chapter 5, Exercise 11).

(14)

Let K be a field and let R be the K-algebra with the basis $\{1, u_0, u_1, u_2, \ldots, e_0, e_1, e_2, \ldots\}$ and the multiplication

$$u_i u_j = 0, \qquad e_i e_j = \delta_{i,j} e_i, \qquad e_i u_j = \delta_{i,j} u_j, \qquad u_i e_j = \delta_{i-1,j} u_i.$$

Show:

(1) For $x = 1k + \sum u_i k_i + \sum e_i h_i \in R$, where $k, k_i, h_i \in K$ we have
 (a) x is left invertible $\Leftrightarrow x$ is right invertible $\Leftrightarrow k \neq 0 \wedge k + h_i \neq 0$ for all $i = 0, 1, 2, \ldots$.
 (b) $x \in \mathrm{Rad}(R) \Leftrightarrow k = 0 = h_i$ for all $i \Leftrightarrow x^2 = 0 \Leftrightarrow x$ is nilpotent.
 (c) $x \in$ centre of $R \Leftrightarrow k_i = 0 = h_i$ for all i.

(2) (a) $(\mathrm{Rad}(R))^2 = 0$.
 (b) $\boldsymbol{r}_R(\mathrm{Rad}(R)) = \mathrm{Rad}(R) = \boldsymbol{l}_R(\mathrm{Rad}(R))$.
 (c) $\mathrm{Soc}({}_R R) = \mathrm{Rad}(R) = \mathrm{Soc}(R_R)$.
 (d) $R/\mathrm{Rad}(R)$ as a ring is commutative and regular.

(3) For the maximal resp. simple ideals of R we have
 (a) The maximal right ideals are precisely

$$\boldsymbol{r}_R(u_0), \qquad \boldsymbol{r}_R(u_1), \qquad \boldsymbol{r}_R(u_2), \qquad \ldots$$

 They are all two-sided ideals and are also precisely all the maximal left ideals.
 (b) The simple right ideals are precisely

$$u_0 R, \qquad u_1 R, \qquad u_2 R, \qquad \ldots$$

They are all two-sided ideals and are also precisely all the simple left ideals.

(c) $u_0R, u_1R, u_2R, \ldots$ is a set of representatives of the simple right R-modules.

(d) $A := \sum_{i=0}^{\infty} e_iR$ is a maximal left ideal in R and R/A, Ru_0, Ru_1, $Ru_2, \ldots$ is a set of representatives of the simple left R-modules. There is no left ideal of R isomorphic to R/A.

(4) For all $i \geq 0$ we have:

(a) e_iRe_i is ring isomorphic to K, in particular e_i is a local idempotent.

(b) u_iR is the unique non-trivial submodule of e_iR and $e_iR/u_iR \cong u_{i+1}R$.

(c) e_iR is injective and consequently R_R is a cogenerator.

(d) R_R is not injective.

(Hint for (c)): If $A \hookrightarrow R_R$, $f \in \mathrm{Hom}_R(A, e_iR)$ and $b \in R$ f may be continued to $A + bR$ if and only if there is a $g \in \mathrm{Hom}_R(bR, e_iR)$ so that f and g coincide on $A \cap bR$. Show that this procedure is also feasible for $b = e_j$, $j \geq 0$ and $b = 1 - e_{i-1} - e_i$ (putting $e_{-1} = 0$).)

(15)

Show for a ring $R : R_R$ is a cogenerator if and only if the injective hull of every finitely cogenerated R-module is projective.

Chapter 13

Quasi-Frobenius Rings

13.1 INTRODUCTION

In the following presentation of *QF-rings* we pursue a direction opposite to that of their historical development. In the historical development there were considered first in the representation theory of finite groups—more or less explicitly—group rings of finite groups with coefficients in a field.

Let $R := GK$ be such a group ring where

$$g_1 = e, g_2, \ldots, g_n$$

are the elements of the group G. Then the mapping

$$\varphi: R \ni \sum_{i=1}^{n} g_i k_i \mapsto k_1 \in K$$

is a K-homomorphism of R into K, i.e. $\varphi \in R^* := \mathrm{Hom}_K(R, K)$. This homomorphism φ has the essential property that $\mathrm{Ker}(\varphi)$ contains no right or left ideal different from 0. By means of this property φ is essentially uniquely determined (i.e. up to multiplication by regular elements from R on the right) and is called the *Frobenius homomorphism*. Since R^* is a right R-module, $\varphi R \hookrightarrow R^*_R$ and for a Frobenius homomorphism it follows in fact that $\varphi R = R^*$. Then

$$\Phi: R_R \ni r \mapsto \varphi r \in \varphi R = R^*_R$$

is an R-isomorphism and conversely every R-isomorphism

$$\Phi: R_R \to R^*_R$$

yields, in the form $\varphi := \Phi(1)$, $1 \in R$, a Frobenius homomorphism $\varphi: R_K \to K_K$. After it had become clearer in the course of the development that many attractive properties of group rings depend only on the existence of a Frobenius homomorphism φ or—what is equivalent—of an isomorphism Φ, the existence of such a φ resp. Φ in regard to a finite-dimensional K-algebra R_K was incorporated in the definition of a Frobenius algebra (even if formulated originally in the context of representation theory, T. Nakayama, 1939).

The next essential step in the development was taken in removing the algebra property. As is easy to see, it follows for a Frobenius algebra by use of φ resp. Φ that the following annihilator equations hold:

$$\boldsymbol{r}_R \boldsymbol{l}_R(A) = A \quad \text{for all} \quad A \hookrightarrow R_R$$

$$\boldsymbol{l}_R \boldsymbol{r}_R(B) = B \quad \text{for all} \quad B \hookrightarrow {}_R R.$$

(Orthogonality relations between a finite-dimensional vector space and its dual space!) By an additional condition on dimensions the Frobenius property of the algebra follows again conversely from the annihilator equations. In these annihilator equations the algebra properties no longer appear.

A two-sided artinian ring, which satisfied the annihilator equations, was then called a *quasi-Frobenius ring* and—with an additional condition—a *Frobenius ring* (T. Nakayama, 1941).

On this basis a plethora of results on quasi-Frobenius and Frobenius rings was established.

An important new impulse stimulated the development with the coming into use of categorical and homological concepts. This led on to to-day's situation in which we have the following results:

A ring is a quasi-Frobenius ring, i.e. is artinian (and hence also noetherian) on both sides and with the annihilator conditions satisfied if and only if it is noetherian on one side and is injective or a cogenerator on one side.

This will be the main theorem of the following analysis. Since accordingly a quasi-Frobenius ring is on both sides an artinian (and noetherian) injective cogenerator, there is at our convenience all of the structure that we have proved up till now for artinian and noetherian modules as well as for rings with perfect duality.

13.2 DEFINITION AND MAIN THEOREM

We prove rather more than is mentioned in the introduction. In place of $\boldsymbol{l}_R$ resp. $\boldsymbol{r}_R$ we write in the following only $\boldsymbol{l}$ resp. $\boldsymbol{r}$.

13.2.1 THEOREM. *Let R_R be noetherian, then we have*:

(a) *The following conditions are equivalent*:

(1) *R_R is injective.*

(2) *R_R is a cogenerator.*

(3) *${}_RR$ is injective.*

(4) *${}_RR$ is a cogenerator.*

(5) $\forall A \hookrightarrow R_R[\boldsymbol{rl}(A)=A] \wedge \forall B \hookrightarrow {}_RR[\boldsymbol{lr}(B)=B]$.

(b) *If the conditions in* (a) *are satisfied then R is artinian on both sides.*

13.2.2 *Definition*

(1) A ring which satisfies the conditions of 13.2.1 is called a *quasi-Frobenius ring.*

(2) A ring R is called a *Frobenius ring* if it is quasi-Frobenius and we have

$$\mathrm{Soc}(R_R) \cong (R/\mathrm{Rad}(R))_R, \qquad \mathrm{Soc}({}_RR) \cong {}_R(R/\mathrm{Rad}(R)).$$

Obviously a ring with perfect duality is accordingly a quasi-Frobenius ring if and only if it is noetherian on one side.

We divide the lengthy proof of 13.2.1 into several propositions, some of which are also of interest in themselves.

13.2.3 PROPOSITION. *If R_R is injective and noetherian then R is artinian on both sides.*

Proof. Since R_R is injective by 12.4.2 we have $\boldsymbol{lr}(C)=C$ for all finitely generated left ideals $C \hookrightarrow {}_RR$. Since R_R is noetherian, then R satisfies the descending chain condition for all finitely generated and in particular for all cyclic left ideals. Consequently by 11.6.3 R_R is perfect. Then 11.6.4 implies that R_R is artinian. Therefore it follows from $\boldsymbol{lr}(C)=C$ that ${}_RR$ satisfies the ascending chain condition for finitely generated left ideals. We reflect that ${}_RR$ is then indeed noetherian. If this were not the case then an ideal $B \hookrightarrow {}_RR$ would have to exist which would not be finitely generated. To every finitely generated subideal of B there is then a proper larger finitely generated subideal. In B there may be defined inductively an infinite properly ascending chain of finitely generated subideals in contradiction to the previous statement. Since R is thus also right artinian and left noetherian it follows from 9.3.12 that ${}_RR$ is also artinian. □

13.2.4 PROPOSITION. *If R_R is noetherian and* (5) *of* 13.2.1 *holds, then R_R is injective and R is artinian on both sides.*

Proof. We wish to apply 12.4.3. For that purpose we have to show for right ideals A and B of R that

$$\boldsymbol{l}(A \cap B) = \boldsymbol{l}(A) + \boldsymbol{l}(B).$$

By (5) we have

$$\boldsymbol{rl}(A \cap B) = A \cap B = \boldsymbol{rl}(A) \cap \boldsymbol{rl}(B) = \boldsymbol{r}(\boldsymbol{l}(A) + \boldsymbol{l}(B)),$$

in which the last equality is easily verified. By application of $\boldsymbol{l}$ it follows therefore that

$$\boldsymbol{l}(A \cap B) = \boldsymbol{lr}(\boldsymbol{l}(A) + \boldsymbol{l}(B)) = \boldsymbol{l}(A) + \boldsymbol{l}(B).$$

From 12.4.3 we deduce then that R_R is injective. The rest follows from 13.2.3. □

13.2.5 PROPOSITION. *If R_R or ${}_RR$ is noetherian and we have*

$$\boldsymbol{rl}(A) = A \quad \text{or} \quad \boldsymbol{lr}(A) = A$$

for every two-sided ideal A of R then Rad(R) *is nilpotent.*

Proof. It suffices to exhibit the proof for the case $\boldsymbol{rl}(A) = A$ since in the other case everything proceeds analogously. Put $N := \text{Rad}(R)$, then $N \hookleftarrow N^2 \hookleftarrow N^3 \hookleftarrow \ldots$ and consequently

$$\boldsymbol{l}(N) \hookrightarrow \boldsymbol{l}(N^2) \hookrightarrow \boldsymbol{l}(N^3) \hookrightarrow \ldots$$

is also a chain of two-sided ideals. Since R_R or ${}_RR$ is noetherian this chain is stationary, i.e. there is an n with

$$\boldsymbol{l}(N^n) = \boldsymbol{l}(N^{n+1}).$$

Therefore we have

$$N^n = \boldsymbol{rl}(N^n) = \boldsymbol{rl}(N^{n+1}) = N^{n+1}.$$

Since R_R resp. ${}_RR$ is noetherian, N^n_R resp. ${}_RN^n$ is finitely generated so that by 9.2.1(d) $N^{n+1} \overset{\circ}{\hookrightarrow} N^n$ follows. The last two relations together imply that $N^n = 0$, which was to be shown. □

13.2.6 PROPOSITION. *If R_R is injective and ${}_RR$ is noetherian then R_R is a cogenerator and R is artinian on both sides.*

Proof. Since ${}_RR$ is noetherian every left ideal of R is finitely generated. From 12.4.2 and 13.2.5 it follows then that Rad(R) is nilpotent. From 9.6.2 we deduce for $Q_R = R_R$ that $\bar{R} := R/\text{Rad}(R)$ is regular. Since ${}_{\bar{R}}\bar{R}$ is noetherian ${}_R\bar{R}$ is noetherian and because of this ${}_{\bar{R}}\bar{R}$ is also noetherian. Then 10.4.9 indicates that every left ideal of $\bar{R}$ is a direct summand of $\bar{R}$, i.e. $\bar{R}$ is semisimple. Consequently by 11.6.3 R is perfect on both sides and by 11.6.4 ${}_RR$ is artinian. Since ${}_RR$ is perfect by 11.6.3 (4), $\mathrm{Soc}(R_R)$ is large in R_R. From 12.5.2 it then follows that R_R is a cogenerator. By 12.4.1 we then have $\boldsymbol{rl}(A)=A$ for all $A \hookrightarrow R_R$. Since ${}_RR$ is noetherian it follows therefore that R_R is artinian. □

Proof of 13.2.1. Since (b) follows from 13.2.4 only (a) has to be shown.

"(1)⇒(2)": By 13.2.3 ${}_RR$ is artinian, thus also noetherian. Then the proposition follows from 13.2.6.

"(2)⇒(5)": By 12.5.3 R_R is injective and by 13.2.3 ${}_RR$ is noetherian. Since R_R is a cogenerator by 12.4.1 we have $\boldsymbol{rl}(A)=A$ for all $A \hookrightarrow R_R$. Since R_R is injective and ${}_RR$ is noetherian by 12.4.2 we have also $\boldsymbol{lr}(B)=B$ for all $B \hookrightarrow {}_RR$, thus (5) holds.

"(5)⇒(1)": By 13.2.4.

"(5)⇒(3)": From (5) and as R_R is noetherian it follows that ${}_RR$ is artinian thus also ${}_RR$ is noetherian. Hence we obtain (3) from 13.2.4.

"(3)⇒(4)": By 13.2.6.

"(4)⇒(5)": From (4) and as R_R is noetherian it follows from 12.4.1 that ${}_RR$ is artinian and hence is also noetherian. Then the proposition follows as "(2)⇒(5)". □

13.3 DUALITY PROPERTIES OF QUASI-FROBENIUS RINGS

The quasi-Frobenius rings can be characterized under noetherian rings by means of the conditions in 12.1.1. To the conditions in 12.1.1 further characterizations can now be added by duality properties. In so doing we take over the notations of Chapter 12.

As a lemma for further considerations we first establish how finiteness conditions carry over to the dual module.

13.3.1 LEMMA. *Let M_R be finitely generated, then we have for $M^* := \mathrm{Hom}_R(M_R, R_R)$:*

(a) *If ${}_RR$ is noetherian then ${}_RM^*$ is noetherian.*

(b) *If ${}_RR$ is artinian then ${}_RM^*$ is of finite length (i.e. artinian and noetherian).*

Proof. (a) If first of all $F := \bigoplus_{i=1}^{n} x_iR$ is a finitely generated free right R-

module with basis $x_1, \ldots, x_n$ then (as in the case of a vector space)

$$F^* = \bigoplus_{i=1}^{n} R\delta_i \quad \text{with} \quad \delta_i(x_j) = \begin{cases} 0 & \text{for } i \neq j \\ 1 & \text{for } i = j \end{cases}$$

is a free left R-module with basis $\delta_1, \ldots, \delta_n$. Since ${}_RR$ is noetherian by 6.1.3 ${}_RF^*$ is also noetherian. Now let

$$M_R = \sum_{i=1}^{n} m_i R$$

and let

$$\eta : F = \bigoplus_{i=1}^{n} x_i R \ni \sum_{i=1}^{n} x_i r_i \mapsto \sum_{i=1}^{n} m_i r_i \in M,$$

then, since η is an epimorphism,

$$\mathrm{Hom}(\eta, 1_R) : M^* \ni \alpha \mapsto \alpha\eta \in F^*$$

is a monomorphism. Since ${}_RF^*$ is noetherian, in consequence, ${}_RM^*$ is also noetherian.

(b) Follows from (a) and 6.1.3. □

If R is artinian on both sides then it follows, from 13.3.1 together with the results of Chapter 6, that for every finitely generated right or left R-module M all submodules and factor modules of M and of M^* are of finite length. Use is made of this in the following without explicit mention. Further recall that $\mathrm{Le}(M)$ is the length of the module M (= length of a composition series of M) (Definition 3.5.4).

13.3.2 THEOREM. *The following are equivalent for a two-sided artinian ring* R:

(1) R *is quasi-Frobenius.*

(2) *Dual modules of simple right and simple left* R*-modules are simple.*

(3) *For every finitely generated right* R*-module and every finitely generated left* R*-module we have*: $\mathrm{Le}(M) = \mathrm{Le}(M^*)$.

Proof. "(1)⇒(2)": Let E_R be simple, then, since R_R is a cogenerator, there is a monomorphism $\mu : E_R \to R_R$; thus $E^* := \mathrm{Hom}_R(E_R, R_R) \neq 0$. Now let $0 \neq \alpha \in E^*$, then we have to show that $E^* = R\alpha$ holds, i.e. that E^* is simple. Since E_R is simple and $\alpha \neq 0$, α must be a monomorphism. Since R_R is injective for every $\xi \in E^*$ there exists a commutative diagram

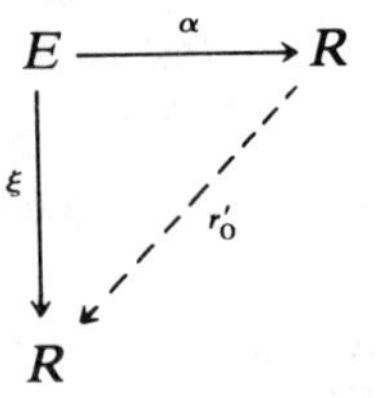

where r_0' is left multiplication by an $r_0 \in R$. Thus $\xi = r_0\alpha$ holds and consequently $E^* = R\alpha$. Analogously for the left side.

"(2)⇒(1)": We show that the annihilator conditions 13.2.1 (5) are satisfied. The proof follows from two steps.

Step 1. *Assertion*: Let $A \hookrightarrow B \hookrightarrow R_R$ and let B/A be simple, then $\boldsymbol{l}(A)/\boldsymbol{l}(B)$ is simple or 0.

Proof. As is easily verified,

$$f : \boldsymbol{l}(A)/\boldsymbol{l}(B) \to (B/A)^*$$

with

$$f(x + \boldsymbol{l}(B))(b + A) := xb, \qquad x \in \boldsymbol{l}(A),\ b \in B$$

defines a monomorphism. Since $(B/A)^*$ is simple by assumption the assertion follows. Evidently the corresponding statement holds also for left ideals.

Step 2. Let now $A \hookrightarrow R_R$. Then there is a composition series of R_R which contains A:

(i) $$0 = A_0 \hookrightarrow \ldots \hookrightarrow A_m = R.$$

In addition consider the series

(ii) $$R = \boldsymbol{l}(0) \hookleftarrow \boldsymbol{l}(A_1) \hookleftarrow \ldots \hookleftarrow \boldsymbol{l}(R) = 0.$$

By the first step it follows that

$$\mathrm{Le}({}_RR) \leqslant \mathrm{Le}(R_R).$$

Since everything is symmetric with regard to sides we also have $\mathrm{Le}(R_R) \leqslant \mathrm{Le}({}_RR)$, thus $\mathrm{Le}({}_RR) = \mathrm{Le}(R_R)$ follows. Consequently (ii) must be a composition series of ${}_RR$. Likewise

$$0 = \boldsymbol{rl}(A_0) \hookrightarrow \ldots \hookrightarrow \boldsymbol{rl}(A_m) = R$$

is then also a composition series of R_R. Since by assumption (i) is a composition series and $A_i \hookrightarrow \boldsymbol{rl}(A_i)$, $i = 1, \ldots, m$, holds it follows that

$A_i = \boldsymbol{rl}(A_i)$, $i = 1, \ldots, m$, from which we conclude that $\boldsymbol{rl}(A) = A$. Analogously $\boldsymbol{lr}(B) = B$ holds for $B \hookrightarrow {}_RR$.

"(1) $\wedge$ (2) $\Rightarrow$ (3)": Induction on $\mathrm{Le}(M_R)$. By (2) the assertion holds for $\mathrm{Le}(M_R) = 1$. Let it hold now for all modules with $\mathrm{Le}(M_R) \leq n$. Then let L_R have $\mathrm{Le}(L_R) = n+1$ and let E be a simple submodule of L. Then by assumption we have $\mathrm{Le}((L/E)^*) = n$. Let, as previously introduced,

$$E^\circ = \{\varphi \mid \varphi \in L^* \wedge \varphi(E) = 0\},$$

then obviously we have $(L/E)^* \cong E^\circ$ and consequently also $\mathrm{Le}(E^\circ) = n$. Since by 12.6.1

$$\psi : L^*/E^\circ \to E^*$$

with $\psi(\varphi + E^\circ) = \varphi \mid E$ is an isomorphism and we have $\mathrm{Le}(E^*) = 1$, it follows that $\mathrm{Le}(L^*) = n+1$.

"(3) $\Rightarrow$ (2)": (2) is a special case of (3). □

13.4 THE CLASSICAL DEFINITION

The characterizations above of quasi-Frobenius rings do not render the classical definition, or further characterizations closely connected with the latter, redundant, these give indeed a good insight into the ideal-theoretic structure of a quasi-Frobenius ring.

The definition of quasi-Frobenius rings goes back to T. Nakayama (1939). In order to be able to present these some notation is needed.

Let R be a two-sided artinian ring with $N := \mathrm{Rad}(R)$. Let

$$\begin{aligned} R &= A_{11} \oplus \ldots \oplus A_{1g_1} \oplus A_{21} \oplus \ldots \oplus A_{2g_2} \oplus \ldots \oplus A_{k1} \oplus \ldots \oplus A_{kg_k} \\ &= e_{11}R \oplus \ldots \oplus e_{1g_1}R \oplus \ldots \oplus e_{k1}R \oplus \ldots \oplus e_{kg_k}R \end{aligned}$$

denote a decomposition into directly indecomposable right ideals $A_{ij} = e_{ij}R$ with orthogonal idempotents $e_{11}, \ldots, e_{kg_k}$; in which the indices are chosen so that $A_{i1}, \ldots, A_{ig_i}$ $(i = 1, 2, \ldots, k)$ are exactly all of the right ideals isomorphic to A_{i1} in the decomposition. For brevity put $A_i := A_{i1}$ and $e_i := e_{i1}$. Let $\bar{R} := R/N$ and $\bar{r} := r + N \in \bar{R}$. In the following let e and e' denote two of the orthogonal idempotents e_{ij}. Then we have by 12.5.1:

$$eR \cong e'R \Leftrightarrow \bar{e}\bar{R} \cong \bar{e}'\bar{R}.$$

Every one of the $\bar{e}_{ij}\bar{R}$ is simple and as well as being a right ideal of $\bar{R}$ is in fact also a right R-module (11.4.3). Further every simple right R-module is isomorphic to one of the $(\bar{e}_{ij}\bar{R})_R$. Summarizing it follows that

$$\bar{e}_1\bar{R}, \ldots, \bar{e}_k\bar{R}$$

is a representative system for the isomorphism classes of simple right R-modules.

13.4.1 *Remark.* Let e and e' be two of the orthogonal idempotents e_{ij}, then we have

$$eR \cong e'R \Leftrightarrow Re \cong Re'.$$

Proof. By 12.5.1 we have

$$eR \cong e'R \Leftrightarrow \bar{e}\bar{R} \cong \bar{e}'\bar{R}.$$

Since $\bar{R}$ as a semisimple ring is two-sided injective it follows by 12.3.1 and 12.3.2 that

$$\bar{e}\bar{R} \cong \bar{e}'\bar{R} \Leftrightarrow \bar{R}\bar{e} = \bar{R}\bar{e}'.$$

Repeated application of 12.5.1 yields the assertion. □

If eR, for an idempotent $e \neq 0$, is directly indecomposable then this means that e cannot be written as the sum $e = e' + e''$ of two orthogonal idempotents $\neq 0$. Therefore it follows that Re is also directly indecomposable. In this regard we recall that e is then called a *primitive idempotent.*

From the right-sided decomposition of R stated at the beginning we obtain the left-sided decomposition

$$R = Re_{11} \oplus \ldots \oplus Re_{1g_1} \oplus \ldots \oplus Re_{k1} \oplus \ldots \oplus Re_{kg_k},$$

which possesses properties corresponding to those on the right side.

The following theorem embraces the original definition of T. Nakayama for quasi-Frobenius rings.

13.4.2 THEOREM. *The following are equivalent for a two-sided artinian ring.*

(a) *R is quasi-Frobenius.*

(b) *For every primitive idempotent e* Soc(eR) *and* Soc(Re) *are simple and in* Soc(R_R) resp. Soc($_RR$) *all simple right* resp. *left R-modules occur up to isomorphism.*

(c) *For every primitive idempotent e* Soc(eR) *and* Soc(Re) *are simple and we have* $\mathrm{Soc}(R_R) = \mathrm{Soc}(_RR)$.

(d) (*Definition of T. Nakayama*). *There exists a permutation π of $\{1, \ldots, k\}$ so that for every $i = 1, \ldots, k$ we have*

$$\mathrm{Soc}(e_iR)_R \cong (\bar{e}_{\pi(i)}\bar{R})_R, \qquad {}_R\mathrm{Soc}(Re_{\pi(i)}) \cong {}_R(\bar{R}\bar{e}_i).$$

Proof. "(a)⇒(b)": Let E be a simple submodule of eR. Since eR as a direct summand of R is injective, eR contains an injective hull of E which is a direct summand of eR. Since eR is directly indecomposable, eR is an injective hull of E. Consequently as $E \overset{*}{\hookrightarrow} eR$ we have $E = \text{Soc}(R)$. Hence the first assertion is proved. Since R_R and $_RR$ are cogenerators, all simple right R-modules resp. left R-modules occur up to isomorphism in $\text{Soc}(R_R)$ resp. $\text{Soc.}(_RR)$.

"(b)⇒(c)": Since in $\text{Soc}(R_R)$ there is contained an ideal isomorphic to $(\bar{e}\bar{R})_R$, we have (as $\bar{e}e = \bar{e}$) $\text{Soc}(R_R)e \neq 0$. Since $\text{Soc}(R_R)$ is a two-sided ideal, $\text{Soc}(R_R)$ consequently contains a subideal $\neq 0$ of Re and hence also $\text{Soc}(Re)$ (since this is simple and large in Re). Thus we have $\text{Soc}(_RR) \hookrightarrow \text{Soc}(R_R)$ (as $\text{Soc}(_RR) = \text{Soc}(\oplus Re_{ij}) = \oplus\, \text{Soc}(Re_{ij})$). Since analogously the reverse inclusion holds the assertion follows.

"(c)⇒(b)": If e is a primitive idempotent, then as $0 \neq \text{Soc}(Re) = \text{Soc}(Re)e$ we have

$$0 \neq \text{Soc}(_RR)e = \text{Soc}(R_R)e.$$

Consequently there is an $x \in \text{Soc}(R_R)$ so that xeR is simple. Hence we then have $xeR \cong \bar{e}\bar{R}$ from which (b) holds.

"(b) ∧ (c)⇒(d)": Since $\text{Soc}(e_iR)$ is simple, to every $i \in \{1, \ldots, k\}$ there is a $\pi(i) \in \{1, \ldots, k\}$ with

$$(*) \qquad \text{Soc}(e_iR) \cong \bar{e}_{\pi(i)}\bar{R}.$$

Since in $\text{Soc}(R_R) = \oplus\, \text{Soc}(e_{ij}R)$ there are contained only simple ideals which are isomorphic to a $\text{Soc}(e_iR)$, $i = 1, \ldots, k$, and from (b) all isomorphism classes of simple right R-modules must be represented, $\{\text{Soc}(e_iR) \mid i = 1, \ldots, k\}$ forms a set of representatives for these isomorphism classes. Since $\{\bar{e}_i\bar{R} \mid i = 1, \ldots, k\}$ is also such a set of representatives, $i \mapsto \pi(i)$ (in the sense of (*)) is a permutation of $\{1, \ldots, k\}$. Let $\text{Soc}(e_iR) = e_ia_iR$ then it follows from $\text{Soc}(e_iR) \cong \bar{e}_{\pi(i)}\bar{R}$ that we have $e_ia_ie_{\pi(i)} \neq 0$, thus $\text{Soc}(e_iR) = e_ia_ie_{\pi(i)}R$. As

$$0 \neq e_ia_ie_{\pi(i)} \in \text{Soc}(R_R) = \text{Soc}(_RR)$$

we have

$$Re_ia_ie_{\pi(i)} \hookrightarrow \text{Soc}(_RR) \cap Re_{\pi(i)} = \text{Soc}(Re_{\pi(i)})$$

and since $\text{Soc}(Re_{\pi(i)})$ is simple it follows that $\text{Soc}(Re_{\pi(i)}) = Re_ia_ie_{\pi(i)}$. Then the epimorphism

$$Re_i \ni re_i \mapsto re_ia_ie_{\pi(i)} \in \text{Soc}(Re_{\pi(i)})$$

yields the isomorphism

$$\bar{R}\bar{e}_i \cong \text{Soc}(Re_{\pi(i)}),$$

by which (d) is proved.

"(d)⇒(b)": By the Krull–Remak–Schmidt Theorem it can be assumed that the idempotent e in (b) is one of the e_i in (d). Then the assertion is immediately clear.

Hence we have demonstrated the equivalence of conditions (b), (c) and (d).

"(b) ∧ (c)⇒(a)": By 13.3.2 it suffices to show that the dual module of every simple right R-module and left R-module is again simple. Since isomorphic modules have isomorphic dual modules it suffices to show that every $\mathrm{Soc}(e_iR)$ and $\mathrm{Soc}(Re_i)$, $i=1,\ldots,k$ has a simple dual module, in which we can by the symmetry confine ourselves to $\mathrm{Soc}(e_iR)$. We show first that every non-zero homomorphism

$$\varphi:\mathrm{Soc}(e_iR)_R\to R_R$$

is induced by multiplication on the left by an element of Re_i. We use, as previously shown, $\mathrm{Soc}(e_iR)=e_ia_ie_{\pi(i)}R$, then it follows that

$$\varphi(e_ia_ie_{\pi(i)}r)=\varphi(e_ia_ie_{\pi(i)})e_{\pi(i)}r, \qquad r\in R.$$

Let $q:=\varphi(e_ia_ie_{\pi(i)})\neq 0$, then $qe_{\pi(i)}R$ is simple and by the same inference as in the proof of (b) ∧ (c)⇒(d) $(0\neq Rqe_{\pi(i)}\hookrightarrow \mathrm{Soc}({}_RR)\cap Re_{\pi(i)}=\mathrm{Soc}(Re_{\pi(i)})\wedge \text{simple } \mathrm{Soc}(Re_{\pi(i)})\Rightarrow Rqe_{\pi(i)}=\mathrm{Soc}(Re_{\pi(i)}))$ it follows that

$$Rqe_{\pi(i)}=\mathrm{Soc}(Re_{\pi(i)})=Re_ia_ia_{\pi(i)}.$$

Thus an $r_0e_i\in Re_i$ exists with

$$qe_{\pi(i)}=r_0e_ia_ie_{\pi(i)}$$

and hence we have

$$\varphi(e_ia_ie_{\pi(i)}r)=qe_{\pi(i)}r=r_0e_ia_ie_{\pi(i)}r.$$

If we write for the left multiplication of $\mathrm{Soc}(e_iR)=e_ia_ie_{\pi(i)}R$ by xe_i, $x\in R$, $(xe_i)^l$, then it follows that $\varphi=(r_0e_i)^l$. Thus the mapping

$$\psi: Re_i\ni xe_i\mapsto (xe_i)^l\in(\mathrm{Soc}(e_iR))^*$$

is an epimorphism. Let $N:=\mathrm{Rad}(R)$; as

$$0=N\,\mathrm{Soc}({}_RR)=N\,\mathrm{Soc}(R_R)$$

we have $Ne_i\hookrightarrow \mathrm{Ker}(\psi)$. Since $\psi\neq 0$ and since by 11.4.3 Ne_i is the unique maximal ideal in Re_i it follows that $\mathrm{Ker}(\psi)=Ne_i$ thus

$$Re_i/Ne_i\cong(\mathrm{Soc}(e_iR))^*$$

and consequently $(\mathrm{Soc}(e_iR))^*$ is simple which was to be shown. □

13.4.3 COROLLARY. *The following are equivalent for a two-sided artinian ring R:*

(1) *R is a Frobenius ring.*

(2) $\mathrm{Soc}(R_R) \cong (R/\mathrm{Rad}(R))_R$ *and* ${}_R\mathrm{Soc}({}_RR) \cong {}_R(R/\mathrm{Rad}(R))$ *hold.*

(3) *R is a quasi-Frobenius ring and either*

$$\mathrm{Soc}(R_R)_R \cong (R/\mathrm{Rad}(R))_R \quad \textit{or} \quad {}_R\mathrm{Soc}({}_RR) \cong {}_R(R/\mathrm{Rad}(R)).$$

hold.

Proof. "(1)⇒(2)": In definition 13.2.2(2) the condition "R is a Frobenius ring" was omitted.

"(2)⇒(1)": Since in $(R/N)_R$ resp. ${}_R(R/N)$ all simple right resp. left R-modules occur up to isomorphism, this holds also for $\mathrm{Soc}(R_R)_R$ resp. ${}_R\mathrm{Soc}({}_RR)$. As

$$\oplus \mathrm{Soc}(e_{ij}R) = \mathrm{Soc}(R_R) \cong (R/N)_R = \oplus \bar{e}_{ij}\bar{R}$$

and since all $\bar{e}_{ij}\bar{R}$ are simple, on the basis of number all $\mathrm{Soc}(e_{ij}R)$ must be simple. Correspondingly this holds for the left-hand side. Hence 13.4.2(b) is satisfied. Consequently "(2)⇒(3)" also holds.

"(3)⇒(1)": Now let $\mathrm{Soc}(R_R)_R \cong (R/N)_R$ be satisfied. By 13.4.2(d) this is evidently equivalent to having $g_i = g_{\pi(i)}$ for every $i = 1, \ldots, k$. Since by 13.4.1 g_i is independent of the side, it follows that ${}_R\mathrm{Soc}({}_RR) \cong {}_R(R/N)$ which was to be proved. □

13.5 QUASI-FROBENIUS ALGEBRAS

The principal aim of the following considerations consists in showing that a quasi-Frobenius ring, resp. a Frobenius ring in the case that it is an algebra over a field, can also be characterized by the classical definition for quasi-Frobenius algebras resp. for Frobenius algebras.

Now let K be a field and let R_K be a unitary K-algebra (see 2.2.5). This implies that R_K is a unitary K-module, i.e. a K-vector space. We call R a finite-dimensional K-algebra if the dimension of R over K (as a vector space) is finite. Let $A \hookrightarrow R_R$, then we have for $a \in A$, $k \in K$,

$$ak = (a1)k = a(1k) \in A,$$

i.e. every right ideal is also a K-subspace of R_K. Now let $B \hookrightarrow {}_RR$, then we have for $b \in B$, $k \in K$,

$$bk = (1b)k = (1k)b \in B,$$

so that left ideals are also K-subspaces. For the K-dimension of a K-

subspace U of R_K we write $\dim_K(U)$. If $\dim_K(R)<\infty$ then for ideals

$$A \subsetneqq B \hookrightarrow R_R \quad \text{resp.} \quad A \subsetneqq B \hookrightarrow {}_R R$$

it follows that

$$\dim_K(A)<\dim_K(B)<\infty.$$

Consequently R is then a two-sided artinian ring for a properly descending chain of ideals must break off after at most $\dim_K(R)$ steps.

We consider now the mapping

$$\kappa: K \ni k \mapsto 1k \in R \qquad (1 \in R).$$

As $1(k_1+k_2)=1k_1+1k_2$ and

$$1(k_1k_2)=(1k_1)k_2=((1k_1)1)k_2=(1k_1)(1k_2)$$

κ is a ring homomorphism. Let e be the unit element of K, then we have $1e=1$, thus κ is not the zero homomorphism and consequently (since K is a field) is a monomorphism. We establish further that $\kappa(K)$ lies in the centre of R:

$$r(1k)=(r1)k=(1r)k=(1k)r, \qquad r \in R,\ k \in K.$$

By virtue of this statement we can and do wish to assume in the following that K is a subfield of the centre of R (i.e. letting $\kappa(K)$ be replaced by K and calling $\kappa(K)$ again K).

Let now $\dim_K(R)=n$. We consider the dual vector space to R_K

$$R^* := \operatorname{Hom}_K(R, K),$$

for which then we likewise have $\dim_K(R^*)=n$ (we notice that now the * refers to K and not as earlier to R!). By putting

$$(\varphi r)(x) := \varphi(rx), \qquad \varphi \in R^*,\ r, x \in R$$

R^* becomes a right R-module. As $n=\dim_K(R)=\dim_K(R^*)$ R and R^* are isomorphic as K-vector spaces. An important question for the following is now whether R and R^* are indeed isomorphic as right R-modules. It will turn out that this is the case if and only if R is a Frobenius ring.

13.5.1 LEMMA. *Let* $\dim_K(R)=n$. *For* $\varphi \in R^*$ *the following are then equivalent*:

(1) $\operatorname{Ker}(\varphi)$ *contains no non-zero right ideal of* R.
(2) $\operatorname{Ker}(\varphi)$ *contains no non-zero left ideal of* R.
(3) $f: R_R \ni r \mapsto \varphi r \in R^*_R$ *is an* R*-isomorphism.*

Proof. "(1)⇒(3)": Let $\varphi r=0$, thus $\varphi(rx)=0$ for all $x \in R$, thus $\varphi(rR)=0$.

By assumption it follows that $r=0$, i.e. f is a monomorphism. As $\dim_K(R) = \dim_K(R^*)$ f is then in fact an isomorphism.

"(3)⇒(1)": From $\varphi(rR)=0$ it follows that $\varphi r=0$ and since f is an isomorphism $r=0$, thus (1) holds.

"(2)⇒(3)": φR is a K-subspace of R^*. Suppose, $\varphi R \neq R^*$ then there is $0 \neq x \in R$ with $\varphi(rx)=0$ for all $r \in R$ (if we take x from the orthogonal complement of φR in R), thus $\varphi(Rx)=0$, contradiction to (2)! Consequently we have $\varphi R = R^*$, i.e. f is epimorphism and thus, on account of dimension, is an isomorphism.

"(3)⇒(2)": For every $0 \neq x \in R$ there is an $\xi \in R^*$ with $\xi(x) \neq 0$. Let $\xi = \varphi r$, then it follows that $\varphi(rx) \neq 0$, thus $\varphi(Rx) \neq 0$, i.e. $Rx \not\subset \mathrm{Ker}(\varphi)$. □

13.5.2 *Definition.* A linear function φ on R_K, which satisfies the conditions of 13.5.1, is called *non-degenerate.*

13.5.3 COROLLARY. *Let* $\dim_K(R) < \infty$, *then the following are equivalent*:

(1) *There exists a non-degenerate function on* R_K.

(2) $R_R \cong R_R^*$.

Proof. "(1)⇒(2)": By 13.5.1.

"(2)⇒(1)": Let f: $R_R \cong R_R^*$ and let $\varphi := f(1)$, then it follows that

$$f(r) = f(1r) = f(1)r = \varphi r,$$

thus f: $R \ni r \mapsto \varphi r \in R^*$ and by 13.5.1 φ is non-degenerate. □

13.5.4 *Definition.* Let $\dim_K(R) < \infty$.

(a) R is called a *Frobenius algebra* :⇔ $R_R \cong R_R^*$.

(b) R is called a *quasi-Frobenius algebra* :⇔ the directly indecomposable direct summands of R_R and R_R^* coincide up to isomorphism and number (i.e. for every directly indecomposable direct summand of R_R there is a corresponding isomorphic copy of R_R^* and conversely).

In this definition we have retained the classical formulation also in order to make the older literature in this area more easily accessible. What this means in modern terms is to be explained immediately. The foundation for everything is the fact that for an arbitrary finite-dimensional algebra R_K the dual space R_R^* as a right R-module is an injective hull of $(R/\mathrm{Rad}(R))_R$, from which it follows immediately that R_R^* is an injective cogenerator. Hence (b) is then equivalent to saying that R_R is also an injective cogenerator thus a quasi-Frobenius ring and (a) implies additionally that

$$\mathrm{Soc}(R_R) \cong \mathrm{Soc}(R_R^*) \cong (R/\mathrm{Rad}(R))_R,$$

from which R is then in fact a Frobenius ring. All of this is now to be explained precisely.

To prove that R_R^* is an injective hull of $(R/\mathrm{Rad}(R))_R$ it must first of all be shown that every finite-dimensional semisimple algebra is a Frobenius algebra. Here an algebra is called *semisimple* if it is semisimple as a ring (see 8.2).

13.5.5 COROLLARIES

(1) *If R_K is a Frobenius algebra, if S_K is a K-algebra and if $R \cong S$ is a K-algebra isomorphism then S_K is also a Frobenius algebra.*

(2) *If R_K is a K-algebra and if*

$$R = A_1 \oplus \ldots \oplus A_m$$

is a direct decomposition into two-sided ideals $\neq 0$, then we have: R is a Frobenius algebra if and only if every A_i, $i = 1, \ldots, m$ is a Frobenius algebra.

(3) *Let L be a skew field which contains K in its centre, for which $\dim_K(L) < \infty$ holds, then the ring of all $n \times n$ square matrices (for $n \in \mathbb{N}$) with coefficients in L is a Frobenius algebra over K.*

(4) *Every finite-dimensional semisimple algebra is a Frobenius algebra.*

(5) *If G is a finite (multiplicative) group and K is a field then the group ring GK is a Frobenius algebra over K.*

Proof. (1) An algebra isomorphism $\rho: R \to S$ is a ring isomorphism for which we have: $\rho(x)k = \rho(xk)$ for all $x \in R$, $k \in K$. Let φ be a non-degenerate linear function on R_K. Then $\varphi\rho^{-1}$ is a non-degenerate linear function on S_K, for from

$$0 = \varphi\rho^{-1}(s_0 S) = \varphi(\rho^{-1}(s_0)\rho^{-1}(S)) = \varphi(\rho^{-1}(s_0)R)$$

it follows that $\rho^{-1}(s_0) = 0$, thus $s_0 = 0$.

(2) Let φ be a non-degenerate linear function on R_K, then $\varphi | A_i$, $i = 1, \ldots, m$ is a non-degenerate linear function on A_i. This implies immediately, if we take note, that we have $A_i A_j = 0$ for $i \neq j$ and consequently $aA_i = aR$ for $a \in A_i$. Conversely let φ_i be a non-degenerate linear function on A_i, $i = 1, \ldots, m$, then $\varphi = (\varphi_1, \ldots, \varphi_m)$ is a non-degenerate linear function on R for from

$$0 = \varphi((a_1 \ldots a_m)R)$$

it follows that $0 = \varphi_i(a_i A_i)$ for all i, thus by assumption $a_i = 0$, $i = 1, \ldots, m$.

(3) Let $w_1, \ldots, w_m$ with $w_1 = 1$ be a basis of L_K over K and let d_{ij} be the matrix with 1 in the (i, j)th place and 0 elsewhere. Then $d_{ij}w_l$ $(i, j = 1, \ldots, n; l = 1, \ldots, m)$ is obviously a basis of the matrix ring L_n over

K. Define $\varphi : L_n \to K$ by

$$\varphi\Big(\sum_{i,j,l} d_{ij} w_l k_{ij}^l\Big) := \sum_i k_{ii}^1$$

then φ is a non-degenerate linear function. Namely let

$$r = \sum_{i,j,l} d_{ij} w_l k_{ij}^l \neq 0,$$

then $k_{i_0 j_0}^{l_0} \neq 0$ exists. Consequently we have also

$$x := \sum_{l=1}^{m} w_l k_{i_0 j_0}^l \neq 0$$

and it follows that

$$\varphi(r d_{j_0 i_0} x^{-1}) = 1.$$

Thus the kernel of φ contains no right ideal $\neq 0$.

(4) On account of 8.2.4 and (2) we can confine ourselves in the proof to a simple finite-dimensional algebra R_K. For this we have 8.3.2 at our disposal. Let E be a simple right ideal of R and let $L := \operatorname{End}(E_R)$, then ${}_L E$ is a finite-dimensional vector space over L. Let $v_1, \ldots, v_n$ be a basis of ${}_L E$, then by 8.3.2 we obtain a ring isomorphism

$$\rho : R \ni r \mapsto (l_{ij}) \in L_n, \qquad l_{ij} \in L$$

in which

$$\begin{pmatrix} v_1 r \\ \vdots \\ v_n r \end{pmatrix} = (l_{ij}) \begin{pmatrix} v_1 \\ \vdots \\ v_n \end{pmatrix}$$

holds. Since K is contained in the centre of R, K is a subfield of L and as $K \subset R$ even a subfield of the centre of L. Thus L_n is also a K-algebra in which $(l_{ij})k = (l_{ij}k)$ for $k \in K$ and $l_{ij}k$ is the multiplication in L. For rk it then follows that

$$v_i(rk) = (v_i r)k = \Big(\sum_{j=1}^{n} l_{ij} v_j\Big)k = \sum_{j=1}^{n} (l_{ij}k) v_j,$$

thus we have $\rho(rk) = \rho(r)k$, i.e. ρ is a K-algebra isomorphism. Then the assertion follows from (1) and (3).

(5) Let $\operatorname{Ord}(G) = n$ and let $G = \{g_1 = e, g_2, \ldots, g_n\}$. Then

$$\varphi : GK \ni \sum_{i=1}^{n} g_i k_i \mapsto k_1 \in K$$

is a non-degenerate linear function for if in $\sum_{i=1}^{n} g_i k_i$ it happens that $k_j \neq 0$, then it follows that

$$\varphi\left(\left(\sum_{i=1}^{n} g_i k_i\right) g_j^{-1}\right) = k_j \neq 0,$$

thus $\operatorname{Ker}(\varphi)$ contains no right ideal $\neq 0$. □

13.5.6 THEOREM. *Let K be a field and let R_K be a K-algebra with $\dim_K(R) < \infty$. Then we have:*

(a) *$R/\operatorname{Rad}(R) \cong \operatorname{Soc}(R_R^*)$ as right R-modules.*
(b) *R_R^* is an injective hull of $(R/\operatorname{Rad}(R))_R$.*
(c) *R_R^* is an injective cogenerator.*

Proof. The proof follows in several steps.

(1) R_R^* is injective. The proof of this follows completely analogously to that of 5.5.2. In place of $\mathbb{Z}$ in 5.5.2 K now appears and K_K now appears in place of $D_{\mathbb{Z}}$. To the $\mathbb{Z}$-injectivity of $D_{\mathbb{Z}}$ corresponds now the K-injectivity of K_K. With these replacements the proof of 5.5.2 can be taken over word for word.

(2) By 9.3.5 we have $\operatorname{Soc}(R_R^*) = \boldsymbol{l}_{R^*}(\operatorname{Rad}(R))$. We claim that $\xi \in \boldsymbol{l}_{R^*}(\operatorname{Rad}(R)) \Leftrightarrow \operatorname{Rad}(R) \hookrightarrow \operatorname{Ker}(\xi)$. To this end let

$$(\xi u)(x) = \xi(ux) = 0$$

for all $u \in \operatorname{Rad}(R)$ and all $x \in R$. For $x = 1$ it follows that $\operatorname{Rad}(R) \hookrightarrow \operatorname{Ker}(\xi)$. Conversely if this is the case then it follows, since $\operatorname{Rad}(R)$ is a right ideal, that

$$0 = \xi(ux) = (\xi u)(x)$$

for all $u \in \operatorname{Rad}(R)$, $x \in R$; thus we have $\xi \in \boldsymbol{l}_{R^*}(\operatorname{Rad}(R))$.

(3) For $\xi \in \operatorname{Soc}(R_R^*)$ let $\bar{\xi}$ be the linear function induced by

$$\bar{\xi}: R/\operatorname{Rad}(R) \ni x + \operatorname{Rad}(R) \mapsto \xi(x) \in K.$$

We claim that

$$\psi: \operatorname{Soc}(R_R^*) \ni \xi \mapsto \bar{\xi} \in \operatorname{Hom}_K(R/\operatorname{Rad}(R), K)$$

is an R-isomorphism. It is clear that this is an R-monomorphism. Let now $g \in \operatorname{Hom}_K(R/\operatorname{Rad}(R), K)$ and let

$$\nu: R \to R/\operatorname{Rad}(R),$$

then it follows that $g\nu \in \operatorname{Soc}(R_R^*)$ and $\overline{g\nu} = g$, thus ψ is an isomorphism.

(4) Since $R/\mathrm{Rad}(R)$ is a finite-dimensional semisimple K-algebra, by 13.5.5 there is an $R/\mathrm{Rad}(R)$-isomorphism

$$\Lambda : \mathrm{Hom}_K(R/\mathrm{Rad}(R), K) \to R/\mathrm{Rad}(R),$$

which can and is to be considered also as an R-isomorphism. In short we have the isomorphism

$$\Lambda\psi : \mathrm{Soc}(R_R^*)_R \to (R/\mathrm{Rad}(R))_R.$$

Let f be the inverse isomorphism. Thus (a) is satisfied.

(5) Let $\iota : \mathrm{Soc}(R_R^*) \to R_R^*$ denote the inclusion. Since $\mathrm{Soc}(R_R^*) \overset{*}{\hookrightarrow} R_R^*$ (because it is artinian) and R_R^* is injective,

$$\iota f : (R/\mathrm{Rad}(R))_R \to R_R^*$$

is an injective hull. Hence we have shown (b).

(6) Since all simple right R-modules occur in $(R/\mathrm{Rad}(R))_R$ up to isomorphism, R_R^* is a cogenerator, thus (c) also holds. □

We come now to the aforementioned characterization.

13.5.7 THEOREM. *Let R_K be a finite-dimensional algebra over the field K. Then we have:*

(1) *R is a quasi-Frobenius algebra if and only if R is a quasi-Frobenius ring.*

(2) *R is a Frobenius algebra if and only if R is a Frobenius ring.*

Proof. (1) In regard to this we recall that a module is a cogenerator if and only if for an injective hull of any simple module it possesses an isomorphic submodule. This is then a directly indecomposable direct summand of the cogenerators. Since R_R^* by 13.5.6 is an (injective) cogenerator, we have consequently: R_R is then also a cogenerator (and then also injective) if and only if R_K is a quasi-Frobenius algebra.

(2) If R_K is a Frobenius algebra then R_K is also a quasi-Frobenius algebra and consequently by (1) a quasi-Frobenius ring. Further by 13.5.6 and as $R_R^* \cong R_R$ we have

$$(R/\mathrm{Rad}(R))_R \cong \mathrm{Soc}(R_R^*) \cong \mathrm{Soc}(R_R),$$

thus by 13.4.3 R is a Frobenius ring. Conversely let R be a Frobenius ring, then R is a quasi-Frobenius ring and by definition we have

$$(R/\mathrm{Rad}(R))_R \cong \mathrm{Soc}(R_R).$$

Consequently the injective hull R_R^* of $(R/\mathrm{Rad}(R))_R$ is isomorphic to the injective hull R_R of $\mathrm{Soc}(R_R)$, thus R_K is a Frobenius algebra. That in fact

R_R is the injective hull of Soc(R_R) follows from the injectivity of R_R and since in an artinian ring $\text{Soc}(R_R) \overset{*}{\hookrightarrow} R_R$. □

13.6 CHARACTERIZATION OF QUASI-FROBENIUS RINGS

In conclusion we return once again to the general case of quasi-Frobenius rings and state an interesting characterization of them. It is particularly of interest for the reason that set-theoretic considerations come essentially here into the proof of algebraic results. For this we need to use some set-theoretic facts, which are not proved here but which however are to be found in any text-book on set theory.

13.6.1 THEOREM (FAITH-WALKER). *The following are equivalent for a ring R:*

(1) *R is quasi-Frobenius.*
(2) *Every projective right R-module is injective.*
(3) *Every injective right R-module is projective.*

Proof. We go through the proof in the following steps: (1)⇒(2), (1)⇒(3), (2)⇒(1), (3)⇒(1), in which the first two implications are easy to prove whereas we must delve deeper for the last two.

"(1)⇒(2)": Since R_R is injective and noetherian, by 6.5.1 every free right R-module is injective and hence also every direct summand of a free right R-module. Consequently every projective right R-module is injective.

"(1)⇒(3)": Let Q_R be an injective R-module, then by 6.6.4 Q is the direct sum of submodules which are injective hulls of simple right R-modules. It suffices therefore for such a module to show that it is projective. Since R_R is a cogenerator then the injective hull of every simple right R-module occurs up to isomorphism as a direct summand in R_R and is therefore projective. □

We preface the rest of the proof by a lemma.

13.6.2 LEMMA. *For an arbitrary ring R and a module M_R we have: If $M^{(\mathbb{N})}$ is injective then R satisfies the ascending chain condition for right ideals of the form $r_R(U)$ with $U \subset M$.*

Proof. Indirect proof. Suppose we have for $U_i \subset M$, $i \in \mathbb{N}$

$$r_R(U_1) \subsetneqq r_R(U_2) \subsetneqq \dots,$$

then it follows (as $\boldsymbol{r}_R\boldsymbol{l}_M\boldsymbol{r}_R(U)=\boldsymbol{r}_R(U)$) that

$$\boldsymbol{l}_M\boldsymbol{r}_R(U_1) \supsetneqq \boldsymbol{l}_M\boldsymbol{r}_R(U_2) \supsetneqq \dots .$$

For every $i \in \mathbb{N}$ let

$$x_i \in \boldsymbol{l}_M\boldsymbol{r}_R(U_i), \qquad x_i \notin \boldsymbol{l}_M\boldsymbol{r}_R(U_{i+1}),$$

then there is an element $a_{i+1} \in \boldsymbol{r}_R(U_{i+1})$ with $x_i a_{i+1} \neq 0$. Let

$$A := \bigcup_{i \in N} \boldsymbol{r}_R(U_i),$$

then we have $A \hookrightarrow R_R$ and for every $a \in A$ there is an $n_a \in \mathbb{N}$ with

$$a \in \boldsymbol{r}_R(U_i) \quad \text{for all} \quad i \geq n_a.$$

Then it follows that

$$x_i a = 0 \quad \text{for all} \quad i \geq n_a,$$

thus for an element $x := (x_1 x_2 x_3 \dots) \in M^{\mathbb{N}}$ we have

$$xa = (x_1 a x_2 a \dots x_{n_a-1} a 000 \dots) \in M^{(\mathbb{N})}.$$

Consequently

$$\varphi_x : A \ni a \mapsto xa \in M^{(\mathbb{N})}$$

is a homomorphism. Since $M^{(\mathbb{N})}$ by assumption is injective a commutative diagram exists:

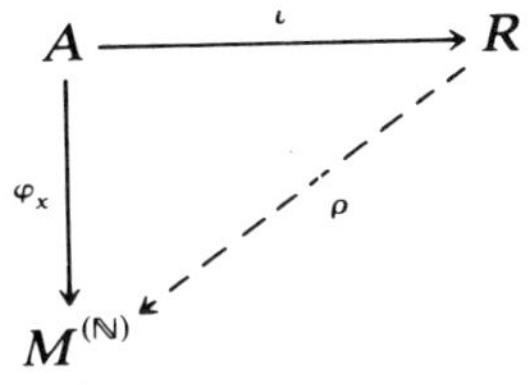

Let $\rho(1) = (z_1 z_2 \dots z_n 000 \dots)$ then it follows for all $a \in A$ that

$$\varphi_x(a) = xa = \rho(a) = \rho(1)a = (z_1 a \dots z_n a \dots z_n a 000 \dots),$$

thus $x_i a = 0$ for all $i > n$ and all $a \in A$, thus in particular $x_i a_{i+1} = 0$ for $i > n$. Contradiction! □

We continue now with the proof of Theorem 3.6.1.

"(2)⇒(1)": Since by assumption $R^{(\mathbb{N})}$ is an injective right R-module, 13.6.2 can be applied in the case $M_R = R_R$. Thus R satisfies the ascending chain condition for ideals of the form $\boldsymbol{r}_R(U)$ with $U \subset R$. Since R_R is injective it follows therefore by 12.4.2 that R satisfies the descending chain condition for finitely generated left ideals and in particular for cyclic left ideals. By 11.6.3 R_R is then perfect.

Let $N := \text{Rad}(R)$, then the chain

$$\boldsymbol{r}_R(N) \hookrightarrow \boldsymbol{r}_R(N^2) \hookrightarrow \boldsymbol{r}_R(N^3) \ldots$$

is stationary, i.e. there is a $t \in \mathbb{N}$ with

$$\boldsymbol{r}_R(N^t) = \boldsymbol{r}_R(N^{t+i}), \qquad i \geqslant 0.$$

Since N^i is a two-sided ideal, $\boldsymbol{r}_R(N^i)$ is a two-sided ideal. Suppose $\boldsymbol{r}_R(N^t) \neq R$, then it follows from 11.6.3 that

$$\text{Soc}({}_R(R/\boldsymbol{r}_R(N^t))) \neq 0.$$

Let $\bar{x}$ be a non-zero element of this socle, then it follows that $x \notin \boldsymbol{r}_R(N^t)$ and, since the socle is semisimple, $N\bar{x} = 0$, thus $Nx \subset \boldsymbol{r}_R(N^t)$. Consequently we have

$$N^t N x = N^{t+1} x = 0,$$

thus $x \in \boldsymbol{r}_R(N^{t+1}) = \boldsymbol{r}_R(N^t)$, contradiction! This contradiction shows that $\boldsymbol{r}_R(N^i) = R$ thus we must have $N^t = 0$, i.e. $N = \text{Rad}(R)$ is nilpotent. Consequently ${}_R R$ is also perfect. By 11.6.3 every right R-module $\neq 0$ then has a non-zero socle. Thus $\text{Soc}(R_R) \overset{*}{\hookrightarrow} R_R$. Therefore 12.5.2 (6) is satisfied and it follows that R_R is a cogenerator. By 12.4.1 we then have $\boldsymbol{r}_R \boldsymbol{l}_R(A) = A$ for every right ideal A of R and consequently R_R is noetherian. Since moreover R_R is injective, by 13.2.1 R is quasi-Frobenius. Hence (2)$\Rightarrow$(1) is shown.

"(3)$\Rightarrow$(1)": Since every injective right R-module is projective every injective module can be mapped monomorphically into a free module. Since every right R-module can be mapped monomorphically into an injective module, by 4.8.2 R_R is a cogenerator. We now show that R_R is noetherian. To this end let Q_R be an injective hull of R_R. Since R_R is a cogenerator Q_R is also a cogenerator. First of all we assume that $Q^{(\mathbb{N})}$ is injective and complete the proof for (3)$\Rightarrow$(1); we put the proof of the injectivity of $Q^{(\mathbb{N})}$ at the end. By 13.6.2 (with $M_R = Q_R$) R satisfies the ascending chain condition for right ideals of the form $\boldsymbol{r}_R(U)$ with $U \subset Q$. Since by 12.4.1 every right ideal of R is of this form, R_R is noetherian, from which the proof is complete up to the injectivity of $Q^{(\mathbb{N})}$.

While the proof was obtained so far in the context of the usual arguments, use must be made in the following of essentially set-theoretic considerations. We formulate separately particular steps of the proof which are of independent interest.

The next aim of our consideration consists in proving the Theorem of Kaplansky which says that every projective module is a direct sum of countably generated submodules. Here "countable" is to include "finite".

13.6.3 LEMMA. *Let R be an arbitrary ring and M an R-module. Suppose we have*

$$M = \bigoplus_{j\in J} M_j = A \oplus B,$$

where every M_j is countably generated. Then to every set $H \neq J$ with the property that for

$$U := \bigoplus_{j\in H} M_j,$$

$U = (A \cap U) \oplus (B \cap U)$ holds we have a set I with $H \subsetneqq I \subset J$ so that for

$$W := \bigoplus_{j\in I} M_j$$

$W = (A \cap W) \oplus (B \cap W)$ holds and $A \cap W = (A \cap U) \oplus C$, where C is a countably generated submodule.

Proof. Let α and β $(= 1_M - \alpha)$ be the projections belonging to the decomposition $M = A \oplus B$. Let $i_0 \in J\backslash H$. Since M_i is countably generated, $\alpha(M_{i_0})$ and $\beta(M_{i_0})$ are countably generated. Hence there is a countable set $I_1 \subset J$ with

$$M_{i_0} \hookrightarrow \alpha(M_{i_0}) + \beta(M_{i_0}) \hookrightarrow \bigoplus_{j\in I_1} M_j.$$

Since every M_j is countably generated and I is a countable set $\bigoplus_{j\in I_1} M_j$ is countably generated. Consequently there is a countable set $I_2 \subset J$ with

$$\bigoplus_{j\in I_1} M_j \hookrightarrow \alpha\Big(\bigoplus_{j\in I_1} M_j\Big) + \beta\Big(\bigoplus_{j\in I_1} M_j\Big) \hookrightarrow \bigoplus_{j\in I_2} M_j.$$

We continue inductively. We obtain therefore a sequence of countable sets

$$I_0 := \{i_0\}, I_1, I_2, \ldots$$

with

$$\bigoplus_{j\in I_n} M_j \hookrightarrow \alpha\Big(\bigoplus_{j\in I_n} M_j\Big) + \beta\Big(\bigoplus_{j\in I_n} M_j\Big) \hookrightarrow \bigoplus_{j\in I_{n+1}} M_j.$$

As $\mathrm{Im}(\alpha) = A$ and $\mathrm{Im}(\beta) = B$ this means that

$$(*)\qquad \bigoplus_{j\in I_n} M_j \hookrightarrow \Big(A \cap \bigoplus_{j\in I_{n+1}} M_j\Big) + \Big(B \cap \bigoplus_{j\in I_{n+1}} M_j\Big).$$

Then both

$$L := \bigcup_{n=0,1,2,\ldots} I_n$$

and $L \backslash H$ are countable sets. Let now $I := H \cup L$ and

$$V := \bigoplus_{j \in L \backslash H} M_j, \qquad W := \bigoplus_{j \in I} M_j = U \oplus V.$$

Then V is also countably generated.

Assertion. $W = (A \cap W) \oplus (B \cap W)$.

For the proof it is first of all clear that $(A \cap W) \oplus (B \cap W) \hookrightarrow W$. For the reverse inclusion we establish that every M_j with $j \in I$ is contained in $(A \cap W) \oplus (B \cap W)$. For $j \in H$ this holds by assumption. Let $j \in L \backslash H$ and let $j \in I_n$, then this holds by $(*)$.

From $W = U \oplus V = (A \cap U) \oplus (B \cap U) \oplus V$

it follows by the modular law that

$$A \cap W = (A \cap U) \oplus C, \qquad B \cap W = (B \cap U) \oplus D,$$

with

$$C := ((B \cap U) \oplus V) \cap A, \qquad D := ((A \cap U) \oplus V) \cap B.$$

Therefore we deduce that

$$W = (A \cap W) \oplus (B \cap W) = (A \cap U) \oplus (B \cap U) \oplus C \oplus D = U \oplus C \oplus D.$$

Since also $W = U \oplus V$ it follows that

$$V \cong W/U \cong C \oplus D.$$

Thus C is an epimorphic image of the countably generated module V and hence is itself countably generated. □

13.6.4 THEOREM. *Let R be an arbitrary ring and M an R-module. If we have*

$$M = \bigoplus_{j \in J} M_j = A \oplus B$$

with countably generated submodules M_j, then A and B are also direct sums of countably generated submodules.

Proof. It suffices evidently to prove the assertion for A. Let $\{A_\lambda \mid \lambda \in \Lambda\}$ be the set of all countably generated submodules of A. Let

$$X := \left\{(H, \Gamma) \mid H \subset J \wedge \Gamma \subset \Lambda \wedge \bigoplus_{j \in H} M_j = \left(A \cap \bigoplus_{j \in H} M_j\right) \oplus \left(B \cap \bigoplus_{j \in H} M_j\right) \wedge \right.$$

$$\left. A \cap \bigoplus_{j \in H} M_j = \bigoplus_{\lambda \in \Gamma} A_\lambda \right\}.$$

Since $(\varnothing, \varnothing) \in X$, $X \neq \varnothing$. Further X is ordered by

$$(H_1, \Gamma_1) \leq (H_2, \Gamma_2) \Leftrightarrow H_1 \subset H_2 \wedge \Gamma_1 \subset \Gamma_2.$$

If $Y \subset X$ is a totally ordered subset then

$$(H', \Gamma') \quad \text{with} \quad H' := \bigcup_{(H,\Gamma) \in Y} H \quad \text{and} \quad \Gamma' := \bigcup_{(H,\Gamma) \in Y} \Gamma$$

is an upper bound of Y in X, as is easily confirmed. Zorn's Lemma ensures then a maximal element $(\bar{H}, \bar{\Gamma}) \in X$. If we now suppose $\bar{H} \neq J$, 13.6.3 yields a properly larger element from X ↯. Thus $\bar{H} = J$ must hold. □

13.6.5 COROLLARY. *For an arbitrary ring R we have: Every projective R-module is a direct sum of countably generated submodules.*

Proof. Since every projective R-module is isomorphic to a direct summand of a free R-module the assertion follows from 13.6.4 in the case $M = R^{(J)}$. □

13.6.6 LEMMA. *Let R be an arbitrary ring and A_R a finitely generated R-module. Then we have: If an injective hull of A is also projective then it is finitely generated.*

Proof. Since all injective hulls of A are isomorphic it can be assumed without loss that A is a submodule of the injective hull Q of A. Since Q is projective there is a monomorphism

$$\mu : Q \to R^{(J)}$$

into a free R-module. Since A is finitely generated there is a finite subset $J_0 \subset J$ with

$$\mu(A) \hookrightarrow R^{(J_0)} \hookrightarrow R^{(J)}.$$

Denote the projection of $R^{(J)}$ onto $R^{(J_0)}$ by π, then it follows that $\pi\mu \mid A$ is a monomorphism. Since $A \overset{*}{\hookrightarrow} Q$ then $\pi\mu$ is also a monomorphism. Consequently $\pi\mu(Q)$ as a direct summand of $R^{(J_0)}$ is finitely generated and hence also is Q. □

13.6.7 COROLLARY. *For an arbitrary ring R we have: Every R-module which is simultaneously projective and injective is a direct sum of finitely generated submodules.*

Proof. By 13.6.5 it suffices to prove the assertion for a countably generated projective and injective R-module M. Let

$$M = \sum_{i \in N} x_i R$$

be one such. Let $Q_1 \hookrightarrow M$ denote an injective hull of $x_1 R$. Since Q_1 is a direct summand of M, Q_1 is also projective and consequently by 13.6.6 finitely generated. Let

$$M = Q_1 \oplus B_1,$$

then B_1 is also projective and injective. Suppose $Q_1, \ldots, Q_n$ and B_n with

$$M = Q_1 \oplus \ldots \oplus Q_n \oplus B_n$$

and

$$x_1, \ldots, x_n \in \bigoplus_{i=1}^{n} Q_i$$

have been inductively determined, then let

$$x_{n+1} = a_{n+1} + b_{n+1} \quad \text{with} \quad a_{n+1} \in \bigoplus_{i=1}^{n} Q_i, \quad b_{n+1} \in B_n$$

and let $Q_{n+1} \hookrightarrow B_n$ be an injective hull of $b_{n+1} R$. For the sequence, so obtained, of finitely generated direct summands

$$Q_1, Q_2, Q_3, \ldots$$

with $x_1, \ldots, x_n \in \bigoplus_{i=1}^{n} Q_i$ we then obviously have

$$M = \bigoplus_{i \in N} Q_i. \qquad \square$$

We wish now to show in the sense of the proof (3)$\Rightarrow$(1) of 13.6.1 that $Q^{(\mathbb{N})}$ is injective, where Q is an injective hull of R_R. By assumption Q is also projective and hence by 13.6.6 finitely generated. Now let τ be an infinite cardinal which is properly bigger than $2^{|R|}$ where $|R|$ is the cardinality of R. If A_R is a finitely generated R-module then τ is bigger than the cardinality of the set of all submodules of A_R for this is a subset of the power set of A_R (for reasons see a book on set theory).

Now let I be a set of cardinality τ (or bigger), then let

$$M := Q^I = \prod_{i \in I} Q_i \quad \text{with} \quad Q_i = Q \quad \text{for all} \quad i \in I.$$

Since Q is injective, M is injective and thus also projective. Let Q'_i be the image of $Q_i = Q$ under the canonical monomorphism $\sigma\eta_i$ (in the sense of 4.1.5). On the other hand by 13.6.7 M is a direct sum of finitely generated submodules:

$$M = \bigoplus_{j \in J} M_j.$$

Now let $i_1 \in I$ be arbitrary. Since Q'_{i_1} is finitely generated there is a finite subset $J_1 \subset J$ with

$$Q'_{i_1} \hookrightarrow \bigoplus_{j \in J_1} M_j.$$

If we now put

$$Q_{(1)} := Q'_{i_1} \quad \text{and} \quad D_1 := \bigoplus_{j \in J_1} M_j,$$

then D_1 is finitely generated and there is a $B_1 \hookrightarrow D_1$ with

$$D_1 = Q_{(1)} \oplus B_1.$$

We now consider the set

$$\{D_1 \cap Q'_i \mid i \in I \wedge i \neq i_1\}.$$

This is a set of submodules of the finitely generated module D. By choice of the cardinality τ of I not all of the $D_1 \cap Q'_i$ can be different from one another (transfinite box principle). Let $i_2, k \in I \backslash \{i_1\}$, $i_2 \neq k$ with $D_1 \cap Q'_{i_2} = D_1 \cap Q'_k$.

As $Q'_{i_2} \cap Q'_k = 0$ it follows that

$$D_1 \cap Q'_{i_2} = D_1 \cap Q'_k = 0.$$

Consequently

$$Q'_{i_2} \xrightarrow{\iota_2} \bigoplus_{j \in J} M_j \xrightarrow{\pi_2} \bigoplus_{j \in J \backslash J_1} M_j$$

is a monomorphism (where ι_2 resp. π_2 is the inclusion resp. the corresponding projection and we have $\mathrm{Ker}(\pi_2\iota_2) = D_1 \cap Q'_{i_2} = 0$ as $\mathrm{Ker}(\pi_2) = D_1$). Since Q'_{i_2} is finitely generated there is a finite set

$$J_2 \subset J \backslash J_1$$

with $\mathrm{Im}(\pi_2\iota_2) \hookrightarrow D_2 := \bigoplus_{j \in J_2} M_j$.

By choice of J_2 we have $J_1 \cap J_2 = \varnothing$. Now let

$$Q_{(2)} := \operatorname{Im}(\pi_2 \iota_2),$$

then it follows that

$$Q_{(2)} \cong Q'_{i_2} \cong Q$$

and there exists a B_2 with

$$D_2 = Q_{(2)} \oplus B_2.$$

Inductively we define Q'_{i_n}, $i_n \in I \backslash \{i_1, \ldots, i_{n-1}\}$ by

$$(D_1 \oplus \ldots \oplus D_{n-1}) \cap Q'_{i_n} = 0$$

and

$$Q'_{i_n} \xrightarrow{\iota_n} \bigoplus_{j \in J} M_j \xrightarrow{\pi_n} \bigoplus_{j \in J \backslash (J_1 \cup \ldots \cup J_{n-1})} M_j$$

as well as $J_n \subset J \backslash (J_1 \cup \ldots \cup J_{n-1})$ with J_n finite and

$$\operatorname{Im}(\pi_n \iota_n) \hookrightarrow D_n := \bigoplus_{j \in J_n} M_j.$$

Further we have

$$J_n \cap (J_1 \cup \ldots \cup J_{n-1}) = \varnothing.$$

If we put

$$Q_{(n)} := \operatorname{Im}(\pi_n \iota_n),$$

then we have again $Q_{(n)} \cong Q$ and there exists a B_n with

$$D_n = Q_{(n)} \oplus B_n.$$

For the inductively resulting sequences

$$\begin{matrix} J_1, & J_2, & J_3 & \ldots, & D_1, & D_1, & D_3, & \ldots \\ Q_{(1)}, & Q_{(2)}, & Q_{(3)}, & \ldots, & B_1, & B_2, & B_3, & \ldots \end{matrix}$$

we then have, if we put

$$H := \bigcup_{i \in \mathbb{N}} J_i$$

$$(*) \qquad Q^{(\mathbb{N})} \cong \bigoplus_{i \in \mathbb{N}} Q_{(i)} \qquad (\text{as } Q_{(i)} \cong Q),$$

$$M = \bigoplus_{j \in J} M_j = \Big(\bigoplus_{j \in H} M_j\Big) \oplus \Big(\bigoplus_{j \in J \backslash H} M_j\Big),$$

$$\bigoplus_{j \in H} M_j = \bigoplus_{i \in \mathbb{N}} \Big(\bigoplus_{j \in J_i} M_j\Big) = \bigoplus_{i \in \mathbb{N}} D_i = \bigoplus_{i \in \mathbb{N}} (Q_{(i)} \oplus B_i),$$

thus

$$M = \Big(\bigoplus_{i\in\mathbb{N}} Q_{(i)}\Big) \oplus \Big(\bigoplus_{i\in\mathbb{N}} B_i\Big) \oplus \Big(\bigoplus_{j\in J\setminus H} M_j\Big).$$

Consequently

$$\bigoplus_{i\in\mathbb{N}} Q_{(i)}$$

is a direct summand of the injective module M and hence in any case injective. By $(*)$ $Q^{(\mathbb{N})}$ is then also injective which was to be shown. Hence the proof $(3)\Rightarrow(1)$ is complete. □

EXERCISES

(1)

Show:

(a) A commutative artinian ring is a quasi-Frobenius ring if and only if it is a direct sum of ideals with simple socle.

(b) Every commutative quasi-Frobenius ring is a Frobenius ring.

(c) If R is a commutative principal ideal domain and $0 \neq A \hookrightarrow R_R$ then R/A is a Frobenius ring.

(2)

Let K be a field and let R be the ring of all matrices of the form

$$\begin{pmatrix} a & b \\ 0 & c \end{pmatrix} \quad \text{with} \quad a, b, c \in K.$$

Show:

(a) For $x = \begin{pmatrix} a & b \\ 0 & c \end{pmatrix} \in R$ we have:

(1) x is left invertible $\Leftrightarrow$ x is right invertible $\Leftrightarrow$ $ac \neq 0$.

(2) x is nilpotent $\Leftrightarrow x^2 = 0 \Leftrightarrow a = c = 0$.

(3) x is an idempotent $\Leftrightarrow x \in \left\{\begin{pmatrix} 0 & 0 \\ 0 & 0 \end{pmatrix}, \begin{pmatrix} 1 & 0 \\ 0 & 1 \end{pmatrix}, \begin{pmatrix} 0 & b \\ 0 & 1 \end{pmatrix}, \begin{pmatrix} 1 & b \\ 0 & 0 \end{pmatrix}\right\}$.

(4) xR is simple $\Leftrightarrow x \neq 0 \wedge a = 0$; Rx is simple $\Leftrightarrow x \neq 0 \wedge c = 0$.

(b) (1) $\mathrm{Rad}(R) = \begin{pmatrix} 0 & K \\ 0 & 0 \end{pmatrix}$, $\quad \mathrm{Soc}(R_R) = \begin{pmatrix} 0 & K \\ 0 & K \end{pmatrix}$,

$\mathrm{Soc}({}_RR) = \begin{pmatrix} K & K \\ 0 & 0 \end{pmatrix}$.

(2) $\boldsymbol{r}_R(\mathrm{Rad}(R)) = \mathrm{Soc}({}_RR)$, $\boldsymbol{l}_R(\mathrm{Rad}(R)) = \mathrm{Soc}(R_R)$, $\boldsymbol{r}_R(\mathrm{Soc}(R_R)) = \mathrm{Soc}({}_RR)$, $\boldsymbol{l}_R(\mathrm{Soc}(R_R)) = 0$, $\boldsymbol{r}_R(\mathrm{Soc}({}_RR)) = 0$, $\boldsymbol{l}_R(\mathrm{Soc}(R_R)) = \mathrm{Soc}(R_R)$.

(3) $\mathrm{Soc}(R_R)$ as a left ideal is a direct summand, however as a right ideal it is not cyclic (thus not a direct summand).

(4) $\mathrm{Soc}({}_RR)$ as a right ideal is a direct summand, however as a left ideal it is not cyclic.

(c) For the determination of the lattice of the right ideals of R show:

(1) $\mathrm{Le}(R_R)=3$.

(2) The maximal right ideals of R are $\mathrm{Soc}(R_R)$ and $\mathrm{Soc}({}_RR)$.

(3) The simple right ideals of R are $\mathrm{Rad}(R)$ and also $E_k := \begin{pmatrix} 0 & k \\ 0 & 1 \end{pmatrix} R$, $k \in K$.

(4) The lattice of the right ideals has the following picture

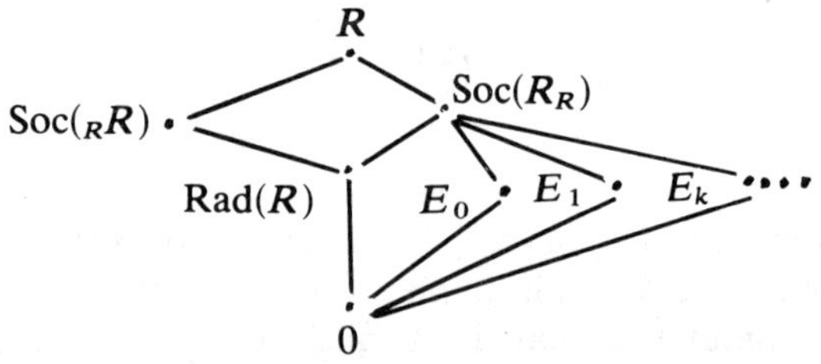

(d) For the determination of the injective hull of R_R show:

(1) For all $k \in K$, $E_k \cong \mathrm{Rad}(R) \cong R/\mathrm{Soc}({}_RR)$ as right R-modules.

(2) The only injective right ideals of R are 0 and $\mathrm{Soc}({}_RR)$.

(3) R is a subring of $S := \begin{pmatrix} K & K \\ K & K \end{pmatrix}$ $(=K_n)$; $R_R \xrightarrow{\iota} S_R$ is an injective hull of R_R.

(3)

Show:

(1) Let R be a quasi-Frobenius ring. If $e \in R$ is an idempotent, for which eR is a two-sided ideal, then e lies in the centre of R.

(2) Let A and B be rings, let ${}_AM_B$ be a bimodule and let $R := \begin{pmatrix} A & M \\ 0 & B \end{pmatrix}$, then we have: R is quasi-Frobenius $\Leftrightarrow$ A and B are quasi-Frobenius and $M=0$.

(Hint for (1): Show that the factor ring R/eR is again a quasi-Frobenius ring and thereby deduce the assertion.)

(4)

Let K be a field and let R be the commutative K-algebra with the basis $1, a, b, c$ and the multiplication $1r = r1 = r$ for $r \in R$, $ab = ba = 0$, $a^2 = b^2 = c$, $ac = ca = bc = cb = c^2 = 0$.

Show:

(a) For $x = 1k_1 + ak_2 + bk_3 + ck_4 \in R \quad (k_i \in K)$ we have

(1) x is invertible $\Leftrightarrow k_1 \neq 0$.

(2) x is nilpotent $\Leftrightarrow x^3 = 0 \Leftrightarrow k_1 = 0 \Leftrightarrow x \in \mathrm{Rad}(R)$.

(3) $x \in \mathrm{Soc}(R) \Leftrightarrow k_1 = k_2 = k_3 = 0$.

(b) Let $N := \mathrm{Rad}(R)$. Then $N^2 = \mathrm{Soc}(R) = cR$ and $0 \hookrightarrow cR \hookrightarrow aR \hookrightarrow N \hookrightarrow R$ is a composition series of R_R. In particular $\mathrm{Soc}(R)$ is simple, thus R is a quasi-Frobenius ring.

(c) If we define $A_k := (ak + b)R$ for every $k \in K$ then we have:

(1) $cR \underset{\neq}{\hookrightarrow} A_k \underset{\neq}{\hookrightarrow} N$; and $A_k \neq A_{k'}$ for $k \neq k'$.

(2) If U is an ideal of R of length 2, then $U = aR$ or $U = A_k$ for a $k \in K$. (Hint: Show first that U is cyclic.)

(3) Determine the lattice of ideals of R.

(d) The factor ring R/N^2 is not a quasi-Frobenius ring.

(5)

Let the ring R be commutative and artinian. Show:

(a) If A is a maximal ideal of R then the injective hull of R/A is finitely generated.

(b) For every finitely generated R-module the injective hull is again finitely generated.

(c) If C is a minimal cogenerator of M_R then the ring $S := \mathrm{Id}(C_R)$, defined in Chapter 12, Exercise 10, is a quasi-Frobenius ring which has a factor ring isomorphic to R.

(Hint for (a): If Q is an injective hull of R/A and if $B_i := \boldsymbol{l}_Q(A^i)$ then show first that B_{i+1}/B_i is finitely generated.)

(6)

Let R_R be noetherian and let every cyclic left R-module be reflexive. Show:

(a) R is artinian on both sides.

(b) Every maximal right ideal B is an annihilator ideal (i.e. $B = \boldsymbol{r}_R\boldsymbol{l}_R(B)$). (Hint: If E is a simple left R-module and if A_R is simple with $A_R^* \hookrightarrow E_R^*$ then it follows that ${}_RA \cong {}_RE$).

(c) If $R_R \hookrightarrow M_R$ and if M/R is simple then R_R is a direct summand of M_R.

(d) R is a quasi-Frobenius ring.

(7)

Show: Every ring with perfect duality, if perfect on one side, is a quasi-Frobenius ring.

(8)

Show: All reflexive modules over a quasi-Frobenius ring are finitely generated.

Note on the Literature

Textbooks on rings and modules (in chronological order)

[1] Artin, E., Nesbitt, C. and Thrall, R., "Rings with Minimum Condition." Ann Arbor, Mich. 1944.
[2] Jacobson, N., Structure of Rings. *Amer. Math. Soc. Coll. Publ.* **37** (1956).
[3] Bourbaki, "Algèbre," Chapter 8. Paris, 1958.
[4] Jans, J. P., "Rings and Homology." New York, 1964.
[5] Lambek, J., "Lectures on Rings and Modules." New York, London, 1966.
[6] Faith, C., Lectures on Injective Modules and Quotient Rings. Springer, Berlin, Heidelberg, New York, 1967. (Lecture Notes in Math. **49**.)
[7] Bourbaki, "Algèbre," Chap. 2. Paris, 1970.
[8] Sharpe, D. W. and Vámos, P., "Injective Modules." London, 1972.
[9] Tachikawa, H., "Quasi-Frobenius Rings and Generalizations, QF-3 and QF-1 Rings." Springer, Berlin, Heidelberg, New York, 1973. (Lectures Notes in Math. **351**.)
[10] Anderson, F. W. and Fuller, K. R., "Rings and Categories of Modules." Springer, Berlin, Heidelberg, New York, 1974.
[11] Stenström, B., "Rings of Quotients." Springer, Berlin, Heidelberg, New York, 1975.

Literature for Chapters 11 to 13 (in chronological order)

(Literature on QF3- and QF1-rings was not recorded; consult the references in the book of Tachikawa, Reference [9].)

[1] Nakayama, T., On Frobenius algebras I. *Ann. Math.* **40** (2) (1939), 611–633.
[2] Nakayama, T., On Frobenius algebras II. *Ann. Math.* **42** (2) (1941), 1–21.
[3] Nakayama, T., On Frobenius algebras III. *Jap. J. Math.* **18** (1942), 49–65.
[4] Kasch, F., Grundlagen einer Theorie der Frobeniuserweiterungen. *Math. Ann* **127** (1954), 453–474.
[5] Eilenberg, S. and Nakayama, T., On the dimension of modules and algebras II (Frobenius algebras and quasi-Frobenius rings). *Nagoya Math. J.* **9** (1956), 1–16.

[6] Dieudonné, J., Remarks on quasi-Frobenius rings. *Illinois J. Math.* **2** (1958), 346–354.
[7] Kaplansky, I., Projective modules. *Ann. Math.* **68** (1958), 372–377.
[8] Morita, K., Duality for modules and its applications to the theory of rings with minimum condition. *Sci. Rep. Tokyo Kyoiku Daigaku* **A6** (1958), 83–142.
[9] Tachikawa, H., Duality theorem of character modules for rings with minimum conditions. *Math. Z.* **68** (1958), 479–487.
[10] Azumaya, G., A duality theory for injective modules. *Amer. J. Math.* **81** (1959), 249–278.
[11] Bass, H., Finitistic dimension and homological generalization of semi-primary rings. *Trans. Amer. Math. Soc.* **95** (1960), 466–488.
[12] Kasch, F., Projektive Frobeniuserweiterungen. *Sitz.-Ber. Heidelberger Akad. Wiss.* (1960/61), 89–109.
[13] Kasch, F., Dualitätseigenschaften von Frobeniuserweiterungen. *Math. Z.* **77** (1961), 219–227.
[14] Kasch, F., Ein Satz über Frobeniuserweiterungen. *Arch. Math.* **12** (1961), 102–104.
[15] Mares, E. A., Semi-perfect modules. *Math. Z.* **82** (1963), 347–360.
[16] Müller, B., Quasi-Frobenius-Erweiterungen. *Math. Z.* **85** (1964), 345–368.
[17] Morita, K., Adjoint pairs of functors and Frobenius extensions. *Sci. Rep. Tokyo Kyoiku Daigaku* **9** (1965), 40–71.
[18] Müller, B., Quasi-Frobenius-Erweiterungen II. *Math. Z.* **88** (1965), 380–409.
[19] Azumaya, G., Completely faithful modules and self-injective rings. *Nagoya Math. J.* **27** (1966), 697–708.
[20] Faith, C., Rings with ascending chain condition on annihilators. *Nagoya Math. J.* **27** (1966), 179–191.
[21] Kasch, F. and Mares, E. A., Eine Kennzeichnung semi-perfekter Moduln. *Nagoya Math. J.* **27** (1966), 525–529.
[22] Miyashita, J., Quasi-projective modules, perfect modules and a theorem for modular lattices. *J. Fac. Sc. Hokkaido Univ.* **XIX** (1966), 86–110.
[23] Morita, K., On S-rings in the sense of F. Kasch. *Nagoya Math. J.* **27** (1966), 687–695.
[24] Osofsky, B. L., A generalization of quasi-Frobenius rings. *J. Algebra* **4** (1966), 373–387.
[25] Rentschler, R., Eine Bemerkung zu Ringen mit Minimalbedingung für Hauptideale. *Arch. Math.* **17** (1966), 298–301.
[26] Faith, C. and Walker, E. A., Direct sum representations of injective modules. *J. Algebra* **5** (1967), 203–221.
[27] Kato, T., Self-injective rings. *Tôhoku Math. J.* **19** (1967), 469–479.
[28] Utumi, Y., Self-injective rings. *J. Algebra* **6** (1967), 56–64.
[29] Kato, T., Torsionless modules. *Tôhoku Math. J.* **20** (1968), 234–243.
[30] Kato, T., Some generalizations of QF-rings. *Proc. Jap. Hc.* **44** (1968), 114–119.
[31] Ondera, T., Über Kogeneratoren. *Arch. Math.* **19** (1968), 402–410.
[32] Björk, I. E., Rings satisfying a minimum condition on principal ideals. *J. Reine Angew. Math.* **236** (1969), 112–119.
[33] Chamard, I. Y., Anneaux semi-parfaits et presque-frobeniusiens. *C. R. Acad. Sci. Paris* **269** (1969), 556–559.
[34] Fuller, K. R., On indecomposable injectives over artinian rings. *Pacific J. Math.* **29** (1969), 115–135.

[35] Kasch, F., Schneider, H.-J. and Stolberg, H. J., On injective modules and cogenerators. *Carnegie-Mellon Univ. Report* **23** (1969), 1–23.
[36] Michler, G. O., Idempotent ideals in perfect rings. *Canadian J. Math.* **21** (1969), 301–309.
[37] Rutter, E. A., Two characterisations of quasi-Frobenius rings. *Pacific J. Math.* **30** (1969), 777–784.
[38] Sandomierski, F. L., On semi-perfect and perfect rings. *Proc. Amer. Math. Soc.* **21** (1969), 205–207.
[39] Jonah, D., Rings with minimum condition for principal right ideals have the maximum condition for principal left ideals. *Math. Z.* **113** (1970), 106–112.
[40] Müller, B. J., On semi-perfect rings. *Illinois J. Math.* **14** (1970), 464–467.
[41] Sandomierski, F. L., Some examples of right self-injective rings which are not left self-injective. *Proc. Amer. Math. Soc.* **26** (1970), 244–245.
[42] Golan, I. S., Quasi perfect modules. *Quat. J. Math. Oxford* **22** (1971), 173–182.
[43] Onodera, T., Eine Bemerkung über Kogeneratoren. *Proc. Jap. Acad.* **47** (1971), 140–142.
[44] Onodera, T., Ein Satz über koendlich erzeugte RZ-Moduln. *Tôhoku Math. J.* **23** (1971), 691–695.
[45] Osofsky, B. L., Loewy length of perfect rings. *Proc. Amer. Math. Soc.* **28** (1971), 352–354.
[46] Rutter, E. A., PF-modules. *Tôhoku Math. J.* **23** (1971), 201–206.
[47] Ware, R., Endomorphism rings of projective modules. *Trans. Amer. Math. Soc.* **155** (1971), 233–256.
[48] Oberst, U. and Schneider, H. J., Die Struktur von projektiven Moduln. *Inventiones Math.* **13** (1971), 295–304.
[49] Anderson, F. W. and Fuller, K. R., Modules with decompositions that complement direct summands. *J. Algebra* **22** (1972), 241–253.
[50] Beck, I., Projective and free modules. *Math. Z.* **129** (1972), 231–234.
[51] Kasch, F. and Pareigis, B., Einfache Untermoduln von Kogeneratoren. Sitz.-Ber. Bay. *Akad. Wiss.* (1972), 45–76.
[52] Onodera, T., Linearly compact modules and cogenerators. *J. Fac. Sci. Hokkaido* **22** (1972), 116–125.
[53] Hannula, A. T., On the construction of quasi-Frobenius rings. *J. Algebra* **25** (1973), 403–414.
[54] Hauger, G. and Zimmermann, W., Quasi-Frobenius-Moduln. *Arch. Math.* **24** (1973), 379–386.
[55] Skornjakov, L. A., Mehr über Quasi-Frobeniusringe (russ.) *Mat. Sbornik* **92** (1973), 518–529.
[56] Azumaya, G., Characterisation of semi-perfect and perfect modules. *Math. Z.* **140** (1974), 95–103.
[57] Cunningham, R. S. and Rutter, E. A., Perfect modules. *Math. Z.* **140** (1974), 105–110.
[58] Ming, R. Y. C., On simple P-injective modules. *Math. Japonicae* **19** (1974), 173–176.
[59] Müller, B. J., The structure of quasi-Frobenius rings. *Can. J. Math.* **XXVI** (1974), 1141–1151.
[60] Zöschinger, H., Komplementierte Moduln über Dedekindringen. *J. Algebra* **29** (1974), 42–56.

[61] Zöschinger, H., Komplemente als direkte Summanden. *Arch. Math.* **25** (1974), 241–253.
[62] Hauger, G., Aufsteigende Kettenbedingungen für zyklische Moduln und perfekte Endomorphismenringe. *Acta Math. Ac. Sci. Hungar.*
[63] Takeuchi, T., The endomorphism ring of a indecomposable module with an Artinian projective cover. *Hokkaido Math. J.* **IV** (1975), 265–267.
[64] Takeuchi, T., On cofinite-dimensional modules. *Hokkaido Math. J.* **V** (1976), 1–43.
[65] Zimmermann, W., Über die aufsteigende Kettenbedingung für Annullatoren. *Arch. Math.* **27** (1976), 261–266.

Index